Wind Propulsion Conference 2023

London, United Kingdom
16-17 February 2023

ISBN: 978-1-7138-9566-4

Printed from e-media with permission by:

Curran Associates, Inc.
57 Morehouse Lane
Red Hook, NY 12571

Some format issues inherent in the e-media version may also appear in this print version.

Printed with permission by Curran Associates, Inc. (2024)

The Royal Institution of Naval Architects
8-9 Northumberland Street
London, WC2N 5DA
United Kingdom

Phone: 020 7235 4622
Fax: 020 7259 5912

publications@rina.org.uk

Additional copies of this publication are available from:

Curran Associates, Inc.
57 Morehouse Lane
Red Hook, NY 12571 USA
Phone: 845-758-0400
Fax: 845-758-2633
Email: curran@proceedings.com
Web: www.proceedings.com

**160 YEARS OF HISTORY,
KNOWLEDGE AND EXPERIENCE**

SALUM ET CARINAE PIGNORA VITAE

THE ROYAL INSTITUTION
OF NAVAL ARCHITECTS

Wind Propulsion Conference

16 - 17 February 2023, London, United Kingdom

Conference venue - International Maritime Organisation

In Association with:

IWSA
International Windship Association

Sponsored by:

YARA

TOGETHER WE WILL SHAPE OUR FUTURE

SALUM ET CARINAE PIGNORA VITAE

THE ROYAL INSTITUTION
OF NAVAL ARCHITECTS

Wind Propulsion Conference

16 – 17 February 2023, London, United Kingdom

Conference venue – International Maritime Organisation

THE ROYAL INSTITUTION OF NAVAL ARCHITECTS
8-9 Northumberland Street
London WC2N 5DA

Telephone: 020 7235 4622
Fax: 020 7259 5912

ISSN No: 2752-6569

CONTENTS

Wind Propulsion Conference
16th – 17th February 2023
London, United Kingdom

AERODYNAMIC OPTIMIZATION OF THE ESAIL, BOUND4BLUE'S SUCTION SAIL FOR WIND-ASSISTED VESSEL PROPULSION

A Llopis Pascual, **G Bailardi**, **B Charrier** and **D Ferrer Desclaux,** Bound 4 Blue S.L., Spain

SUMMARY

bound4blue (B4B) is a WAPS (Wind-Assisted Propulsion System) provider which has developed the eSAIL, a fully autonomous suction sail. This paper showcases bound4blue's journey in using CFD to design the eSAIL, from initial modelling all the way to complex simulations of realistic conditions that the sail operate in.

CFD studies were carried to initially validate the modelling technique and later to conduct parametric studies in 2D and 3D to optimise the design of the eSAIL. The results were verified and validated against a wind tunnel campaign, showing very good correlation in aerodynamic forces and pressure distributions. This results in an eSAIL which generates a CL 20% higher for the same coefficient of power (Ca) when compared to the Turbovoile, achieving a lift coefficient of up to 8.2.

Following the design optimisation, results are shown for a sail-to-sail interference case study. The results of this study are used to provide a custom control algorithm to maximise the savings offered by the eSAIL.

NOMENCLATURE

AoA	Angle of Attack (degrees)
AWA	Apparent Wind Angle (degrees)
Ca	Aerodynamic Power Coefficient (-)
CD	Drag Coefficient (-)
CL	Lift Coefficient (-)
Cp	Pressure Coefficient (-)
Cq	Suction Coefficient (-)

1. INTRODUCTION

The world is facing one of its ever-major challenges: global warming and climate change – which will transform the way we all live. Shipping industry is not alien to it, driven by the IMO environmental regulations and increasing fuel price. In the 20th-century, diesel engines revolutionized the industry substituting sails. A new revolution is now happening: new environmentally-friendly technologies will be implemented on new-era vessels. Funny fact: sails are back to help on this mission.

bound4blue is a European WAPS (Wind-Assisted Propulsion System) provider which has developed the eSAIL, an autonomous suction sail that combines the positive features of a conventional wingsail (simple operation, excellent capacity to sail upwind) but takes profit of active boundary layer control techniques (based on suction) to prevent flow detachment (stall), achieving extraordinarily high values of lift coefficients (x6-7 compared to regular sails), resulting in smaller size and lower weight sails for equivalent savings. Bound4Blue offers a portfolio with different eSAIL sizes, so that it is suitable for the different types and sizes of merchant vessels. Full-scale eSAIL installations have already been executed on 2 vessels, 3 additional installations are ongoing, and additional full scale realizations are in preparation.

R&D and constant improvement are part of the DNA of bound4blue and its team. This paper summarizes the results obtained thanks the work and effort of the aerodynamics' department team.

2. ORIGIN/CONCEPT

2.1 TURBOVOILE

In the early 1980's, Captain Jacque-Yves Cousteau, oceanographer, and explorer, wanted to build a new ship to replace the *Calypso*. Aware of the depletion of the natural resources and of the impact of humankind development on the nature, Captain Cousteau wished an ecological exploratory ship with wind assisted

propulsion. The Cousteau Foundation mobilized a Research Team under the direction of Prof. Lucien Malavard and PhD. Bertrand Charrier to propose wind assisted propulsion systems that are true wind engines, efficient from an aerodynamic point of view in order to limit as much space as possible on the ship's deck, easily automated, requiring low maintenance, meeting all safety criteria.
The Aerodynamic Team identified the Flettner Rotor, implemented in 1933 on Barbara ship, as one system that responded to most of the criteria requested for *Calypso II*. For Captain Cousteau, the device had an important disadvantage, the high cylinder turned. For him, a very large tower rotating at more than 200 rpm posed questions of safety for the navigating personnel, of the impact of the gyroscopic effect on vessel stability and above all the absence of possible improvements in the aerodynamic and energy efficiency of the wind propeller.

The challenge launched by Captain Cousteau to the Research Team was to identify another system with aerodynamic performances equivalent or superior to those of the Flettner rotor, namely, a coefficient of lift as high as possible, a coefficient of drag as low as possible, a minimum expenditure of energy to create the aerodynamic phenomenon, and non-rotating.

Research and analysis led to studies of boundary layer control systems. Aeronautical research centers had developed multiple systems to control boundary layer detachment either through suction or blowing, but always on thin wing profiles. To obtain very high lift coefficients as requested for the maritime transport, the laws of aerodynamics dictate that very thick or even circular profiles should be used.

More than 350 wind tunnel tests were carried out on models with a chord of 10 to 30 cm, a wingspan of 50 to 150 cm, in infinite and finite aspect ratio, for Reynolds numbers from 40,000 to 500,000, with different suction and blowing systems to control the boundary layer separation made up with one to three fans, with suction zones of variable permeability, with or without flap at the rear of the profile. Naturally, different profiles were tested from the circular cylinder, to the parabola, to the ellipse and others. Theoretical studies of pressure field modeling were carried out and compared to wind tunnel tests [2].

Within the scientific validation of the performances of the suction sail during the four wind tunnel campaigns, a full-scale prototype of 12 m high was installed on a catamaran, *Moulin à Vent*, 20.75 m length. More than 150 sea trials were carried out under very different wind conditions. After the completion of trials on *Moulin à Vent*, the construction of *Alcyone* was decided, a boat designed with two suction sails, named Turbovoile, foreshadowing the future *Calypso II*.

With the drop in oil prices of 1986-87, research into wind propulsion for ships was abandoned for over thirty years. This parenthesis is now over.

2.2 BOUND4BLUE'S ESAIL

Taking the lessons learnt from the development of the Turbovoile, bound4blue set out to improve the performance of the suction sail. The culmination of this work is the eSAIL: a fully autonomous suction sail delivering an increase of lift-to-power ratio of up to 20% when compared to the Turbovoile. The autonomous eSAIL consists of the following parts:

Pedestal. This is the structure connecting the vessel foundations and the eSAIL itself, with two important interfaces:

- At the lower end, the vessel connecting flange, offering a flange-bolted connection to the vessel.
- At the top end, the eSAIL connecting flange, offering a flange-bolted connection between the pedestal and the eSAIL rotation system.

eSAIL main body. This is the main structure of the eSAIL, generating the desired aerofoil shape and containing/holding all eSAIL parts and elements. The eSAIL body contains the three actuators of the eSAIL (rotation motors, flap positioning motors and suction fans) and the eSAIL Cabinet of the control system.

Movable flap. This is a structure located at the rear part of the eSAIL body and extending along the whole eSAIL height. It can position on one side or the other of the eSAIL body, generating the desired aerofoil asymmetry both for port and starboard winds. The movement is done by electric motors geared to the flap.

Suction system. The suction is in charge of ensuring the extraordinary aerodynamic characteristics of the eSAIL and is performed by means of electric axial fans that blow air from the inside of the eSAIL main body upwards. The fans are located on top of the eSAIL main body secured by means of a bolted flange connection. This position is selected to ensure maximum fan efficiency, maximum safety for the crew (no moving parts or strong airblow next to the deck) and to reduce potential noise.

Orientation system. This system oversees the rotation/ orientation of the eSAIL to adjust its position with the existing wind direction. It is located on the pedestal, connected to the eSAIL-connecting flange. It includes a slew-bearing, which is geared to an electric motor delivering the required torque to rotate the eSAIL and the required brake to hold its position under any external situation.

Control system. The operation of any Wind Assisted Propulsion System shall be fully autonomous with no crew intervention. The eSAIL autonomous control system trims, for each sailing condition, the suction power, eSAIL body rotation and flap positioning in an autonomous way, to ensure maximum efficiency of the system. Additionally, it ensures safety in case of any malfunction or risk. An interesting advantage of the eSAIL is that it is inherently safe, as failure modes cause the stop of the suction fan, removing all aerodynamic forces and taking the eSAIL to a safe rest position. It is, thus, fail-safe.

The system can be monitored from two different locations. The main one is via the touchscreen control panel located at vessel bridge. Additionally, a portable tablet, offering a remote desktop of the main touchscreen is available for line-of-sight eSAIL monitoring. Both locations also have an emergency stop push button to force an immediate stop of the system.

Figure 1: eSAIL Model 1 – 17m

3. TECHNOLOGY IMPROVEMENT

To achieve the improvement in performance described above, bound4blue has conducted extensive work to develop the necessary tools for aerodynamic optimisation of the eSAIL and their validation. Numerical tools such as CFD allow the exploration of the design space. The first step was to validate the CFD modelling techniques using experimental data of the Turbovoile. This was followed by 2D and 3D parametric studies of the suction sail to explore the design space and identify the configuration which provides the best lift-to-power ratio. Finally, a wind tunnel campaign was conducted to validate the results seen in the CFD 3D parametric study.

3.1 CFD VALIDATION

CFD tools are useless if they cannot be adjusted and calibrated to perfectly capture the reality. To do so, the results must be validated against experimental data.

In order to simulate properly all the components and the phenomena involving the eSAIL during its operation, several modelisation techniques were evaluated. The main goal was to validate the results against wind tunnel campaigns run in the past and improve the understanding of the influence of different parameters. Since the high angles of attack and wide range of apparent wind speeds, multiple types of turbulence models were evaluated along with the steadiness of the results. Turbulence was computed from two to seven equations models as well as laminar to turbulence transition was simulated in an unsteady setup. All the results were compared against the experiments. Moreover, the proper modelling of the suction area has been integral part of the CFD studies such that a good matching with pressure distribution was achieved. The study considered 2D simulations where the grid was fully meshed and different fidelity models where the pressure and the shear distribution effects on the aspiration zone were considered.

The final choice of best CFD model setup came from a deep understanding of the eSAIL variables sensitivity on its operational performances. The geometry definition comprising the flap dimensions and the flap angle as well as the Reynolds number effect on the flow coefficient have been considered. Comparisons have been made with previous results at scale model obtained by Charrier et. Al [1].

As can be seen from the image below, on the left, a snapshot is presented of the complex flow field that is created in such a high lift pressure distribution at a certain AoA. On the right, the different simulation conditions/hypothesis creates quite different polar curves, represented by the different line's colours in the plot. The square markers represent the experimental points of the geometry in study.

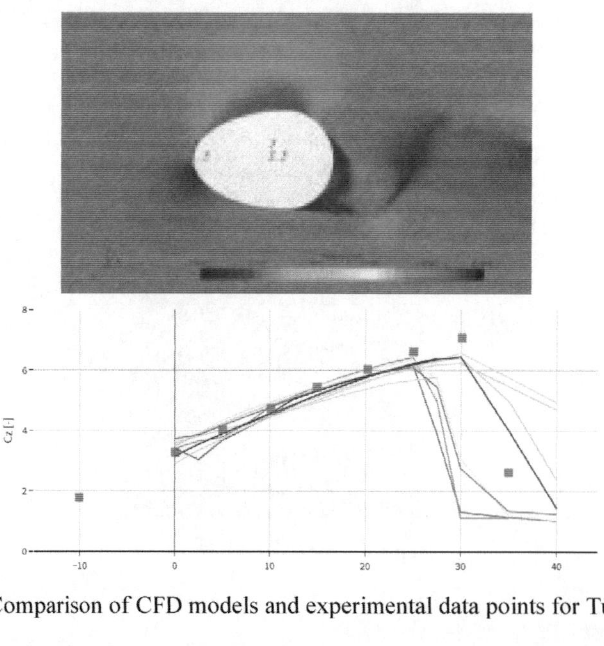

Figure 2: Comparison of CFD models and experimental data points for Turbovoile.

The best setup was selected as the one where the physical quantities are well predicted (lift curve slope and pressure distributions) but underpredicts the stall angle. This means that the CFD results will be conservative, providing lower results in CLmax for a fixed Ca. The conservative results means that performance estimation (and control and operation) of the eSAIL will be on the "safe side" and could realistically be higher.

3.2 PARAMETRIC CFD STUDY

Once the CFD setup was validated, a parametric design of experiments (DOE) was performed that included all the geometric and physics variables that can improve the eSAIL performance.

At first, a fully parametric geometry is created that could be correctly meshed and that would simulate rapidly in two dimensions. Then, parametrized the physics is parametrised input conditions such as the flow coefficient, Cq, and the suction area characteristics. Particular attention has been made in ensuring a robust level of details and a full post-processing comprised of wall shear stress and pressure distribution in all the different conditions.
The DOE allowed to create surrogate models of the eSAIL 2D performances and to evaluate top candidates according to operational criteria.

The optimum design variables found during the 2D study were then used as starting point to simulate a complete eSAIL in 3D. The goal of the 3D simulation was to add/consider important features like the winglets, the flap shape and position, the atmospheric boundary layer, the Reynolds number influence, the fan characteristics and all the local disturbances due to mechanical/structural limitations.

3.2 (a) 2D CFD Parametric Study

The 2D design of exploration has been done using 5 to 7 variables and permutating them with the Latin Hypercube Sampling method. The variables explored are listed below:

- Suction area size and characteristics
- Flap geometry and position
- eSAIL aerofoil shape
- Flow coefficient (Cq) / Power coefficient (Ca)
- Angle of Attack (AoA)

In the figure below, an example of a scatter matrix plot (SPLOM) of the DOE can be seen. By having as rows and columns the variables and the objectives, it is possible to identify trends and area where the eSail is performing well.

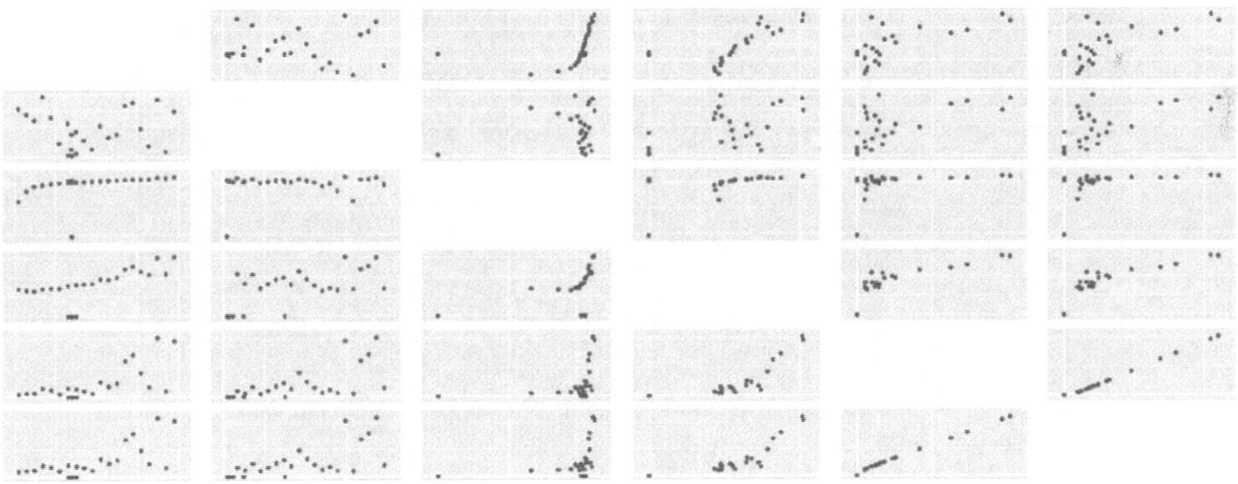

Figure 3: SPLOM of design variables and objectives

The simulations are also compared one to each other looking at the polars and the pressure distribution at specific AoA. An example of the influence of the suction area position can be seen from the image below.

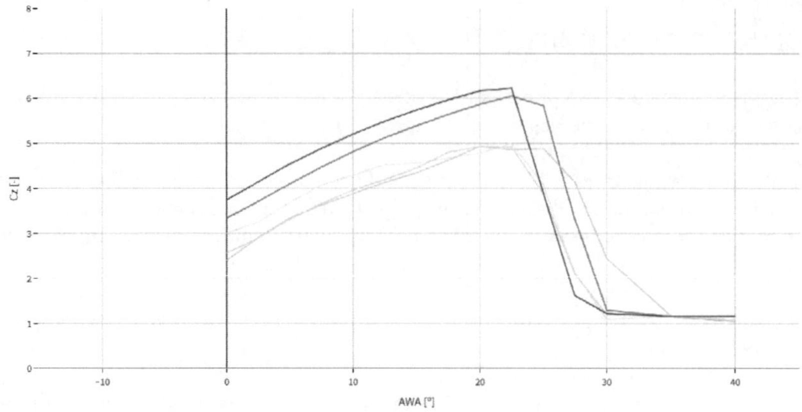

Figure 4: Effect of varying suction area position in 2D CFD.

3.2 (b) 3D CFD Parametric Study

The 3D studies focused on adding complexity to the model and understanding the 3D aerodynamic effects of some components, including:

Fan Selection: Based on the results obtained in the 2D simulations, a fan is selected that will allow enough suction of the profile boundary layer and, therefore, guarantee very high lift conditions. The fan variables are identified as fan diameter, number of blades, pitch of the blade and number of poles. Each fan has a performance curve that is usually provided as volumetric flow versus pressure jump. The setup allows to evaluate the output power needed by the eSAIL to effectively produce energy and driving force.

The atmospheric boundary layer, ABL, is a model that takes into account the true wind speed that has a logarithmic profile due to the wind fetch on the sea and the height of the wind sensor. Along with the ship speed, the ship heading and the true wind angle, it is possible to introduce the ABL profile as velocity inlet condition such that the 3D simulation considers the apparent wind twist effect.

Local disturbances. One of the most innovative and challenging evaluation that can be done with the 3D simulation concerns the local disturbances in the suction area and the internal structure. These local disturbances are mainly generated by structure-related elements. This allows to perform an aerodynamic-structure design optimization of the eSAIL.

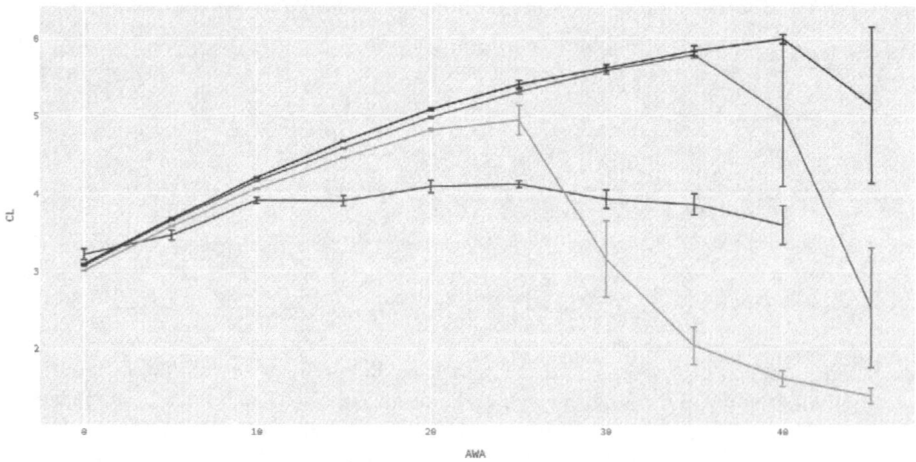

Figure 5: Effect of local disturbances on lift curve slope.

Winglet. Also, an important feature such as the winglet shape and positions were extensively simulated. By improving the design of this eSAIL component, the maximum lift coefficient and its angle of attack can both be increased considerably, with variations of more than 30% in CL.

3.3 WIND TUNNEL VALIDATION

The CFD methodology had been verified against existing Turbovoile data. A new wind tunnel testing campaign was conducted to provide more validation points and to confirm the findings of the CFD optimisation study. The wind tunnel testing campaign was conducted in the S10 wind tunnel at Institut Aerotechnique, at Saint Cyr, France in September 2022. The S10 is an atmospheric double return tunnel capable of wind speeds up to 55 m/s and a test section of 3m x 5m x 10m (height, width, length).

The model consists of a main body with a chord of 400mm and a span of 2m with interchangeable components: flap, porous plate, and winglets. The model was attached to a turntable, used to vary the angle of attack of the model. Forces generated by the model were measured using a 6-component force balance attached to the turn table. Suction was achieved by the installation of fans on the top of the model, exhausting the air vertically into the test section. Control of suction was achieved through control of the fan RPM, calibrated separately. The model was instrumented with 96 pressure taps to obtain internal and external pressure distribution and chordwise pressure distributions. Figure 6 shows an example configuration of the model installed in the wind tunnel.

Figure 6: Wind tunnel model in a test configuration with large winglets.

The test plan covered the same parametric studies conducted in the CFD optimisation, with the aim of validating the results obtained, including:

- Reynolds number (up to 800.000)
- Suction area porosity
- Suction area position
- Winglets
- Effect of local disturbances in suction area
- Range of suction coefficients, Cq.

3.3 (a) Wind Tunnel - CFD Validation

The data obtained from the wind tunnel testing campaign was used to validate the CFD modelling technique in two ways: comparison of the lift curve slope and a comparison of the pressure distributions. To conduct a fair comparison, the wind tunnel data is compared to the CFD of an eSAIL with the same aspect ratio and configuration.

Figure 7 a) and b) shows a comparison of the lift polars of the wind tunnel and CFD for various Ca values. Both show a good match between the data with varying results in predicting stall, this would suggest a dependency in Ca. As indicated in section 3.1, stall angle is underpredicted by CFD, providing conservative results of maximum lift coefficient achievable.

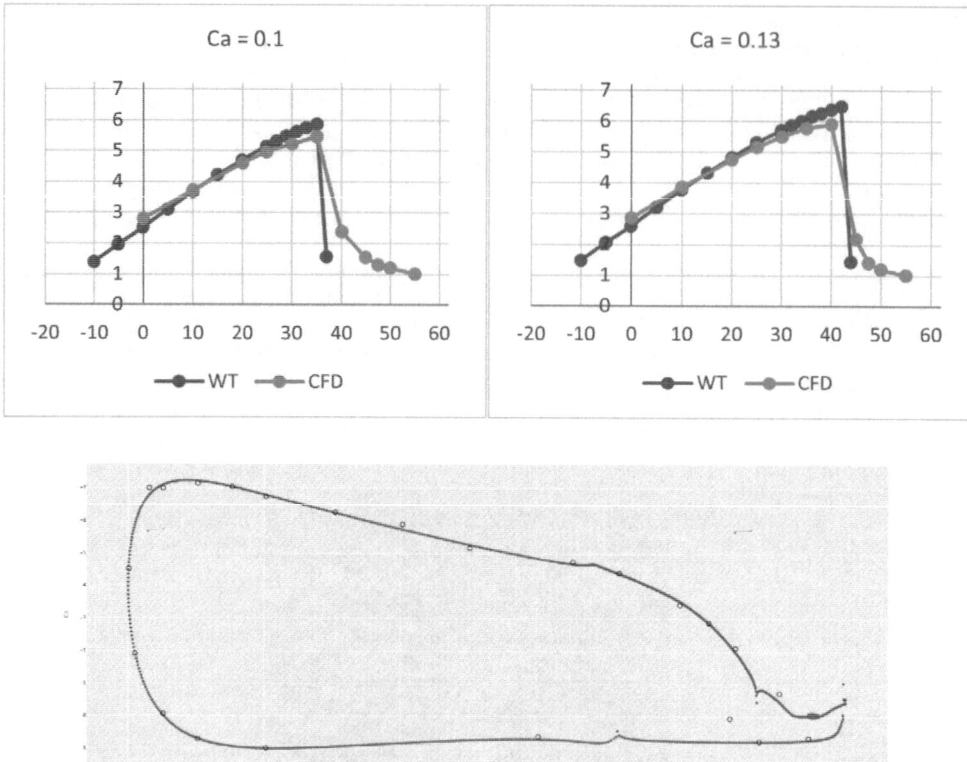

Figure 7: Comparison of wind tunnel and CFD lift polars for aspect ratio 5 eSAIL a) and b) show a comparison of the lift polar at different values of Ca and c) shows the Cp distribution at an example angle of attack.

Figure 7 c) shows an example of the experimental (markers) versus the CFD (line) chord pressure distribution for a given configuration at one example angle of attack. The experimental data presented here is at the mid-span location on the model. An excellent correlation between the numerical and experimental distributions is observed, re-validating that the modelling technique is accurate.

3.4 ESAIL PERFORMANCE

The strong match between the experimental and numerical models have been achieved for an eSAIL with an aspect ratio 5, inviting to a high confidence in the numerical modelling techniques. These can now be unleashed to carry out more design optimisation and crucially, to provide the aerodynamic performance of the eSAIL.

The performance of a larger and preferred eSAIL with an aspect ratio of 6 can be examined. Figure 8 shows the polars obtained for an optimised eSail from CFD analysis compared to the data from the Turbovoile, where an increase in CL of up to 20% for the same aerodynamic power coefficient, Ca, can be seen. Note that these improvements are considering the "conservative" results provided by CFD. The Turbovoile data is taken from the figure in page 1.47 from Malavard [3].

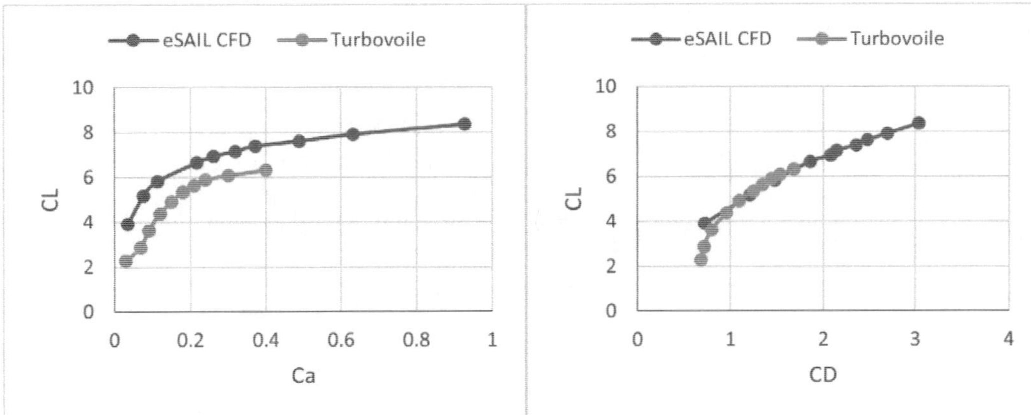

Figure 8: Polars for an eSAIL with an aspect ratio of 6

4. DETAILED STUDIES FOR EACH ESAIL INSTALLATION

The high confidence in the numerical modelling techniques means that it is now possible to add complexity to understand the effects of realistic conditions on the performance, instead of idealised conditions. Examples include: installation height, the effect of the atmospheric boundary layer, sail to sail interaction and ship-sail interaction. This work allows to fully predict the conditions experienced by the sail on each ship and therefore provide a custom control algorithm to maximise the savings offered by the eSAIL. Some examples of the studies that have been carried out are as follows.

The onset flow onto a sail on a ship is not uniform. The flow has a vertical speed gradient, since the apparent wind speed and direction changes with height. Using real ship data, it is possible to model the atmospheric boundary layer for the operation conditions that the eSAIL will encounter. This allows to examine the effects on the performance of the eSAIL and therefore make improvements to the design.

For specific installations it is possible to analyse the sail-to-sail interaction. In this example, two Model 1 eSAILs are fitted at the stern of a ship (Amasus Eems Traveller) separated at a distance of 7.6m (centre to centre). Since both eSAILs are quite close to each other, a study was conducted to understand the aerodynamic effect the sails had on each other as a function of apparent wind angle (AWA). The images in Figure 9 show the pressure distribution at different AWA. The images are aligned with the ship, where the bow is to the left and the wind is simulated at different AWA.

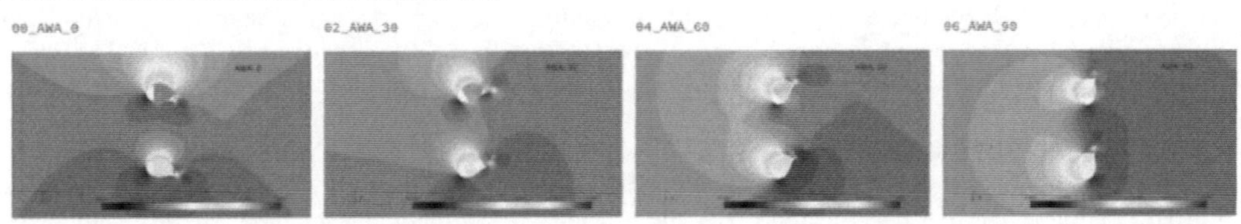

Figure 9: Pressure coefficient flow field of two eSAILs in proximity at different apparent wind angles

Figure 10 shows an example CFD simulation of a full installation for Amasus Eems Traveller. This kind of simulation is the culmination of all the results above, allowing to fully predict the performance in this specific case. A detailed analysis of these results allows to understand the influence each eSAIL has on the other and use that information to customise the control algorithm to maximise the performance of the systems for this specific installation.

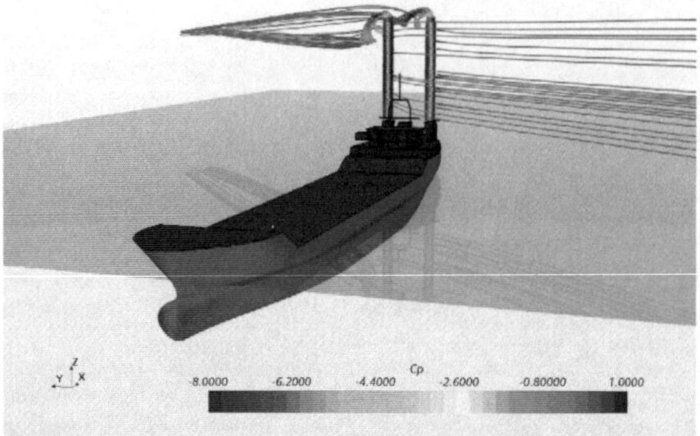

Figure 10: Example simulation of a full installation case study of two eSAILs on a ship.

5. FUTURE DEVELOPMENTS

The above sections show a glimpse of the tools and analysis bound4blue has developed to understand the physics of a suction sail and to accurately predict the performance of the eSAIL.

Section 4 gives an insight into the understanding of real-life conditions that the eSAIL is expected to operate in. Further CFD studies will be used to explores this further and improve the autonomous control algorithm and on a ship-by-ship case.

Further wind tunnel testing campaigns are planned to provide some additional validation points for CFD modelling. These include extending the CL vs Ca curve to the very high range of Ca points and interaction between eSAILs.

6. CONCLUSIONS

bound4blue is a WAPS (Wind-Assisted Propulsion System) provider which has developed the eSAIL, a fully autonomous suction sail. This paper showcases bound4blue's journey in using CFD to design the eSAIL, showing details at every step of the way from initial modelling all the way to complex simulations of realistic conditions that the sail is expected to operate in.

The first step was to validate the correct CFD modelling technique with available experimental data for Turbovoile. Changing the modelling technique leads to significant differences in data agreement. Finally, a model was selected where there is a very close match in the aerodynamic forces generated by the suction sail.

Once the CFD model was validated, extensive 2D and 3D parametric studies were carried out. Through these studies an optimum design was selected, showing an increase in performance of up to 20% when compared to the Turbovoile. To further validate and verify these results, a wind tunnel test campaign was conducted.

With the high confidence in the numerical modelling techniques, bound4blue is able to add complexity to understand the effects of realistic conditions on the performance, instead of idealised conditions. Including atmospheric boundary layer effects and sail to sail interaction. This allows bound4blue to predict the conditions experienced by the sail on each ship and therefore provide a custom control algorithm to maximise the savings offered by the eSAIL.

7. ACKNOWLEDGEMENTS

The authors would like to thank the team at IAT for their invaluable help in the wind tunnel test campaign of September 2022.

8. REFERENCES

1. CHARRIER, B. et al, 'Foundation Cousteau and Windship Propulsion', Journal of Wind Engineering and Industrial Aerodynamics, 1985.
2. MALAVARD, L., 'Un nouveau Propulseur Eolien de Navires', Comptes Rendus de l'Académie des Sciences (CRAS), Tome 1, n°1, 1984
3. MALAVARD, L., CHARRIER, B., 'Etude aerodynamique de dispositif a tres haute portance pour la propulsion eolinenne des navires', Fondation Cousteau, 1981

9. AUTHORS BIOGRAPHY

Alberto Llopis Pascual holds the current position of Lead Aerodynamics Engineer at bound4blue. He is responsible for aerodynamic development of bound4blue's technology, both numerical and experimental. Alberto holds a PhD in Aerospace Engineering in the field of experimental aerodynamics (University of Manchester, 2017). Previously, he spent 6 years working with University of Manchester, BAE Systems and a NATO task group researching and developing fluidic control effectors for aircraft, designing, and building the first aircraft to fully integrate and demonstrate this technology. Later, he also spent 2 years carrying out aerodynamic design of medical delivery drones. Alberto is co-author in papers presented at aerodynamics conferences related to active flow control and the impacts on system sizing and cost.

Giovanni Bailardi holds the current position of fluid dynamicist at Bailardi Engineering. He is responsible for hydro- and aero-dynamic simulation, design optimization and HPC orchestration. His previous experience includes America's Cup CFD simulation of sails, foils and hulls, fast planning yacht and rotating machinery. Lately is focusing on design optimization, data visualization and UX/UI development. He has published several papers in the marine and the aeronautic field, recently focus on fluid structure interaction of submerged wings (NAV2022 and HIES2022).

Bertrand Charrier, PhD in Physics, integrates in 2009 the French Development Agency (AFD) in charge of Innovations for Development. Prior to this, Bertrand Charrier worked for 15 years with Captain Jacques Cousteau, explorer and oceanographer. Charrier is the co-inventor of the Suction Sail, named Turbosail as commercial name, with Pr. Lucien Malavard and Jacques-Yves Cousteau. The Suction Sail is an efficient system for Wind Assisted Ship Propulsion (WASP) based of the cylinder-flap wing with boundary layer suction. Alcyone, ship of Cousteau Foundation was successfully equipped with two SS. Bertrand Charrier is founder of BC Consulting, specialized in Sciences and Environmental Policy is contributing to the research and development of WASP. Since 2021, Bertrand Charrier is partner of bound4blue Company, for the research and development of the eSail, name of the Suction Sail for commercial ships.

David Ferrer Desclaux is co-founder and CTO at bound4blue. He is responsible for the lead and management of the team in charge of the development of bound4blue's technology. He is co-author of three research papers related to the use of the wingsail technology for various applications and co-inventor of three bound4blue patents. David holds a MSc in Aerospace Engineering, with a specialisation in aircraft design and performances (Polytechnic University of Catalonia – Supaero, 2009); and a MSc in Project Management (La Salle – Ramon Llull University, 2016). In 2011 he attended the Space Studies Program (ISU) and has also attended several seminars on start-up management at IE Business School. Previously, he spent 6 years at zero2infinity, right after its foundation, acting as CTO and Project Manager, where he designed, built and tested Near-Space capsules and launched high-altitude balloons from a Spanish Air Force base. He is also co-author in papers related to microgravity's impact on humans and near-space manned flight.

PERFO: METHODOLOGY BENCHMARK FOR WIND ASSISTED PROPULSION SHIP PERFORMANCE ESTIMATION

J Bataille, C Blayo and **P Sergent** Bureau Veritas Solutions Marine & Offshore, France

SUMMARY

The maritime sector has seen a significant development of Wind Propulsion Assistance Systems (WASP) to act towards reducing the emissions induced by goods transportation at sea.
The installation of a WASP decreases the propeller required thrust but introduces significant side forces, modifying the attitude of the vessel and adding drift-induced resistance. It is unsuitable to account for the gains as a simple reduction of effective power.
The PERFO project develops a methodology to estimate wind-assisted vessel's performance. Its robustness is insured by CFD simulations-based validation process and a literature review. The project aims at providing ship owners with ship performance predictions.
It balances the hull's hydrodynamic response to the transverse force generated by the WASP, inducing a couple of drift and rudder angles. The interaction between the superstructure of the vessels and the WASP is also accounted for.
The open-source benchmark vessel ONRT has been used as a study case.

NOMENCLATURE

AWA	Apparent Wind Angle (deg)
AWS	Apparent Wind Speed (m s^{-1})
CFD	Computational Fluid Dynamic
IMO	International Maritime Organisation
LPP	Length between Perpendicular
ONRT	Office of the Naval Research Tumblehome
PPP	Power Prediction Program
TWA	True Wind Angle (deg)
TWS	True Wind Speed (m s^{-1})
VPP	Velocity Prediction Program
WASP	Wind Assisted Ship Propulsion

1. INTRODUCTION

The maritime industry has seen a significant development of the Wind-Assisted Ship Propulsion (WASP) to act toward reducing the global emissions induced by shipping.

Wind-propelled vessels were historically used for global maritime transport. But sailing ships have been replaced by steamships powered by coal and finally by large merchant ships consuming fuel. Hence, hulls are no longer designed to support side forces and the culture of planning routes depending on the most favourable wind conditions has almost disappeared.

Nowadays, while facing an oil crisis, both for saving fuel and to comply with the restrictive IMO indicators, ship brokers are forced to lower considerably the emissions of their vessels. Thus, ships have started to slow down and adding a WASP on board emerges as a promising solution.

Nevertheless, it should be stressed that introducing a modern sail on a large merchant ship is not equivalent to adding an engine: along with the thrust, WASP induces side forces which need to be balanced by a properly appended hull. Many other constraints (such as air draft, restricted visibility, weight, position of the centre of gravity, optimal wind window, limited space on deck, safety) make it complex to select the most suitable technology. Therefore, assessing the vessel power gain on the vessel's operating profile depending on the selected WASP selected and its aerodynamic characteristics has become a strategic issue.

While routing the course of a wind-assisted vessel, the adjustable variable is the power delivered by the engine. This solving approach is referred to as Power Prediction Program (PPP). This kind of tools provides weather routing optimisation process with ship performance in order to help shipbrokers' decisions.

Based on two existing PPP tools, xWASP and SEECAT, the PERFO project aims at developing an extensive methodology to assess the wind-assisted vessel's performance on its operating route.

PERFO is a collaborative R&D project between the French Oceanographic centre Ifremer, the naval architecture firm Stirling Design International and the engineering office Bureau Veritas Solutions Marine & Offshore. It enables to create the ideal configuration where the ship owner in need to modernize its fleet, meets the shared effort of the naval architect in close relation with WASPs developers and the engineering consultants with their broad spectrum expertise.

The extensive methodology developed here, aims at helping the shipbroker at any step of its WASP project progress. Extensive means as accurately as possible without mobilizing unnecessarily expensive computational resources. The accuracy of the performances delivered by a PPP tool mainly comes from the assessment of the loads acting on the vessel, provided as input data. Here, xWASP as a state-of-the-art tool, using fully empirical rapid methods is compared to the in-house SEECAT advanced solver provided with loads from CFD modelling. Both approaches will allow to characterise the vessel's response. They will be compared: qualitatively through forces polar curves and quantitatively in terms of performance gain averaged over a standard operating route.

2. PERFO, AN EXTENSIVE METHODOLOGY

PERFO aims at developing a methodology that helps the shipbroker at any step of its WASP project progress. Two PPP solvers have been benchmarked. On one hand, xWASP as an open-source state-of-the-art solver quickly applicable is using semi-empirical formulations. On the other hand, SEECAT as an advanced in-house tool is provided with loads from CFD modelling.

2.1 STUDIED CASE

2.1 (a) Ship description

The open-source vessel ONRT (Office of the Naval Research Tumblehome) is here used to build a reference study case. She is an academic benchmark vessel which has been introduced at the 2015 Tokyo CFD Workshop. This vessel presents a 154 m length, an 18.78 m width and a displacement of 8507 tons.

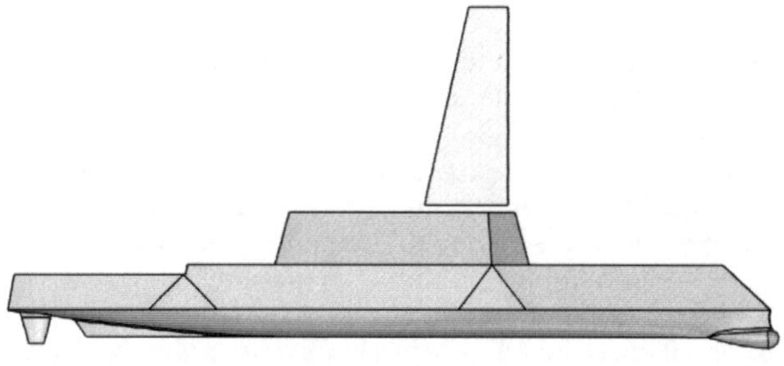

Figure 1: The ONRT with its sail

This choice has been made because of the extensive literature available. This vessel is documented in the workshop with many appendages. Here, the hull is considered with the skeg and the rudders. The parts below the waterline are referred to as "underwater hull".

The upper parts of the vessel including the superstructures and the hull above the waterline are referred to as "deadworks" in this paper.

2.1 (b) Traditional sail

A 500m² "traditional sail" is here used to build a reference study case. The sail's characteristics are taken as for the main sail from the ORC 2022 VPP report [2].

The location of the sail on-board the vessel was defined above the wheelhouse and as close as possible to the CLR (Center of Lateral Rotation) to minimise the yaw moment induced by the sail.

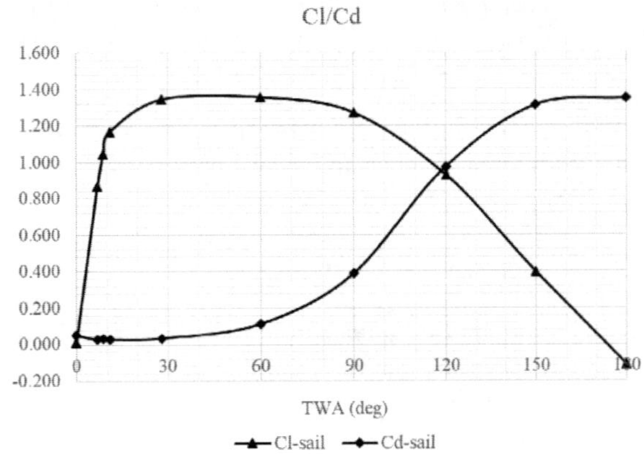

Figure 2: Aerodynamic coefficient of the sail

2.1 (c) Convention

The wind is assumed to come from the port side. The drift and the rudder angles are considered positive according to the following convention.

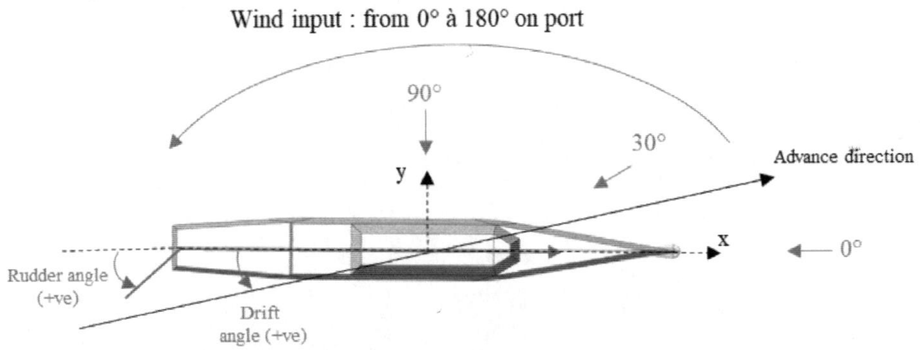

Figure 3: Coordinate system convention

2.2 ASSUMPTIONS AND LIMITS

At this stage of the methodology development, the following assumptions have been done:

- The forward speed is fixed and constant;
- The load balance is solved in static;
- Only horizontal planar motion is considered (the heel effect is ignored);
- Calm water hypothesis is made (no wave taken into account);
- The propeller efficiency modified by the load taken by the sail is ignored.

2.3 METHODOLOGY OVERVIEW

The methodology's global outcome is the gain of performance induced by having the WASP on board.

To begin, external forces, i.e. hydrodynamic and aerodynamic loads acting on the vessel, have to be carefully assessed. Depending on the level of confidence required at the project's current stage, it is possible to use quick and non-expensive empirical methods with xWASP, or an advanced model based on CFD simulations results, with SEECAT.

In this study, wind is accounted for. At this stage, there is no sea state. The aerodynamic load induced by wind on the sail and the deadworks is counteracted by the hydrodynamic response of the underwater hull. Under horizontal planar motion assumption, the balance of the side forces (Fy) and yaw moments (Mz) expressed in the vessel coordinate system enables to derive the angles of drift and rudder. These angles are used to assess the added resistance in leeway.

The resulting forces are transposed in the forward direction to derive the vessel's performance. Thus, for each specified operating condition, the part of the sail aerodynamic force contributing to the propulsion of the vessel in its equilibrium attitude is finally deduced.

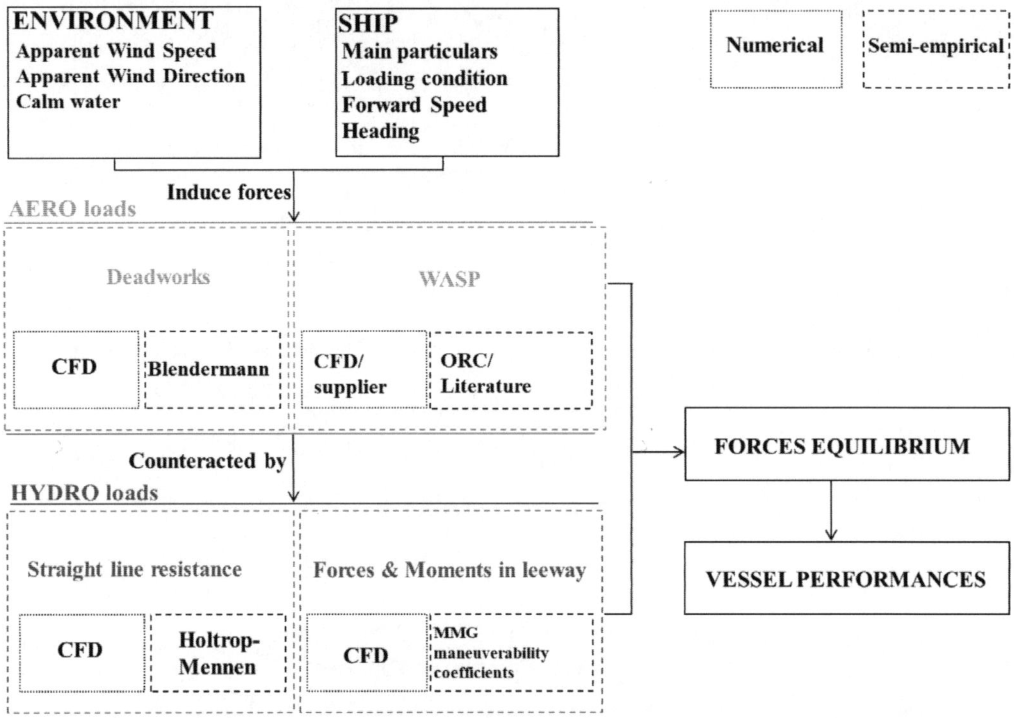

Figure 4: PERFO methodology

2.4 MECHANICAL SOLVERS DESCRIPTION

In the current application case, mechanical solvers are used in static mode. Wind is the only environmental load. The horizontal planar forces equilibrium is solved to balance the external loads on the wind-assisted vessel and deduce the 'sail' contribution to the propulsion.

2.4 (a) xWASP

xWASP is an open-source code developed during the PhD of Moran Charlou at the LHEEA (Laboratoire de recherche en Hydrodynamique, Énergétique et Environnement Atmosphérique, Ecole Centrale de Nantes).

It is a 6 DoFs Dynamic Velocity/Power Prediction Program dedicated to Wind-Assisted Ship Propulsion. It uses the xdyn ship simulator developed by SIREHNA for force computation and time integration. xWASP has statics and dynamics capabilities and is generally used to compute a ship's response to wind and sea conditions.

2.4 (b) SEECAT

SEECAT (Ship Energy Efficiency Calculation and Analysis Tool) is developed within Bureau Veritas Marine & Offshore since 2011.
It is a ship modelling platform (based on SimulationX) aiming at assessing the global efficiency of ships for realistic operational profiles and weather conditions. SEECAT is a modular tool capable of modelling any kind of marine propulsion architecture, thanks to an extensive library of models coded in the Modelica language. It is continuously developed to answer the shipping industry's need. Hence, it has been upgraded with a ship model to deal with wind assistance applications.

2.4 (c) Difference

Benchmarking both solutions is relevant here because xWASP is tested as the state-of-the-art quick tool based on available semi-empirical methods, whereas SEECAT is used as an advanced in-house tool easily customizable with CFD-based data.

xWASP only requires vessel's main particulars and is very cost efficient. SEECAT enables to take into account more physics as the entry data come from CFD simulations. However, these are more demanding in terms of validation process and computational cost.

The main differences lie in the underwater hull hydrodynamic response assessment: xWASP uses unsteady manoeuvrability coefficients while SEECAT uses steady CFD database.

3. LOADS ASSESSMENT

This study dedicates a particular effort to load assessment. The loads acting on the ONRT are provided in xWASP or SEECAT, as blocks entry data: the block characterising the sail, the one characterising the deadworks etc. Their number is not limited. This way, the model can account for more or less physics.

In this section, the methods used to get the block's entry data are presented. In most cases, a CFD-based approach is presented in parallel to an empirical one.

3.1 SEMI-EMPIRICAL APPROACH

Existing semi-empirical models enable to quickly approximate the external loads applying on vessel. Hypotheses and applicability of these models are major concerns of this study.

3.1 (a) Underwater hull (calm water) resistance in straight line

The ship resistance in a straight line can be empirically estimated using Holtrop and Mennen's method [2]. It is one of the most common methods when it comes to displacement ship resistance predictions. It is a semi-empirical method based on the regression analysis of a vast range of model tests and trial data. Within this method, the total resistance is split into several components: frictional resistance, appendages resistance, wave resistance due to the bulbous bow near the water surface, and pressure resistance due to the immersed transom. Its accuracy varies significantly depending on the similarity between the hull assessed and the hulls used for the regressions. It is generally estimated to be between 0 and 20% accuracy.

3.1 (b) Underwater hull hydrodynamic loads in leeway

An additional force model is needed to account for the hydrodynamic loads applying on the underwater hull in leeway. As a quick approach, manoeuvrability coefficients derived from semi-empirical models can be used. The MMG (Manoeuvring Modelling Group) standard method [3] is usually taken as reference.

3.1 (c) Deadworks aerodynamic loads

The vessel's deadworks are usually not modelled for calm water resistance assessment. However, its windage area induces a significant additional drag/lift with the wind characteristics at stake. Blendermann built an extensive database of aerodynamic coefficients extracted from wind tunnel tests and suitable for many types of vessels [4]. Blendermann's approach is used here to model the deadworks loads without requiring to numerical models.

3.2 UNDERWATER HULL HYDRODYNAMIC CFD DATABASE

CFD calculations on underwater appended hull have been performed to build a hydrodynamic database.

3.2 (a) Validation in straight line

Prior to modelling the vessel at full scale, a model-scale CFD model has been set up with STAR-CCM+ software in order to calculate the underwater hull calm water resistance in a straight line. The RANSE (Reynolds-Averaged Navier-Stokes Equations) model is set up as follows:

- k-ω SST turbulence model;
- 15 million cells unstructured mesh with a boundary layer;
- unsteady computations;
- free heave and pitch;
- volume of fluid method at water-air interface;
- model-scale boundary layer target Y+ = 0.5
- full-scale boundary layer target Y+ = 200

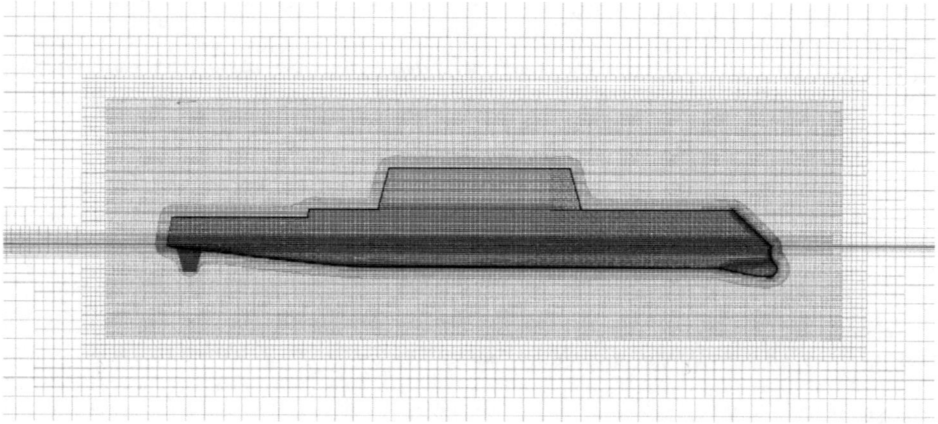

Figure 5: Hydrodynamic set up mesh

The model has been calibrated at model scale, with Lpp = 3.147 m, which corresponds to a scale factor of 48.94. The vessel's speeds tested are 1.11 and 1.67 m/s, respectively corresponding to Froude 0.2 and 0.3.

The literature [5] highlights discrepancies between EFD & CFD Benchmarks. These deviations could be explained by some towing tank experimental set-up features that were not documented in [5], such as basin water temperature, degree of turbulence measured in the towing tank or model roughness. Therefore, it has been chosen to focus the comparison of the developed model setup with the CFD results available in [5]. The agreement with these being satisfying, the above-numerical set-up has been used for full scale hydrodynamic database computation.

3.2 (b) Full scale underwater hull hydrodynamic loads in leeway

Drift and rudder angles are introduced in the previously described CFD model to extract the hydrodynamic loads on the underwater hull in leeway and are presented in Table 1.

Table 1: Underwater hull hydrodynamic loads.

Vessel Speed	Drift Angle	Rudder Angle	Fx	Fy	Mz
knots	deg	deg	kN	kN	kNm
12.0	0.0	0.0	-146.3	-7.8	-364.0
	0.0	5.0	-151.8	186.9	-12 715.7
	5.0	0.0	-137.1	419.6	7 779.8
	5.0	5.0	-144.8	626.9	-5 781.5
	5.0	-5.0	-141.8	209.5	21 146.0

Vessel centreline symmetry is used to symmetrize the forces and moments. Thus, the resulting matrix of 3 drift angles and 3 rudder angles loads enables to build a response surface of force and moment applied on the underwater hull in order to:

- derive the drift and rudder angles that balance the hydrodynamic Fy/Mz couple induced by wind load on vessel;
- deduce the resistance added by these drift and rudder angles.

3.3 AERODYNAMIC CFD DATABASE

With a WASP installation on board, there is a real added value to accurately capture the aerodynamic loads.

Additional loads due to the deadworks windage and the degradation of the wind profile in the vicinity of the WASP are studied to quantify their influence on the sail's performances.

3.3 (a) Deadworks aerodynamic loads

A full scale CFD model has been set-up to accurately assess the aerodynamic loads induced by wind on the deadworks. The wind is modelled by a logarithmic profile (corresponding to the 1-hour mean wind speed from NPD spectrum), accounting for the speed reduction due to skin friction at the free surface. The reference speed is 30 knots at 10m above the free surface. Wind incidences are set from 0° to 180°, every 30°.

The RANSE (Reynolds-Averaged Navier-Stokes Equations) model is set up as follows:

- k-ω SST turbulence model;
- 18 million cells unstructured mesh with boundary layer;
- unsteady computations;
- 0-DoFs motion.

The resulting forces and moments are used to build polar coefficients to assess additional aerodynamic loads due to deadworks. Table 3 gathers aerodynamic coefficients derived from Blendermann formulations and from CFD modelling.

Table 2: Deadworks Aerodynamic coefficients

	Empirical: Blendermann			Numerical: CFD		
TWA	Cx	Cy	Cmz	Cx	Cy	Cmz
0	-0.600	0.000	0.000	-0.255	-0.001	0.000
30	-0.560	-0.408	-0.096	-0.020	-0.622	-0.101
60	-0.382	-0.738	-0.104	-0.115	-0.830	-0.094
90	0.000	-0.871	-0.040	-0.383	-0.770	-0.022
120	0.382	-0.738	0.036	-0.202	-0.800	0.060
150	0.560	-0.408	0.058	0.425	-0.492	0.056
180	0.600	0.000	0.000	0.520	0.014	0.001

3.3 (b) Degradation of wind profile in the deadworks's wake

WASP technologies performance is usually characterized by wind tunnel tests, i.e. under an infinite constant speed. Nevertheless, it is well known that an air wake develops in the vicinity of a ship moving forward, degrading the wind profile seen by the WASP onboard.

Wind speed profiles calculated at the WASP location has been extracted from the CFD aerodynamic model, for each wind direction. Variation between degraded and theoretical wind speed profile is plotted in percentage versus the wind incidence in Figure 5. Corrected aerodynamic drag and lift coefficients are also derived.

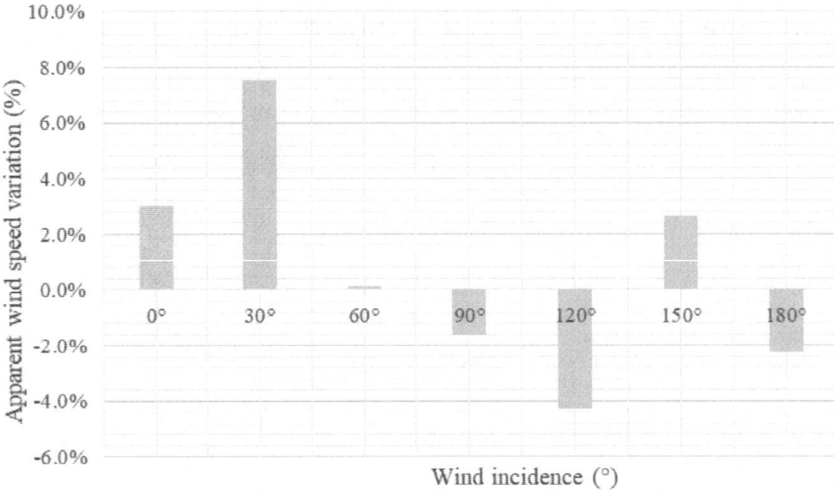

Figure 6: Percentage of wind profile degradation

To go further, it would be worthwhile to model properly sail-deadworks interaction by including the sail in the CFD modelling.

4. RESULTS

The studied case aims at applying this methodology to a first reference case.

First, from a comparative point of view, forces acting on the vessel are plotted depending on the TWA, in the vessel's coordinate system. Focus is made on aerodynamic side force Fy (along with associated drift angle or yaw moment Mz) and underwater hull longitudinal response Fx.

Finally, to quantify these influences, the sail's aerodynamic contribution to the propulsive load needed in the forward direction is calculated for each test. All the tests are summarized in terms of performance gain averaged over a standard operating route.

4.1 Solvers comparison

The ONRT is set to sail in 30 knots wind, at 12 knots, with wind coming from port starting at 0° (headwinds) to 180° (tailwinds) and a sail area of 500 m^2.
To make a relevant initial comparison, both solvers are provided with:

- the same sail coefficients from ORC,
- the same resistance curve issued from the CFD analysis (instead of Holtrop-Mennen for xWASP),, at 12 knots, the corresponding value is presented in Table 1 for 0 degree rudder and drift angles.

The only difference in the entry data are the following ones:

- xWASP uses the manoeuvrability coefficients (grey curves);
- SEECAT uses the matrix of forces and moments issued from CFD computations (black curves).

On the left-hand side graph, the side forces induced by the sail are plotted, with the associated drift angles. As the input sail coefficients are the same, rebuilt aero forces are identical. However, the resulting drift angles are significantly different.

On the right-hand side, the plot shows the forces to be compensated by the propeller at each wind angle.

$$Fx_{Hydro} = Fx_{sail} + Fx_{deadworks} + Fx_{resistance\ (12kn)} + Fx_{added\_DriftAndRudder}$$

A difference in Fx_{Hydro} is observed for TWA up to 100°, which is linked with the difference in the resultant drift angles. It can thus be assumed that xWASP underestimates the added resistance in X direction. At TWA higher than 100°, the drift angles become very small and hence so does the difference between the resulting hydrodynamic forces.

In comparison, xWASP underestimates the drift angle and overestimates the added resistance. This is most likely because these are calculated based on manoeuvrability coefficient. Manoeuvrability coefficients are known to have poor consistency for small drift angles. This will have to be confirmed by completing a CFD matrix with smaller drift angles (or with another case resulting in higher drift angles).

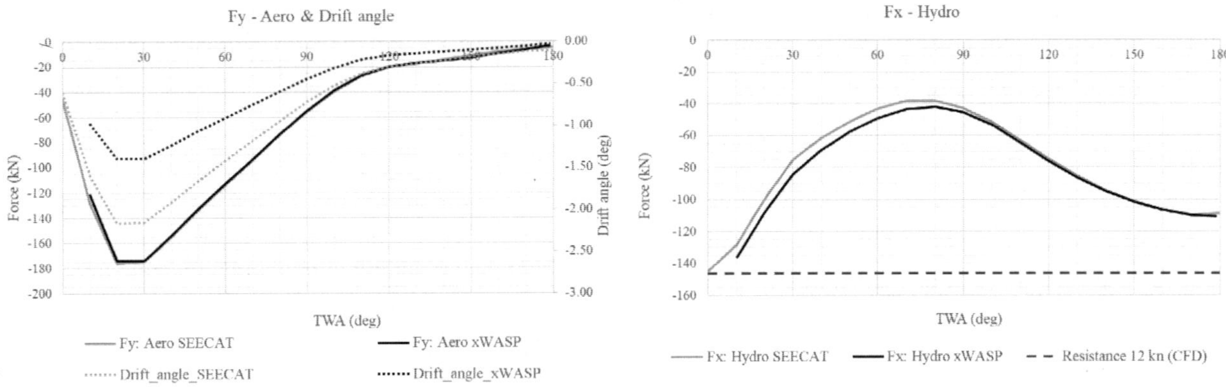

Figure 7: Efforts in the vessel's X and Y axis - no deadworks

4.2 Accounting for the deadworks aerodynamic loads

Depending on the project progress, it might be convenient to neglect the efforts induced by the deadworks's windage. The graphs below show the impact of accounting for deadworks on the resulting forces and moments, without and with the sail on board.

In addition, it was mentioned that the deadworks air wake affects the wind profile seen by the sail. Below, the modified sail characteristics are presented under "sail with degraded profile" notation. On the left, both Fy and its associated Mz moments are represented. Fy in black and Mz in grey.

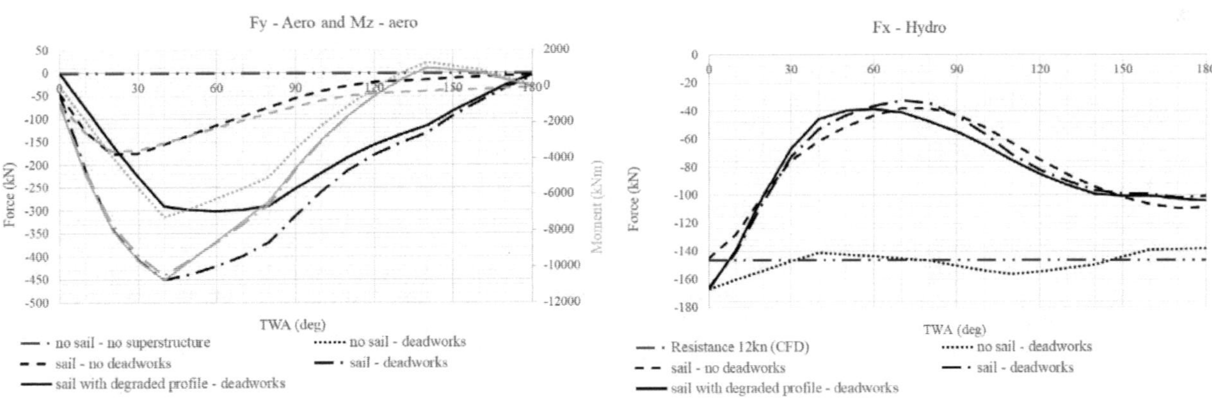

Figure 8: Efforts in the vessel's X and Y axis

The deadworks have a significant effect on the side forces and yaw moment. Hence deadworks-induced efforts cannot be neglected, not even at early-stage performance analysis.

Concerning the impact of the wind profile degradation, the yaw moment and side forces are barely affected in this case. As for the forces along the X-axis Fx_{Hydro}, for TWA higher than 60°, the degraded profile case shows a higher magnitude. This means that the drive force generated by the sail is lower, as expected.

4.3 Empirical approaches compared to numerical modelling data

4.3 (a) Aerodynamic of the deadworks

As the deadworks loads cannot be neglected, it might be useful to assess if an empirical method enables to reach a satisfying approximation. Aerodynamic coefficients have been derived from Blendermann's method.

The model provided with Blendermann coefficients (grey curves) is compared to the one provided with coefficients from CFD aerodynamic database (black curves) in the graphs below.

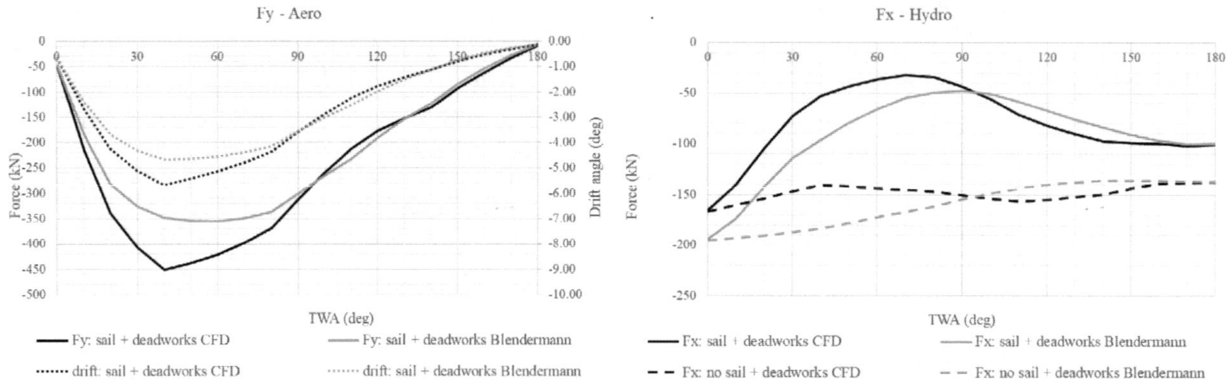

Figure 9: Efforts in the vessel's X and Y axis - empirical vs numerical deadworks loads

If we take the results obtained from aerodynamic CFD model as the reference, it appears that Blendermann's approach well captures the tendency of the side forces (Fy) but underestimates their magnitude at low TWA.

In the X direction, for TWA up to 90°, Blendermann's coefficients overestimate the added resistance due to the deadworks.

4.3 (b) Calm water resistance

In a similar way, the ship resistance can be approximated empirically using Holtrop-Mennen's method. At 12 knots, it results in a 156.1 kN resistance, i.e. 7% higher than the one obtained in CFD (146.3 kN).

4.4 Gains provided by the WASP

Finally, a WASP benefit will be assessed on the propulsive gain over the vessel's operating route.

The MEPC (Maritime Environment Protection Committee) has established wind occurrence probabilities on the main maritime routes [7]. For every combination of TWS/TWA, the associated occurrence probabilities have been used to derive statistics of gain based on standard wind conditions. To remain realistic, an arbitrary 40 knots operational limit of apparent wind was set.

Both solvers balance the loads according to the X and Y axis linked to the vessel; but here, the forces equilibrium is expressed in the flow direction. Hence, the "gain" accounts for the sail aerodynamic contribution to reduce the power that the mechanical propulsion must deliver.

Table 3: Overall gains assessment

Case	Solver	Resistance	Response to side forces	Deadworks	Degradation wind profile	Gain
Case 1 : Fully Empirical	xWASP	H-M	Coef. Manoe.	Blendermann	No	9.0%
Case 2 : Fully Numerical	SEECAT	CFD	CFD	CFD	Yes	7.5%
Case 3	SEECAT	CFD	CFD	CFD	No	7.8%
Case 4	SEECAT	CFD	CFD	No	No	11.2%

Case 1 is the fully empirical case. When compared to the fully numerical Case 2, it appears less conservative, even if the gain in % is affected by the 7% overestimation of resistance. Case 2 to 4 shows the effects of accounting for each parameter on the overall gain. It appears that accounting for more physics tends to lower the gain.

5. CONCLUSIONS

In this paper, a methodology is described that assesses WASP performances.

The results of a first reference case results have been presented. The following conclusions can already be drawn:

- While working on Wind Assisted Ship performance predictions, accounting for the deadworks is not negligible;
- Blendermann's approach allows a satisfying approximation of the deadworks aerodynamic loads at early-stage study;
- Concerning the response to side forces, the manoeuvrability coefficients calculated empirically are a very rough approach. Manoeuvrability coefficients from sea trials or CFD modelling would be more accurate.
- The resistance estimated using Holtrop-Mennen method is reasonable for an early-stage estimation. However, it is only applicable to "traditional" hulls. In the case of a modern hull, it would be recommended to at least perform one point of CFD resistance calculation to calibrate the Holtrop-Mennen results;
- Accounting for wind profile degradation is the first stage of a more global ship-to-sail interactions model. Further developments on that topic are currently being undertaken and could not be disclosed within that paper.

These conclusions are intended to be consolidated through further work within PERFO project and thanks to other more realistic study cases.

6. ACKNOWLEDGEMENTS

This study was supported by the region Pays de la Loire in France. The authors would like to thank the PERFO project partners: Ifremer and Stirling Design International.

7. REFERENCES

1. ORC, 'VPP Documentation 2022', *Offshore Racing Congress*, 2022.

2. HOLTROP, J., MENNEN, G.G.J., 'An Approximate Power Prediction Method', *International Shipbuilding Progress*, 1982.

3. TAIMURI et al., 'A 6-DoF manoeuvring model for the rapid estimation of hydrodynamic actions in deep and shallow waters', *Ocean Engineering*, 2020.

4. BLENDERMANN, W., 'Wind loading of ships: Collected data from wind tunnel tests in uniform flow', *Inst. für Schiffbau*, 1996.

5. DELEN, C., BAL, S., 'Uncertainty analysis of numerical and experimental resistance tests for ONR Tumblehome', *Sustainable Development and Innovations in Marine Technologies,* 2019.

6. ITTC – Recommended Procedures and Guidelines, 'Resistance and Propulsion Test and Performance Prediction with Skin Frictional Drag Reduction Techniques', *International Towing Tank Conference*, 2017.

7. IMO, 'MEPC 76/6/2 – Energy Efficiency of ships', *International Maritime Organization*, 2013.

8. AUTHORS BIOGRAPHY

Juliette Bataille holds the current position of hydrodynamic engineer at Bureau Veritas Solutions M&O. She is responsible for the development of methods to improve performance prediction of wind-assisted vessels. Her previous experience includes ship design and modern sails integration on-board vessels.

Camille Blayo holds the current position of senior hydrodynamic engineer at Bureau Veritas Solutions M&O. She is responsible for supervising the development of methods to improve performance prediction of wind-assisted vessels.

Pierrick Sergent holds the current position of Head of Fluid Dynamics Department at Bureau Veritas Solutions M&O. As CFD & Ship performance assessment specialist, he is also responsible for supervising the development of methods to improve performance prediction of wind-assisted vessels.

BUILDING TRUST IN THRUST

J. Kuuskoski and **V. Paakkari**, Norsepower Oy Ltd, Finland

SUMMARY

This paper discussed the status of the emerging wind propulsion market and describes Norsepower's experiences of entering the shipping market with a modernized Flettner rotor, the Norsepower Rotor Sail™. Recommendations and proposals for collaboration are made to accelerate the growth of the wind propulsion market. The benefits, features and performance predictions of various available technologies should be marketed with realistic expectations and claims for performance and preferably backed by third-party verified measurement results.

1. STATUS OF WIND PROPULSION MARKET DEVELOPMENT

Interest in wind propulsion technology is growing rapidly and clear signs of increased activity are visible on the market. This paper outlines some key topics and hurdles related wind propulsion projects and technology acceptance which are encountered in discussions with customers, authorities and other stakeholders. The wind propulsion market includes various suppliers offering different sail technologies. This variety and the relative novelty of the technologies leads to questions related to ship design, fuel saving performance, reliability, safety, and many other operational and regulatory issues. Technology acceptance is a must for wider application of wind propulsion for ships and suppliers have a common interest in facilitating the positive development.

2. MAIN PRODUCT: NORSEPOWER ROTOR SAILS™

Norsepower was established in 2012 with the purpose of modernising the Flettner rotor concept. The result of the modernised mechanical sail is the Norsepower Rotor Sail™, a modern auxiliary wind propulsion system for ships. The target of developing a low-maintenance, easy to use, safe and reliable mechanical sail was achieved. The Rotor Sail provides a ship with propulsive thrust in favorable wind conditions enabling fuel savings and emissions reduction. Norsepower's first Rotor Sail has been in operation since 2014, proving the technical solutions to be reliable and efficient.

Extensive measurement campaigns verified by third-party specialists have built confidence in the fuel saving performance and convinced a number of customers to install Norsepower Rotor Sails on their vessels. Information of references and the results of the fuel saving performance are available for interested parties.

Demand for Norsepower Rotor Sails has increased and we are experiencing strong growth with installation projects ongoing for various ship types in Asia and Europe.

The discussions with ship owners, ship designers, shipyards and other stakeholders have highlighted the need for consistent communication related to wind propulsion systems, their performance and operational characteristics. The relative novelty of modern wind propulsion systems is still causing hesitation and uncertainty with customers. To overcome such hesitation, positive customer experiences and verified performance of systems in actual operating conditions is required.

3. NORSEPOWER'S KEY REFERENCES

Norsepower's experience with the ship owners and shipyards as well as many other stakeholders in the shipping and shipbuilding industries is that there is a strong demand for proven solutions to improve energy efficiency and reduce emissions from ships. Norsepower has learned and continues to learn from our customers.

The following are short descriptions of ships which have been installed with Norsepower's Rotor Sails. The fuel saving performance is noted indicating the third-party verifier where applicable.

RoRo ship M/V Estraden

The first installation of model 18x3 (rotor height of 18 m and diameter of 3 m) Rotor Sails on the roro-vessel MV Estraden (Fig. 1) has been in operation since 2014. Fuel saving 6.1% on average, verified by NAPA.

Figure 1: Bore's roro vessel MV Estraden.

Cruise ferry Viking Grace

The installation of one model 24x4 Rotor Sail on cruise ferry Viking Grace (Fig. 2) was completed in April 2018 and the three year test period ended in 2021. Annual fuel saving approximately 300 tons, verified by ABB and NAPA.

Figure 2: Viking Line's cruise ferry Viking Grace.

LR2 tanker Maersk Pelican

The installation of two model 30x5 Rotor Sails was completed in the end of August 2018 on the Maersk Pelican (Fig. 3), a 109 000 DWT LR2 tanker owned at the time by Maersk Tankers. Fuel saving 8.2% on average, verified by LR.

Figure 3: LR2 tanker Maersk Pelican.

Hybrid ferries M/V Copenhagen and M/V Berlin

The installation on the M/V Copenhagen (Fig. 4), a hybrid ferry owned by Scandlines, was completed in the end of May 2020. The same model 30x5 Rotor Sail was installed on the sister vessel M/V Berlin in May 2022. Fuel saving 4% on average, verified by SSPA.

Figure 4: Scandlines hybrid ferry M/V Copenhagen.

Ro-Ro vessel SC Connector

Two model 35x5 Rotor Sails, fitted on tilting foundations, were installed in Dec. 2020 on the Ro-Ro vessel SC Connector (Fig. 5), owned by Sea-Cargo. The ship's fuel consumption and emissions are estimated to be reduced by 25% in its operation in the North Sea region.

Figure 5: Sea-Cargo's roro ship SC Connector.

Very Large Ore Carrier (VLOC) Sea Zhoushan

Five model 24x4 Rotor Sails, fitted on tilting foundations, were installed in April 2021 on the newbuilding VLOC Sea Zhoushan (Fig. 6) delivered from New Times Shipbuilding Co., Ltd. The ship is owned by Pan Ocean Ship Management and chartered by Vale. The ship's estimated average fuel savings are 8% and annual reduction of CO_2 is 3400 tons.

Figure 6: Pan Ocean's VLOC Sea Zhoushan.

4. SUGGESTED GUIDELINES FOR TECHNOLOGY PROVIDERS

This paper is intended to raise awareness of the main topics and questions encountered when preparing ship designs to apply wind propulsion as an auxiliary or main propulsion system. In addition, some general guidelines and recommendations are introduced for the shipping industry stakeholders to accelerate the application of wind propulsion in the global shipping fleet.

Examples of topics where collaboration and guidelines are needed:

- Technology provider collaboration to open the market for wind propulsion
- Collaboration for developing ship design practices and classification rules
- Realistic performance predictions and measurements
- Performance prediction methods and tools
- Safety related matters
- Highlighting the market potential in communication

The existing rules and regulations for ship design and operation are in development to accommodate features specific for various wind propulsion technologies. Such work has been ongoing already for several years and will continue as more data is collected from operating installations. There is a strong regulatory push to improve energy efficiency and reduce the emissions of shipping. A well designed wind propulsion installation saves fuel, reduces emissions and is an economically profitable investment. Sails are very likely enablers for alternative fuels which compete with traditional fossil fuels. Active collaboration of technology providers to promote development of safe and efficient regulations, rules and operational practices is a benefit for the growing wind propulsion market and a great benefit for our planet.

5. TECHNOLOGY PROVIDER COLLABORATION TO OPEN THE MARKET FOR WIND PROPULSION

Wind propulsion is one of the few technologies that has potential to provide shipping with two-digit emission reductions. The first installations are gaining more and more running hours, academia is producing more research, technology providers are finetuning their products.

At Norsepower alone, the past 12 months have shown rapid developments: The current order backlog of Norsepower includes six ship installations, which will double the number of ships sailing with Norsepower's Rotor Sails. An accelerated ramp up of production and first repeat orders signal growing confidence and acknowledgement of the benefits of the product.

Significant increase in inquiries for installation projects indicates growing acceptance of wind propulsion technology among ship owners, charterers and shipyards. With the increasing demand, also the competition between technologies and technology providers is expected to become more active. However, at the same time, it is important to recognize the fact that the technologies are still relatively new and there are many common challenges. This paper outlines Norsepower's perspective on the common goals and suggests some common ground, on which providers could collaborate. The maritime industry needs a reliable, safe and economically sustainable transition for decarbonising shipping. Wide adoption of wind propulsion plays an important role in this transition.

6. COLLABORATION FOR DEVELOPING SHIP DESIGN PRACTICES AND CLASSIFICATION RULES

The emerging market of wind propulsion on modern commercial ships requires adjustments and development of existing rules and regulations. Much of this work has been started in IMO and on national levels. IMO has updated the regulation on EEDI and EEXI to better accommodate and incentivize wind propulsion. Another example is the FuelEU Maritime legislation which also explicitly considers wind propulsion. At the same time, for example the benefit gained from wind propulsion is not fully accounted for FuelEU Maritime legislation: ships having 30% of the propulsion power produced from wind gain only 5% reward factor in emissions calculation.

Improving and refining the environmental regulations and how they accommodate wind propulsion should be a joint effort for all sail providers. When the rules consider the effect of wind in the energy production of the ship accurately and transparently, the design and efficiency of implemented wind propulsion systems can be optimized for fuel efficiency and emissions reduction.

In addition to the environmental regulations, there is significant amount of work to do in various operational and design regulations. In general, it can be noted that all wind propulsion technologies will affect the navigational light, radar and visibility arrangements. When the current rules were written, there were no requirements to accommodate tall but relatively narrow structures on the ships deck, apart from the cargo handling cranes. When it comes to operational and navigational safety regulation, it is noted that the current industry standard of applying for project specific exemptions is not considered as a sustainable way. At least on the following aspects are required to be considered to allow various methods and solutions for achieving compliance:

- Bridge visibility
- Navigational light visibility
- Radar visibility

All of the above-mentioned rules are such that the input from all technology providers is required so that the rules accommodate varying properties and challenges of different wind propulsion technologies.

The possibility of establishing a working group focused on developing approved solutions for complying with above mentioned requirements would facilitate the application of wind propulsion on ships. Current requirement for exemption applications consume time from many parties and present a repeating process dealing with almost identical issues. Such a working group could include flag states, classification societies and wind propulsion suppliers.

7. REALISTIC PERFORMANCE PREDICTIONS AND MEASUREMENTS

The overall fuel saving potential of wind propulsion is vast and can be further increased by integration with the power plant and propulsion control systems as well as technologies such as route planning and voyage optimization. When discussing the fuel savings and emission reduction potential of wind propulsion, there are couple of important aspects that should be addressed.

First one is the use of transparent performance indicators. Recent developments such as the workshops organized by SSPA on common Key Performance Indicators, WiSP & WiSP2 joint development projects, revised EEDI/EEXI calculation guidelines are steps to right direction. However, it must be stressed, that in the end, the wind propulsion providers are responsible for indicating the performance of their systems in a transparent and reliable way.

Trust is created by realistic performance predictions and over optimistic fuel saving estimates must be avoided. Misleading information slows down the uptake of wind propulsion in general. Wind propulsion has enough potential even without exaggeration.

Shipping is a conservative industry and there is good reason for that. Seamen rely on ship designers, shipyards and technology providers to create a safe working environment thousands of kilometers away from the closest shore. Wind propulsion providers need to be worthy of that trust.

Finally, it is noted that communicating the overall potential and the fact that the transition is already happening has capability to accelerate this transition. Emphasis on the wider market potential in messaging should be encouraged.

8. PERFORMANCE PREDICTION METHODS AND TOOLS

Most wind propulsion providers have built their own, or have access to, a performance prediction tool. Accurate performance prediction is considered an essential part of the project feasibility analysis and assessing the business case of a sail system on a particular ship and operational profile. Still, relatively little material is provided about the validation of these tools. To improve shipping community's trust in wind propulsion technologies, trustworthy performance prediction should be based on a foundation of verified performance data. Over the years, ships installed with Norsepower Rotor Sails have gone through extensive measurement campaigns by various independent verifiers. Methods for reliable and accurate measurement of thrust and fuel saving performance are an area where continuous development is ongoing.

9. SAFETY RELATED MATTERS

Safety is the highest priority in shipping. Ship owners, captains and crews as well as ship designers and shipyards must be convinced of the safety of the wind propulsion systems they work with. Also other parties, such as pilots, port and terminal authorities, flag states which might not be working with sails on a daily basis must be confident of safety aspects. Wind propulsion suppliers are responsible for designing their products and operational instructions to comply with high safety standards.

Active involvement of technology providers together with operators and regulatory authorities is a crucial component in ensuring safe operations of wind assisted ships. Norsepower is committed to ensuring safety of our product and improving the safety of the windy propulsion industry.

Training and operational support for superintendents, ship officers and crews will become increasingly important as more wind propulsion installations are installed on ships. Crew rotation and new employees need training to familiarize themselves with the relatively new technology of wind propulsion. Introducing modern sail technology in the training programs for seagoing personnel will also be necessary. Wind propulsion suppliers have a supporting role to provide and assist with training programs by providing material and opportunities for visiting training sites.

10. HIGHLIGHTING THE MARKET POTENTIAL IN COMMUNICATION

Finally, to expand the market further, technology providers are invited to a common effort of communicating about the vast potential of wind propulsion in general.

The shipping industry is facing one of the most difficult challenges, as it is striving towards decarbonization. Shipping is already one of the least carbon intensive method of transport for goods and reducing its carbon footprint requires a global approach. Unlike some other zero emission technologies, wind propulsion is ready for scale up and at a feasible cost level. Communicating this opportunity is a common effort for all sail providers.

Furthermore, it should be highlighted that this transition to utilize wind propulsion is happening already. The increasing number of successful installations builds confidence and trust and accelerates the uptake of wind propulsion. The frontrunners of ship owners and charterers are actively making plans to install Norsepower Rotor Sails on their fleet and evaluating the best routes and ships to lead the way. Ship designers and shipyards must keep up with the requirements related to structural and machinery design as well as production process development. When major shipyards are comfortable with wind propulsion installation processes, competitiveness of wind propulsion improves and becomes another standard part of a modern ship.

11. ROLE OF THE IWSA

The active and professional work of IWSA, led by secretary general Gavin Allwright, is widely recognized and appreciated. IWSA is involved in many initiatives and participates in the development of eg. IMO regulations and building awareness of wind propulsion within the shipping community and general public. Events and publications by IWSA are of key importance in communicating the benefits of wind propulsion to the shipping market and wider public. Norsepower is a founding member of the IWSA and continues to strongly support the work with a growing number of members.

12. CONCLUSIONS

Wind propulsion suppliers have the main responsibility to communicate the features, benefits, technical and economical aspects relevant for customers and other stakeholders in the value chain. Ship owners and charterers have new issues to discuss when contracts including wind propulsion systems are negotiated. How to agree the potential sharing of investments, operating costs and benefits. There are already charter contracts in place where the financial impact of wind propulsion installation is considered. It will take time before such terms become more common and widely accepted for use in charter contracts. However, the recognition of all parties in shipping industry of how wind propulsion benefits their business, and the planet, is the key to successfully installing thousands of wind propulsion applications on ships.

Norsepower urges the shipping industry stakeholders to embrace new technology and applications with an open and collaborative mind set.

13. AUTHORS BIOGRAPHY

Ville Paakkari holds the current position of Sales Engineer responsible for performance modelling and analysis of Rotor Sail systems at Norsepower.

Jukka Kuuskoski holds the current position of Chief Sales Officer at Norsepower.

Wind Propulsion Conference
16th – 17th February 2023
London, United Kingdom

VALIDATION OF PERFORMANCE PREDICTION PROGRAM AGAINST CYBER-PHYSICAL MODEL TESTS

K Eide, **A Östman** and **A Alterskjær,** SINTEF Ocean, Norway

SUMMARY

The present paper describes a validation of a numerical performance prediction program against model tests. The performance prediction program solves the motion of the vessel in three degrees of freedom (surge, sway and yaw), where the applied manoeuvring coefficients, resistance curve, wind propulsor model and propeller characteristics, can be obtained experimentally using model tests or numerically by CFD simulations or by other faster numerical tools. The SINTEF Ocean bulk carrier (SOBC-1) was chosen in the validation study due to recently performed cyber– physical empirical model tests [1], where the effect of the wind propulsors were included using a novel hybrid testing technique. In the experiments, the SOBC-1 model was running at self-propulsion, where propulsion was delivered from four numerically modelled Flettner rotors based on published CFD simulations results [2], in addition to a single screw conventional propeller. The forces generated by the wind propulsors were applied to the model using a system of wires, connecting the model to actuators, providing forces in 5 degrees of freedom according to prescribed wind conditions. In the present validation, the computed ship speed, drift angles and rudder angles can be directly compared against the model test results. The drift-angle induced added resistance of the vessel is indirectly compared by evaluating the propulsion point of the single screw propeller and compared against measured RPM, torque and power during model tests. The validation is conducted for a set of true wind directions ranging from 30 to 150 degrees, while true wind speed varies from 10 to 20m/s. The results show a good correlation between model tests and simulations, capturing the essential physical effects of operating at different points of sail. Further refinements to the numerical models are expected to provide a tool with increasing accuracy and flexibility for future performance assessments of wind-assisted ships.

1. INTRODUCTION

With the increasing focus on reducing emissions and reducing dependence on fossil fuels, wind propulsion is again establishing itself as a promising technology for the maritime industry. In order to evaluate business cases to support uptake of the technology, improved performance prediction methods are developed by both academia, research institutes and industry with the intention to provide sufficiently accurate and reliable estimates of the benefits of using wind as a propulsion source (see for instance [3], [4]). In general, the precision of numerical performance predictions will depend on a) the physical effects included in the method, and b) the quality of the input modelling data. Since carrying out route-and operational studies requires computations of many different combinations of speeds, wind speeds, headings, loading conditions etc, the numerical performance prediction analysis needs to be fast. This inevitably requires simplifications or neglections of some physical effects. Since a ship under wind propulsion represents a complex problem, several studies have looked into the consequences of various simplifications [3] and neglections [5] to the validity of the predictions. Even if the performance prediction method itself has the capacity to represent the physical phenomena in sufficient detail, the quality of the input data needs also to be accurate. Typically for wind-assisted vessels the inputs to a performance prediction program will be aerodynamic sail models, hydrodynamic force models (e.g. manoeuvring model) including propeller and rudder, and machinery models – and preferably also including a model of the control system for the ship to be studied.

Numerical methods for hydrodynamic phenomena are typically validated towards experimental results. In the case of performance predictions for wind assisted vessels, sail models can be validated by (or generated from) e.g. wind tunnel tests, hydrodynamic force models towards for instance PMM (Planar Motion Mechanism) tests, propeller and rudder models towards dedicated propulsor tests, and so on. In this paper, performance predictions from a numerical method are compared to model tests where wind propulsion forces are applied by a hybrid test technique, allowing also to study how well the interaction between the different physical aspects are captured by the numerical method – bringing the validation process for performance predictions for ships with wind propulsion one step closer to what is typically done for tools for conventional ships.

2. DESCRIPTION OF TEST CASE

2.1 SINTEF OCEAN BULK CARRIER (SOBC-1)

The present validation study is conducted based on experimental results of the SINTEF Ocean Bulk Carrier (SOBC-1). The geometry of the vessel is open-source, and allows for benchmarking and validation of tools and methods applied on a state-of-the-art efficient hull design. One of the main intentions of the vessel is to serve as a testbed for evaluation of various techniques to reduce energy consumption, such as air lubrication, energy saving devices and wind propulsion. The vessel is single-screw, equipped with a 4-bladed controllable pitch propeller together with a high-efficiency rudder design. A 3D-CAD illustration of the SOBC-1 vessel is shown in Figure 1. The length, LPP, and breadth of the full-scale vessel is 190 m and 32.2 m respectively, with a low block coefficient, CBLW = 0.7018. The vessel is in the present study equipped with 4 wind propulsors of Flettner rotor type. All rotors are located on the ship's centerline, equally spaced by 35 m, where the aftmost rotor is located 30 m ahead of the aft perpendicular. The main particulars of the vessel at design water line are summarized in Table 1.

Table 1: SOBC-1 main particulars at design water line (DWL).

Parameter and symbol	Value
Length overall, LOA	200 m
Length betw. perp., LPP	190 m
Length on waterline, LWL	196.942 m
Breadth waterline, BWL	32.201 m
Draught at FP/AP and LPP/2, T	11 m
Wetted surface	8485.24 m^2
Volume displacement ∇	48957 m^3
Block coefficient (based on LWL), CBLW	0.7018
Propeller diameter, D	6.75 m
Pitch ratio at r/R = 0.7	0.970 –
Blade area ratio, A_E/A_0	0.483 –
Hub diameter ratio, d/D	0.252 –
Number of rotor sails, n_{rotors}	4
Rotor sail height, H	35 m
Rotor sail diameter, d	5 m
Rotor sail end-plate diameter, d_e	6 m
Rotor sail rotation velocity	180 rpm

Figure 1: SINTEF Ocean Bulk Carrier (SOBC-1).

2.2 EXPERIMENTAL SETUP

Free running model tests of the SOBC-1 vessel at a scale λ = 1/32 were conducted in the SINTEF Ocean towing tank [1]. The experimental work utilises a cyber–physical empirical method which allows for forces to be applied on the vessel through a system of wires and actuators. The applied forces simulate the wind forces acting on the real wind propulsors. This technique has been developed at SINTEF Ocean over several years and has also been successfully applied in model tests on offshore floating wind turbines [6-8].

The ship model was placed in the seakeeping carriage (Figure 2). A stiff frame is mounted on the ship's deck, and wires connected to the frame apply forces in surge and sway on the model. The positioning of the wire system also allows moments in roll, pitch and yaw to be transferred. The wires are connected to responsive actuators that allows the tension in each line to be updated at a rate of 200 Hz. The reader is referred to [1] for further details. In the test case, the applied aerodynamic force is based on the systematic CFD study with parametric variations of the Flettner rotor geometry, as presented in [2], but the underlying model can be easily changed during the same model test setup to represent e.g. wing sails or a different configuration of the sails. The same aerodynamic model was applied in the numerical calculations, and a brief description of the model is presented in Section 3.2.

The model tests were run with a constant propeller power, controlled with a PID regulator. A thrust corresponding to a delivered power of 3078 kW, which corresponded to the required power to achieve 12.25 kn (full-scale) without wind nor sails, was applied. A set of constant, as well as varying, wind conditions were tested, with wind speed ranging from 10 to 20 m/s, and wind direction ranging from 30 to 150 degrees. Tests were also carried out for different wind conditions in seakeeping in head waves and zig-zag tests. However, in the present validation study, only constant wind conditions in calm water are considered. Depending on the applied wind forces, different rudder angles are needed to maintain a course which is aligned with the direction of the towing tank. This was solved by an autopilot controller connected to the rudder. The resulting yaw angle was monitored during the tests. As a consequence of the outlined setup, the speed at which the vessel travels will depend on the wind condition during the test, thus, for the carriage to follow the model, the speed of the carriage was automatically adjusted during the tests. Also, a correction force due to Reynolds number dependence on skin friction coefficient was applied on the vessel through the wire system.

Figure 2: Model of SOBC-1 in the towing tank. The thin lines which transfer forces to the model are visible in the picture, from [1].

3. NUMERICAL METHOD

3.1 PERFORMANCE PREDICTION PROGRAM

The SteadySail performance prediction program performs a static simulation of hydrodynamic and aerodynamic loads acting on the vessel, where the target is to simulate a vessel sailing on a steady course. For the simulation to reach a steady state, all forces acting on the vessel must be in balance, in other words, the sum of all forces must be zero:

$$\tau_{visc} + \tau_{wind} + \tau_{wave} + \tau_{propulsion} = 0$$

Where τ_{visc} are lift and drag forces on the underwater hull due to relative speed through water, τ_{wind} are wind forces on the above-water hull due to relative speed through air, τ_{wave} are mean second-order wave drift forces, and $\tau_{propulsion}$ are forces generated by propulsion and steering devices. The methodology for calculating each of these force contributions, dependent on ship velocity, attitude, propulsive parameters, and environmental conditions, is largely based on SINTEF Ocean's VeSim time domain simulator [9]. More details are provided in sections 3.2 through 3.4. The part of the hybrid model tests considered for this study were run without waves, therefore a description of the model for wave forces is omitted here. Also not considered in the model tests were the wind forces on the hull above water due to simulated wind. These forces have therefore also been omitted from the simulations.

The SteadySail program currently considers three degrees of freedom; surge, sway, and yaw. For a given vessel speed, the drift angle of the vessel is varied iteratively until the sum of all forces acting on the vessel is zero. An outer loop may then analyse the propulsive performance, and then repeat the process with a different vessel speed until the target propulsive power is obtained.

3.2 FLETTNER ROTOR MODEL

The numerical model of the aerodynamic forces acting on a Flettner rotor is based on the work done by De Marco et al. [2]. They performed a comprehensive set of CFD simulations with variations of rotor aspect ratio and size of end plate (also denoted Thom disc). The geometrises were tested at four spin ratios ranging from SR=1 to SR=3. The results from the simulations were compiled into multivariable polynomial expressions for lift (C_L) and drag (C_D) coefficients as functions of aspect ratio (AR), end-plate diameter ratio (d_e/d) and spin ratio (SR):

$$C_L = \sum_{i=1}^{4} \sum_{j=1}^{4} \sum_{k=1}^{3} a_{ijk}(SR)^{i-1}(AR)^{j-1}(d_e/d)^{k-1}$$

$$C_D = \sum_{i=1}^{4} \sum_{j=1}^{4} \sum_{k=1}^{3} b_{ijk}(SR)^{i-1}(AR)^{j-1}(d_e/d)^{k-1}$$

where the polynomial coefficients a_{ijk} and b_{ijk} are given in Table 2.

The lift and drag calculated using the expressions for C_L and C_D assumes uniform wind conditions. In reality, both the direction and speed of the apparent wind changes with height over sea surface. In order to take this effect into account, a strip theory inspired method is implemented, where each rotor is divided in sections with constant height. In our case 11 uniformly divided sections is chosen, resulting in a $\Delta H = 3.18$ m vertical thickness of each section. The apparent speed and direction are considered to be constant at each section. The aerodynamic lift and drag force at a given section can be calculated as

$$f_{L,D} = \frac{1}{2}\rho_{air} \cdot C_{L,D}(AR, SR, d_e/d) \cdot d \cdot \Delta H \cdot (AWS)^2$$

where $\rho_{air} = 1$ kg/m^3 and AWS is the apparent speed. Using the apparent wind direction for each section, the local $f_{L,D}$ force can be decomposed in surge and sway force components. And, by summing over all sections for each rotor the total force acting on the complete rotor is obtained. Each rotor is treated individually, not considering interaction effects from the other rotors.

The AWS is calculated based on the vector sum of ship velocity and the undisturbed free-stream wind velocity, where the free-stream wind velocity $U_w(h)$ at height h is given by

$$\frac{U_w(h)}{U_w(h_0)} = \left(\frac{h}{h_0}\right)^{1/9}$$

and the reference height is chosen as h_0 = 20 m.

Table 2: Coefficients a_{ijk} and b_{ijk} used in the expressions for C_L and C_D

i	j	a_{ij1}	a_{ij2}	a_{ij3}	b_{ij1}	b_{ij2}	b_{ij3}
1	1	46.75579	−78.30137816	20.28158	−79.91876737	70.07144	−15.26803135
1	2	−37.84627709	60.05841	−15.30145164	60.03045	−50.611248	11.01285
1	3	8.096156	−11.87436219	2.987145	−11.64851051	9.810299	−2.141092131
1	4	−0.460229167	0.639485	−0.160067313	0.668844	−0.557905642	0.121478
2	1	−89.78524934	140.01	−35.67163964	140.7181	−125.0475503	27.37696
2	2	77.60674	−112.2043683	27.97673	−104.7031285	89.13657	−19.29174369
2	3	−16.46077427	22.03782	−5.408808016	20.21478	−17.09768028	3.697994
2	4	0.915301	−1.152095456	0.279901	−1.155096291	0.964007	−0.20726945
3	1	38.54701	−62.54475786	16.26618	−76.5057191	69.27083	−15.18862485
3	2	−37.55599168	54.80146	−13.70360517	57.9508	−49.79267524	10.70937
3	3	8.15294	−10.75328967	2.627145	−11.1802424	9.493153	−2.034099935
3	4	−0.439901094	0.533755	−0.128120357	0.636172	−0.53087876	0.112735
4	1	−4.895453074	8.642221	−2.29418742	11.79866	−10.53866427	2.289727
4	2	5.189632	−7.946352738	2.016174	−9.411835208	8.06482	−1.716103541
4	3	−1.153622261	1.555612	−0.383202934	1.830524	−1.54716509	0.327419
4	4	0.060046	−0.072807988	0.017516	−0.103856573	0.08591	−0.017961236

3.3 CALM WATER MANOEUVRING MODEL

The calm water manoeuvring model describes lift and drag forces acting on the underwater hull due to relative speed through water. The model is based on precalculated hydrodynamic coefficients, and can take input from various sources:

- Resistance curve
- PMM model tests
- PMM CFD simulations
- Current coefficients
- HullVisc [10]

As a basis for straight-line resistance force, a resistance curve derived from ship model resistance tests performed for SOBC-1 has been applied. For calculation of lift and drag effects due to varying drift angle, HullVisc [10] has been chosen to provide simulation input for this study. It is easy to use, calculates results quickly, and requires input only in the form of ship geometry. HullVisc provides manoeuvring coefficients in three degrees of freedom; surge, sway, and yaw. The coefficients are calculated based on slender body strip theory with corrections from empirical methods [11]. The HullVisc software provides a fully dynamic model suitable for time domain simulation. For the simulations performed by SteadySail, only a subset of the model is active, providing forces and moments in the static context of the simulation.

3.4 PROPULSION

Numerical models describing conventional propulsion and steering devices are based on SINTEF Ocean's propulsion library. This is a software library containing detailed models of multiple types of propellers and rudders based on numeric methods and model tests. Such a model is typically configured by providing input in the form of propeller and rudder geometry, and optionally open water curves and speed-dependent propulsive coefficients.

For this study, a model describing an open propeller in front of a conventional rudder was selected. Propeller geometry was input according to the numbers presented in Table 1. Rudder geometry was input according to the data presented in Table 3. Additionally, open water curves describing thrust and torque coefficients as function of advance ratio, and propulsive coefficients describing wake fraction, thrust deduction and relative rotative efficiency were input based on data derived from open water and propulsion model tests performed for SOBC-1.

Table 3: Rudder geometry for SOBC-1

Parameter	Value
Headbox height	1.667 m
Headbox chord	4.75 m
Headbox area	7.885 m^2
Rudder height	7.8 m
Rudder chord	4.75 m
Rudder area	37.05 m^2

4. RESULTS

This section presents a comparison of results from the hybrid model tests and results from SteadySail. Simulations were performed for the same test matrix used in the hybrid model tests:

- True wind directions ranging from 30 to 150 degrees in 30-degree steps.
- True wind speeds of 10, 15 and 20 m/s.
- Constant propeller power of 3078 kW.

The results from the hybrid model tests are taken as mean values over a time window with relatively stable behaviour and subsequently converted to full scale equivalents using Froude scaling and appropriate corrections according to standard ITTC procedures. The hybrid model test results for 30 degrees TWD (True Wind Direction) and 20 m/s TWS (True Wind Speed) were discarded, and as such simulation results for that condition have been omitted. As is seen in Figure 3, there is a surprisingly good match in obtained vessel speed, especially for the 15 m/s and 20 m/s TWS conditions. For the conditions with 10 m/s TWS the differences are more pronounced for certain wind directions. The drift angle comparison shows a good match for broad reach conditions, with a slight divergence as the drift angle increases. As we approach close hauled conditions, the difference in drift angle generally seems to decrease. An overall good match is seen for the rudder angle. The wind thrust fraction plots depict how much of the forward thrust force is generated by the rotor sails in relation to the total longitudinal force generated by both the propeller and sails. For the 30-degree TWA conditions this value is negative, showing that the rotors sails have a detrimental effect on the forward propulsion of the vessel. For the other TWA conditions, the simulations show a higher wind thrust fraction. A further explanation follows towards the end of the results section.

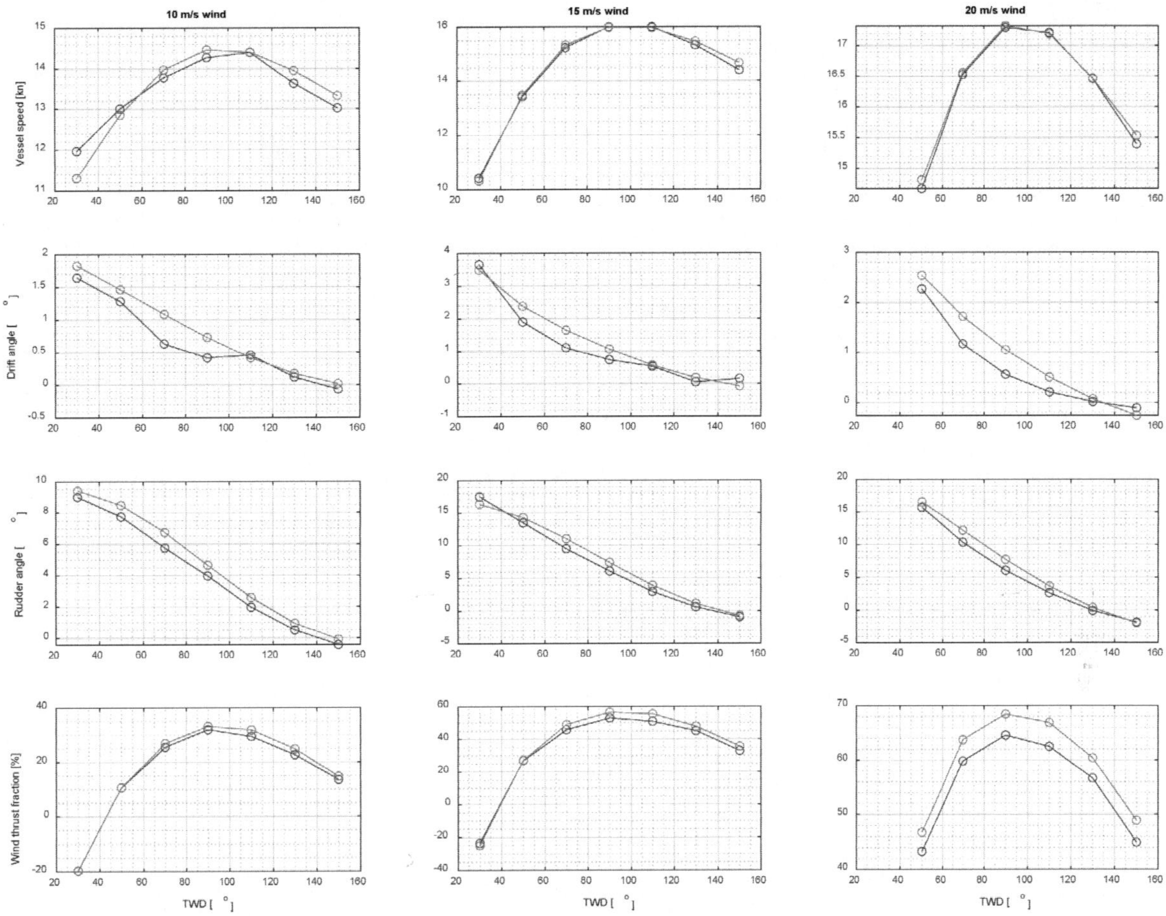

Figure 3: Comparison of obtained key vessel parameters as a function of true wind direction (TWD) and true wind speed (TWS). Results from the hybrid model tests are presented in blue while results from SteadySail are presented in red.

Figure 4 presents a comparison of forces and moments generated by the rotor sails. As the exact same mathematical model has been applied for both the hybrid model tests and the SteadySail simulations, it should not come as a surprise that the comparison shows a good match. The two results diverge for the 30-degree TWA and 10 m/s TWS condition, which is where we also see the largest difference in vessel speed, affecting the apparent wind speed seen by the rotor sails. As observed for the wind thrust fraction plots, the longitudinal sail force becomes negative at 30 degrees TWA.

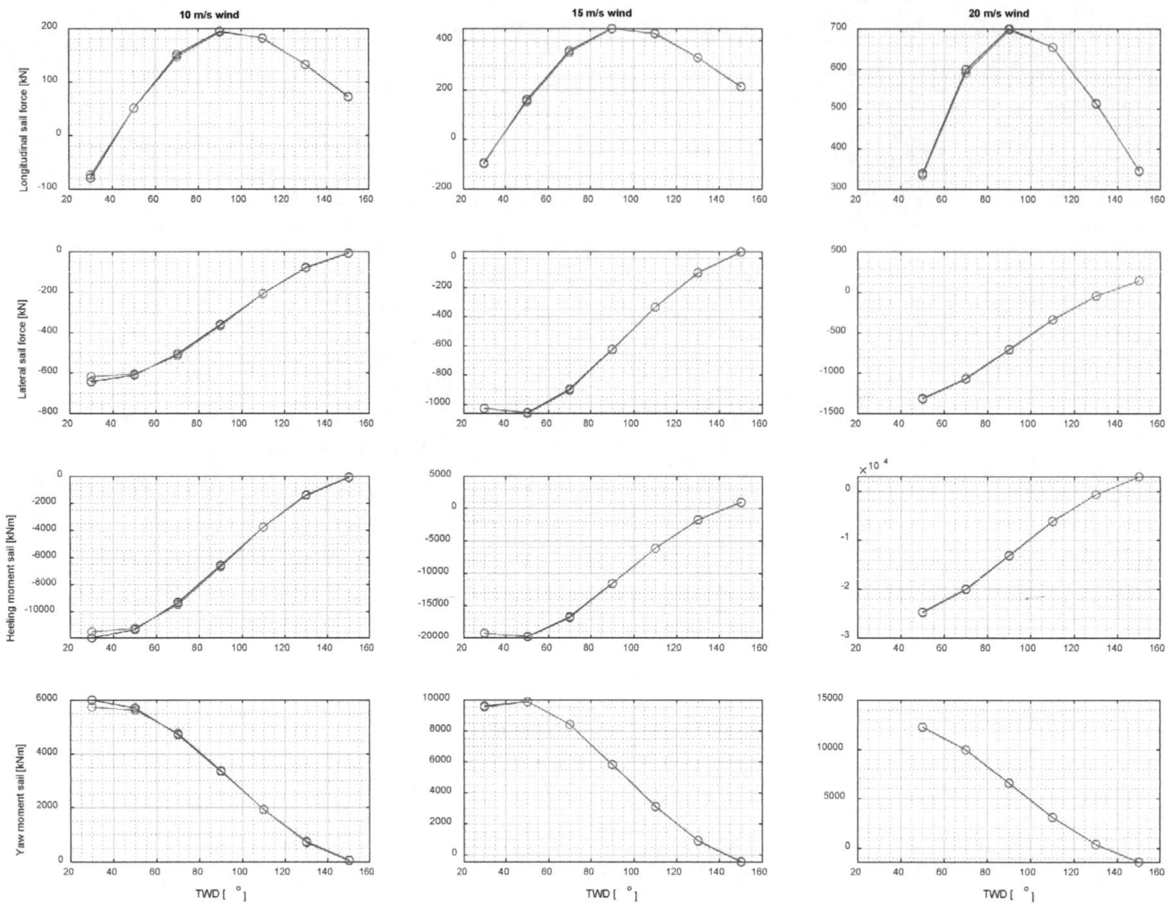

Figure 4: Comparison of generated sail forces and moments as function of true wind direction (TWD) and true wind speed (TWS). Results from the hybrid model tests are presented in blue while results from SteadySail are presented in red.

A comparison of propulsive performance is given in Figure 5. The propeller RPM results seem to generally follow the same trend, with higher values in beam reach conditions. Observe that since the numerical simulations and the model tests are carried out for a constant power, the propeller RPM and torque curves are inversely proportional to each other. For each condition a different vessel speed is obtained, which leads to a difference in advance speed between conditions. Following the relationship between torque, rate of revolution and power, as the torque coefficient changes with advance ratio, a higher vessel speed will require a higher rate of revolution in order to maintain the same propeller power. This is perhaps more easily seen when comparing the RPM plot curves to the vessel speed curves, which especially for the simulation results follow the same shape.

The propeller thrust plots show a significant difference between simulations and model tests, where the propeller thrust in the simulations is generally smaller than in the model tests, causing the difference in wind thrust fraction as seen in Figure 3. An interesting comparison is the condition for 15 m/s TWS and 110 degrees TWA. Here the vessel speed, drift angle, rudder angle and rotor sail forces match up very well. The propeller RPM and torque figures also show a good match, while there is a 17 %, or 69 kN, difference in propeller thrust. This indicates that the simulation model underpredicts the hydrodynamic resistance of the vessel in this condition. On the other hand, it also underpredicts the propulsive efficiency, and these two effects nearly cancel each other. In [1], an increase in effective wake fraction was observed for the conditions run with rotor sails compared to runs without rotor sails, increasing with increasing drift angle. Assuming constant thrust deduction fraction, this infers increased hull efficiency η_H. Further, the reduction in propeller loading for the runs with rotor sails also resulted in a higher open water efficiency. The effect of drift angle on wake fraction has not been modelled in the simulations and is likely a contribution to the observed differences.

As the SteadySail simulation is a planar 3 degree of freedom analysis, the effects of heel and trim are not included. Although the values recorded during the model tests were small, this may have consequences for the modelling of lift and drag forces on the hull. Any changes in thrust deduction, wake and rotative efficiency due to heel, trim and drift angle are also not captured by the simulations.

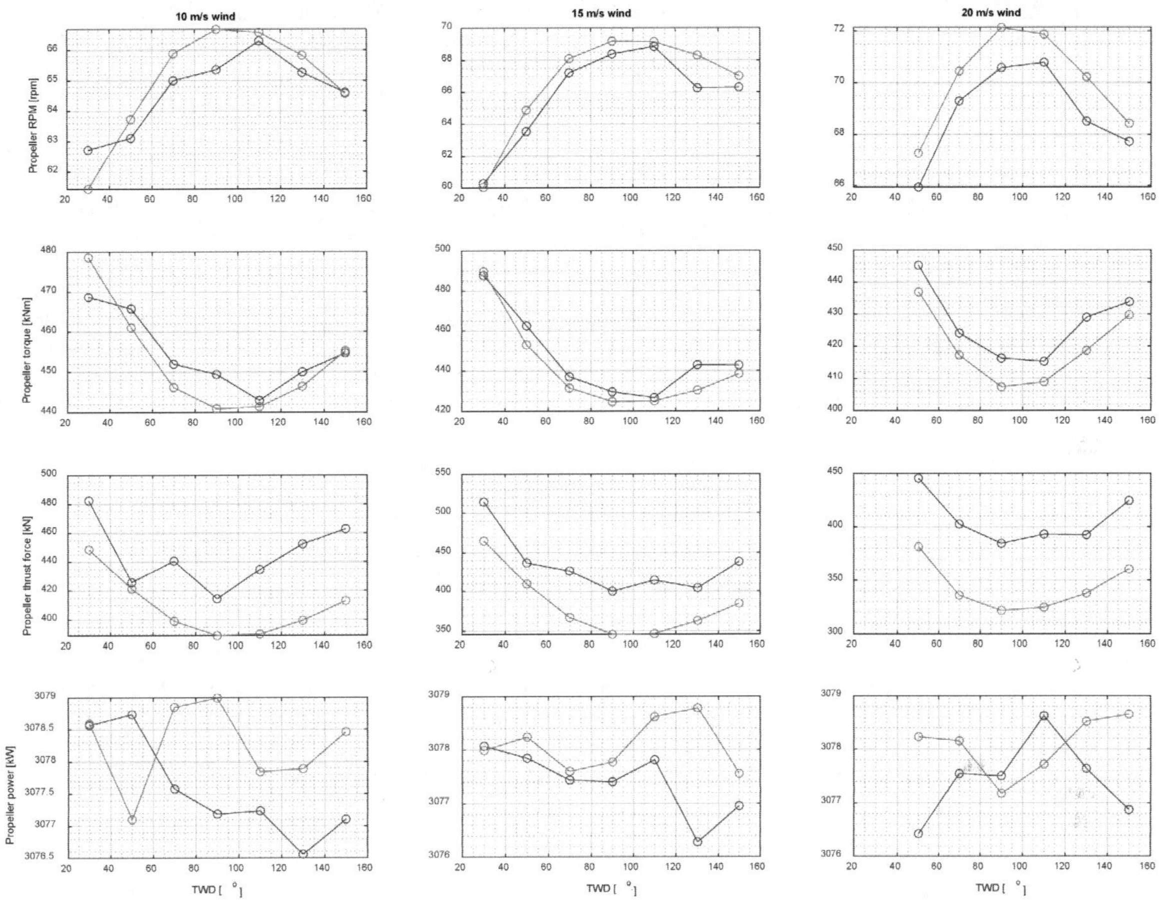

Figure 5: Comparison of propulsion parameters as function of true wind direction (TWD) and true wind speed (TWS). Results from the hybrid model tests are presented in blue while results from SteadySail are presented in red.

5. CONCLUSION

This paper presents a comparison of results from a newly developed steady state numerical simulation method against results from a hybrid model test of a wind-assisted cargo ship. For both model tests and simulations, the same model of a wind propulsion device was applied. While the results show some differences between the hybrid model tests and the simulations performed with SteadySail, it is shown that the numerical method captures the same physical behaviour when varying wind speed and direction. A good match is seen for vessel velocity and drift angle, while a significant difference is seen for propeller thrust force. There will always be differences between numerical models and the real world, and finding the cause of these differences as well as tuning numerical models are interesting topics for future work.

An initial uncertainty was the HullVisc model used to describe the lift and drag forces acting on the vessel when subject to forward speed and a drift angle. This model is based on empirical data and may not always provide a suitable match for the vessel at hand. Using for example CFD simulations or model tests, it should be possible to derive a more tailored model to describe the manoeuvring behaviour of the vessel. The simulation results do however show a good match for both vessel speed and drift angle, making it interesting to see which improvements a more refined manoeuvring model could provide.

In the search of causes for the differences in results found in this study, SteadySail can be run in different modes to realize what-if scenarios. Instead of aiming for a target propeller power, the simulations can be run with for example constant propeller RPM, constant propeller thrust, or constant vessel speed. While not presented in this paper, simulations were repeated with a constant vessel speed, targeting the same values as obtained in the model tests. When comparing output propeller power, the results were generally within ± 10 % of the original target of 3078 kW, while the difference in propeller thrust was similar to what has been presented in Section 4, indicating that further investigation into the effects of drift angle on propulsion performance and resistance is needed.

While the applied wind propulsion model lacks certain physical effects such as wind shear, interaction effects between sails, and interaction effects between sails and hull, it is shown that the same model can be applied in both model tests and numerical studies. More complete and involved wind propulsion models can be utilized by both methods.

The presented numerical method shows great promise in becoming a tool for designers and researchers for analysis and optimisation of wind propulsion device layout and operational parameters. In addition to analysing single conditions, the simulation tool may be employed in routing simulations to provide insight into potential fuel savings for a wider set of conditions encountered during a voyage. Many opportunities for further refinements of the numerical models lay ahead, and topics for future work include:

- Comparison of sailing performance with an improved manoeuvring model from CFD simulations.
- Comparison of simulations to hybrid model tests performed in waves.
- Comparison of zigzag manoeuvring model tests to time domain simulations.
- Investigation into the effects of dynamic heel, trim and drift angle on the propeller wake.

6. ACKNOWLEDGEMENTS

The development of the SteadySail performance prediction program has been funded by the Industry Project CruiZero (Norwegian Research Council project number Project Number: 310075 - IPNÆRINGSLIV19) and SFI Smart Maritime (https://www.smartmaritime.no/). The authors would like to thank Andrew Ross, Edvard Ringen and Thomas Sauder for help in development of the numerical models used for this study.

7. REFERENCES

1. SAUDER, T., ALTERSKJÆR, S.A., 'Hydrodynamic testing of wind-assisted cargo ships using a cyber–physical method', *Journal of Ocean Engineering*, Volume 243, 2022

2. DE MARCO, A., MANCINI, S., PENSA, C., CALISE, G., DE LUCA, F., 'Flettner rotor concept for marine applications: a systematic study', *Int. J. Rotating Machinery*, 2016.

3. F. THIES, J W RINGSBERG 'Performance Prediction and Design of Wind-Assisted Propulsion Systems', *Proceedings of RINA 2021 International Conference on Wind Propulsion*, 2021

4. A. PERSSON, D.-Q. LI, F. OLSSON, S. WERNER, U. Dhomé, 'Performance predictions of wind propulsion systems using 3D CFD and route simulation ', *Proceedings of RINA 2019 International Conference on Wind Propulsion*, 2019

5. KRAMER, J. V.; STEEN, S., 'Sail-induced resistance on a wind-powered cargo ship ', *Journal of Ocean Engineering 2022 ; Volume 261*, 2022

6. SAUDER, T., CHABAUD, V., THYS, M., BACHYNSKI, E.E., SÆTHER, L.O., 'Real-time hybrid model testing of a braceless semi-submersible wind turbine. part I: the hybrid approach.', *Proceedings of the ASME 2016 35th International Conference on Ocean, Offshore and Arctic Engineering*, 2016.

7. THYS, M., CHABAUD, V., SAUDER, T., ELIASSEN, L., SÆTHER, L.O., MAGNUSSEN, Ø.B., 'Real-time hybrid model testing of a semi-submersible 10 MW floating wind turbine, and advances in the test method.', *1st International Offshore Wind Technical Conference*, ASME, 2018.

8. THYS, M., SOUZA, C., SAUDER, T., FONSECA, N., BERTHELSEN, P.A., ENGEBRETSEN, E., 'Experimental investigation of the coupling between aero- and hydrodynamical loads on a 12 mw semi-submersible floating wind turbine', *Proceedings of the ASME 2021 40th International Conference on Ocean, Offshore and Arctic Engineering*, 2021.

9. VeSim, SINTEF Ocean time domain vessel simulator. https://www.sintef.no/en/software/vesim/

10. ROSS, A., FATHI, D., 'Manoeuvring validation analysis of the M/F Landegode', *Proceedings of the ASME 2017 36th International Conference on Ocean, Offshore and Arctic Engineering*, 2021.

11. MARTINUSSEN, K., RINGEN, E., 'Manoeuvring Prediction During Design Stage', *Presented at the International Workshop on Ship Manoeuvrability at the Hamburg Ship Model Basin*, HSVA, 2000.

8. AUTHORS BIOGRAPHY

Kristoffer Eide holds the current position of Senior System Developer at SINTEF Ocean. He is the main developer behind the SteadySail performance prediction program, as well as being responsible for maintenance and development of hydrodynamic analysis software and time domain simulation tools and models. His previous experience includes hydrodynamic modelling, control engineering and development of ship performance assessment software.

Anders Östman holds the current position of Senior Research Scientist at SINTEF Ocean. He is responsible for CFD simulations of maritime and offshore related topics. He is currently involved in the implementation of a fast lifting-line numerical tool for calculation of forces acting on wind propulsion devices. His previous experience includes a wide range of CFD studies, for example, analysis of ship hull performance and propulsion systems, simulating extreme wave impacts on offshore structures, green water analysis, implementation of dedicated hull roughness wall functions, analysis of aquaculture systems, analysis of air-lubrication systems, etc.

S. Anders Alterskjær holds the current position of Research Manager for Ship Hydrodynamics at SINTEF Ocean. He is responsible for research activities with ship hydrodynamics in SINTEF Ocean's laboratories and numerical software. His previous experience includes a wide range of numerical and experimental work, for example ship hull performance and propulsion systems, seakeeping and manoeuvring, performance and dynamic stability of high speed vessels along with increasing activity on wind assisted propulsion through performance assessments, route studies and hybrid model testing. He is currently a member of ITTC's Specialist Committee on Wind Powered and Wind Assisted Ships.

EMPIRICAL METHODS FOR DEVELOPING WASP PERFORMANCE INDICATORS AND HEURISTICS-BASED DECISION MAKING UNDER UNCERTAINTY

K M Fakiolas, Naval Architect & Marine Engineer, CEO Finocean Ltd, Finland

SUMMARY

Wind-assisted ship propulsion is recently perceived positively by Ship Operators as one of a potentially promising energy saving and emission reduction investment choice, however the variability in wind conditions, the uncertainty about trade routing and sometimes even complicated performance prediction analysis, create a lack understanding of the main dynamic functional principles of WASP leading to reluctance in decision making. Ship Operators need to have independent indicators of empirical nature to use at hand for achieving a fast technical and business evaluation on the 'ship-systems-trade' WASP applicability so that a common understanding is established on expectations about performance and target settings.

NOMENCLATURE

AWA	Apparent Wind Angle (deg)
AWS	Apparent Wind Speed (deg)
TWS	True Wind Angle (deg)
TWA	True Wind Angle (deg)
V_S	Ship speed (knots)
ρ	Air density (kg/m^3)
C_L	Lift coefficient
C_D	Drag coefficient
S	Projected surface area (m^2)
L	Lift force (kN)
D	Drag Force (kN)
T_F	Forward Thrust (kN)
CF_{WASP}	Capacity Factor for WASP
VR	Velocity Ratio
$P_{D\ AVERAGE}$	Average Delivered Net propulsion power (kW)
$P_{WASP\ max}$	Max delivered power from WASP (kW)

1. INTRODUCTION

Scientific methods are heuristic in nature. Heuristics are simplifying, incomplete, underdetermined, and fallible problem-solving rules that can nevertheless serve certain goals in certain contexts better than truth-preserving algorithms [1]. In other words, Heuristic is a simple procedure that helps find adequate, though often imperfect, answers to difficult questions, with the root word being same as 'eureka' [2], the ancient Greek word used by Archimedes to express his excitement for the spontaneous discovery of the volume displacement of a water immersed body, a crucial fundamental for ship design.

Mechanical sails for Wind-assisted ship propulsion are a novel application, currently perceived positively by ship operators but the variability in wind conditions, the uncertainty about trade routing, the complicated performance prediction analysis coupled with over-confident marketing figures, create a mistrust primarily due to lack of understanding the main dynamic functional principles of WASP on the ship. This paper attempts to provide such a first set of heuristic, empirical performance indicators, sourcing from the fundamental working principles of wind-assisted ship propulsion, related to the ship type, WASP system type, the wind conditions and routing.

2. FORMULAS AND DATA

Formulas frequently used for the calculations performed in this paper noted as AWS, AWA, L, D, T_F :

$$AWS = \sqrt{V_S^2 + TWS^2 + 2 \cdot V_S \cdot TWS \cos(TWA)}$$

$$AWA = \arccos\left(\frac{TWS \cdot \cos(TWA) + V_S}{AWS}\right)$$

$$L = \frac{1}{2}\rho c_L AWS^2 S$$

$$D = \frac{1}{2}\rho c_D AWS^2 S$$

$$T_F = L \cdot \sin\,(AWA) - D \cdot \cos\,(AWA)$$

For developing empirical and heuristic approaches for WASP performance assessment, hypothetical aerodynamic performance figures will be chosen for a theoretical Rotor Sail with dimensions of 30m x 5m and a theoretical Wing Sail of 40m x 12m (Figure 1). These aerodynamic figures do not represent or reflect the actual performance figures of any specific technology provider; however, they are in close fit with the scale, range and variance of aerodynamic coefficients as observed and measured in experiments and research of both types of mechanical sails.

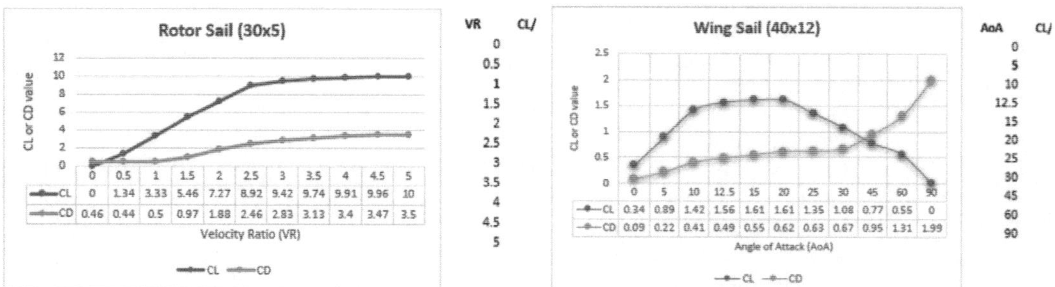

Figure 1: Theoretical C_L-C_D curves for an exemplary Rotor Sail and Wing Sail.

It should be hereby noted that full scale trials and improved designs of mechanical sails have sometimes demonstrated better C_L, C_D values, hence the aerodynamic performance figures used in this paper are considered to be conservative.

3. AVAILABLE WINDS IN THE OCEANS

One prime aspect - not overall well apprehended within the shipping industry - is the awareness of the realistically available winds through the oceans. Usually ship operators pay attention to observe and detect wind intensities above BF7-8 which become a dangerous environment for ship navigation and cargo safety, while largely affecting ETA and fuel consumption agreed with the charterers. For the average (BF 4-7) and low wind conditions (BF 0-3), there is no current incentive to explore wind distribution patterns during various voyages, even if ample wind datasets are recorded from anemometers and potentially included in noon reports and performance monitoring systems.

In principle, ship Captains and Operations Department already know that the ocean winds - at any given time - are generated by continuously moving air masses between the dynamically forming Barometric Highs and Lows. This physical pattern resembles of an 'air mass gearbox' where each 'pinion-gear' couple is a rotating and simultaneously a moving (transposed) air mass that excites the movement of another adjacent air mass, while the Earth's motion contributes the Coriolis effect in the whole system motion. Two methods can be used to determine these moving air mass generated wind speed distribution: 1) the time series method and 2) the statistical analysis method [3]. The result obtained by the time series method maybe more accurate but when there is a lack of long-term wind speed data, it is more feasible to use statistical analysis to explain the behaviour and characteristics of historical wind speed data. The most widely used probability density function for historical wind speed data in wind speed modelling is the Weibull distribution.

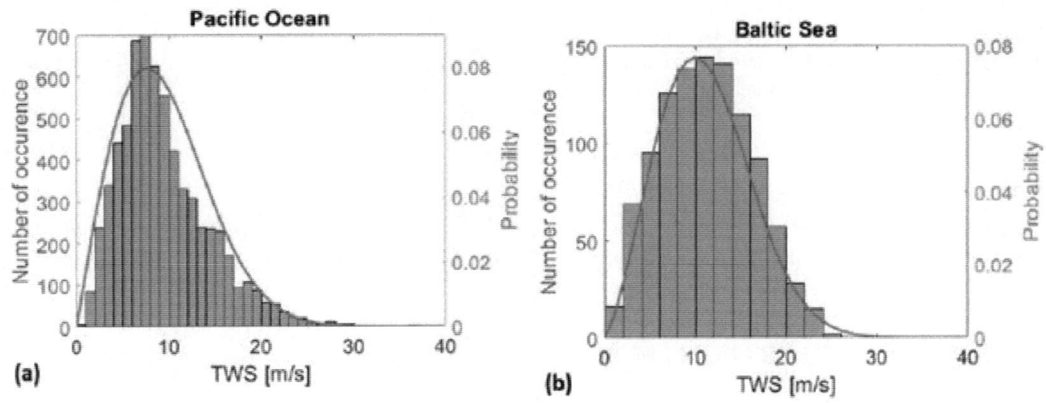

Figure 2: TWS probability density for the Pacific Ocean & Baltic Sea [4].

In Figure 2 a Weibull probability density is shown by means of TWS occurrences in two different ocean environments, the Pacific Ocean, and the Baltic Sea. It can be observed that the Pacific Ocean presents a concentration of the highest occurrences in a very specific TWS domain, and this is neither the too low one (i.e., below 5 m/sec = BF 2-3) nor the too high one (i.e., above 14 m/sec ≥ BF 7). The pattern repeats in the Baltic Sea, but with wider spread of occurrences on the high probability density area (i.e., 5-18 m/sec) compared with the Pacific Ocean.

IMO MEPC 77 adopted the Global Wind statistics matrix [5], where a similar Weibull probability density distribution is used based on a statistical analysis of the ocean winds distribution related to a complex network of global cargo ship movements [6]. In this study, the year's 2007 itineraries (490 517 nonstop journeys) of 16,363 cargo ships (5 498 bulk dry carriers, 3 100 container ships and 2 628 oil tankers), sizes larger than 10, 000 GT (gross tonnage) - which make up close to 90% of the world's total capacity for oceangoing ship cargo transport - were analysed from AIS sources to construct a network of links between 951 ports [7]. For these routes a total of 106 wind condition charts were analysed based on 868.500 individual wind data from WetterWelt (2011), and the below wind probability chart is calculated related to the wind angles and wind speeds [5] as in Figure 3.

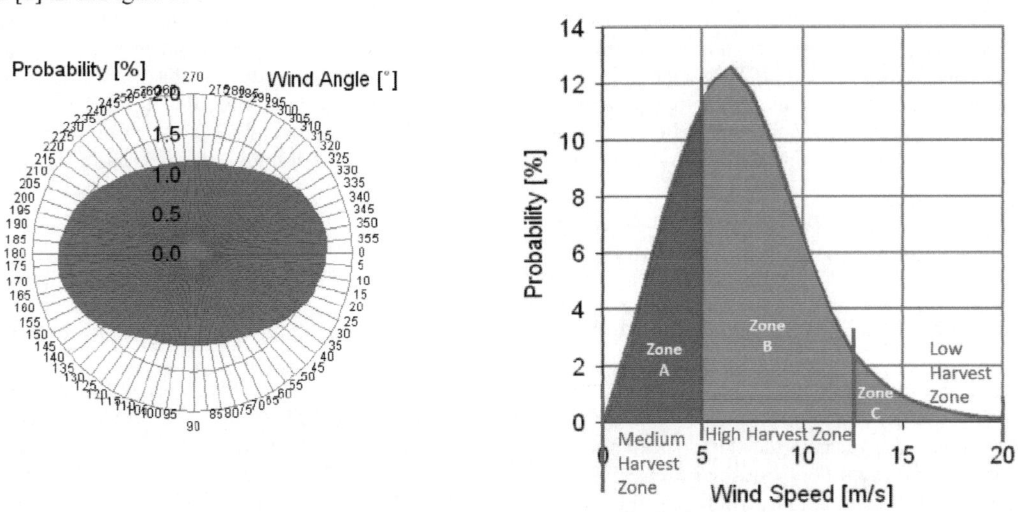

Figure 3: Global wind probability charts adopted for EEDI calculation with WASP [5] and indications of the suggested Harvest Zones in the TWS probability density curve [6].

For these specific routes, the statistical analysis demonstrate that tailwinds (broad reach regions) and headwinds (close haul regions) tend to occur a bit more often than beam reaching winds. The probability density function follows a similar to Figure 2 Weibull curve shape, with a large probability of wind speed occurrence within BF 2-6 (4-11 m/sec), while the directional probability can vary a lot, with the wind potentially coming from any direction on very close equality, with a slightly higher propensity for tail and head winds. This is a useful first empirical observation bringing awareness on what kind of winds the ship expects to face during mechanical sail operation.

Heuristic No 1:

The most available ocean winds at a given route, present a Weibull probability distribution mostly concentrated around an average intensity range (BF2-BF6, i.e.,6-20 knots), statistically potential to blow almost equally from any direction, with 1-2% higher propensity as head or tail winds.

4. USEFUL WINDS FOR PROPULSION

By observing any Weibull-type wind probability density distribution, for any part of the ocean seas three (3) main wind energy Harvesting Zones are identified which can be associated with the ship's speed range, thus reflecting the quality (and quantity) of the useable apparent wind, the one that the wind propulsion system installed on board is 'feeling' and operating [6]. Depicted in Figure 3, three wind energy Harvest Zones are distinguished as follows:

4.1 ZONE A: MEDIUM HARVEST ZONE (BF 0-3)

In this region the probabilities of encounter sum up to abt 30% for a TWS below 5m/sec (equal to 9.72 knots, being within BF 3 and lower), hence the speed of deep-sea cargo vessels over 5000 GRT will be at lowest around 10 knots, exceeding the true wind speed, by bringing most of the true wind as an apparent closed haul/headwind. Even minor variations of wind routing for improving L/D delivery from a given WASP system will be potentially inadequate to make large difference in performance for such relatively low AWS, occasional exceptions could exist if the voyage is very long and continuous. This means that the T_F values delivered by WASP systems will be relatively small and mostly favoured by those able to deliver the highest C_L/C_D during operation in a tight close haul region. This Zone can be characterized as of a 'medium' harvest, due to the good aggregate probability density, despite the low intensities.

4.2 ZONE B: HIGH HARVEST ZONE (BF 4-6)

This region represents the best working area for a WASP system, since the TWS will tend to exceed the ship's sailing speed range (especially for cargo ships sailing below 16 knots), going up to 12.5m/sec (24.2 knots). The total aggregate probability density is higher, of 53-55%, and at the same time there is a higher T_F potential coming from the WASP systems. There will be apparent winds that provide high T_F for high C_L WASP systems and stronger close-haul winds that will favour energy harvesting from wing sails with better C_L/C_D values. Minor adjustments to the ship course to bring the incoming wind more beamy will also support the performance savings without large deviations from the ship's course, even for relatively short routes of 5-10 days, while the wind induced wave height in this region is around 1-3m, hence manageable for navigation and ship safety for such ship sizes with over 5-6m drafts (laden or ballast).

It is the most desired and workable wind potential to be exploited on the ship for any WASP system configuration, and - thankfully to the nature - it is also the most probable to happen.

4.3 ZONE C: LOW HARVEST ZONE (BF 7-12)

This region represents a paradox: the wind intensities are quite high - which is very favourable for the stand-alone aerodynamic performance of a wind propulsion system and can become quite beamy and broad reached since the wind speed is quite higher than the ship's speed range, favouring the T_F development even by both the L and D components, however, they might not be desired for the overall ship operation. In such wind conditions, wave and wind resistance is quite high, the main engine is already utilizing a good part of the sea margin, while hull fouling might make things practically worse for achieving the required ship speed. The ship needs good manoeuvring properties and can be susceptible to dynamic rolling/pitching combination where the excess aerodynamic components of the WASP systems may contribute negatively to course instability, a fact observed already in broad-reach wind tunnel tests [8]. Ships with imposed EEXI engine limitations (i.e., highly EPL limited ships) might have issues when operating in the low RPM zone of the engine. Both beamy and broad-reach high waves may impact dynamic stability and parametric rolling of the ship, especially for unfavourable GM ship conditions.

Furthermore, the specific wind intensity probability is the lowest aggregate of the other harvest zones (less than 15%), much less frequent to appear and at the same time exploited under certain WASP system limitations. Thus, despite if they bring the best of the WASP systems, these winds are not potentially useful for operations all the time – for these reasons characterized as a Low Harvest Zone. For the deliberations of the heuristics approach, we can accept that the MEPC 77 global wind matrix directions (true winds) can almost be equal in probability of encounter.

The above attempted analysis brings many similarities in mind with the Benz Law that applies in the land-based wind turbines, according to which, no wind turbine can capture more than 59.3% of the wind's kinetic energy, however for different reasons related to the Newtonian nature of the winds and the turbine geometry limits [9].

Heuristic No 2:

The most Useable wind intensities for the cargo shipping fleet exist within BF 4-6, occurring abt 53% of the time.

5. VOYAGE DEPENDED AVAILABLE WIND ENERGY

By taking the global wind statistics benchmark of MEPC77, and by using the Apparent wind formulas for a ship speed of 12 knots, it can be shown how the AWA probability distribution will be affected (Figure 4):

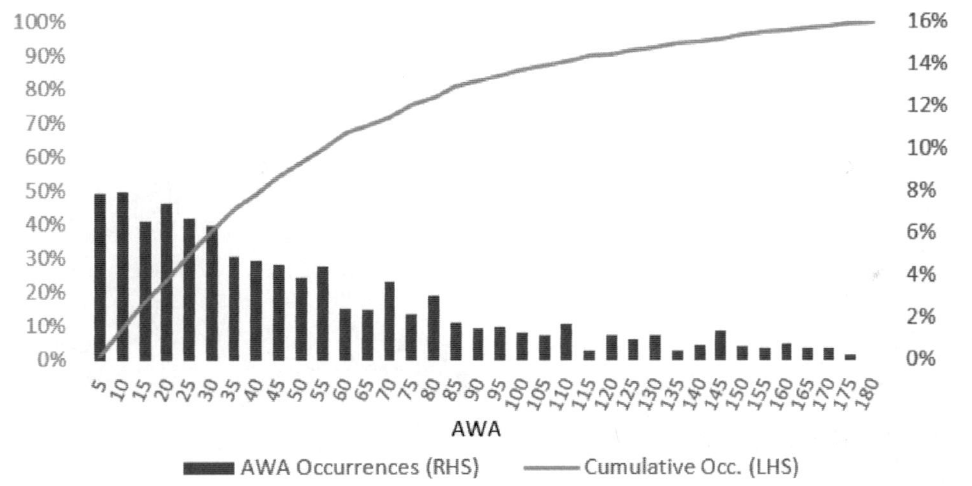

Figure 4: Global wind statistics AWA probability of occurrence for ship speed of 12 knots *(Courtesy: Ayro).*

It can be observed that abt 80% of the probable Apparent wind directions will occur on the ship within a region of 0° - 90° (including both Port and Stbd directions), with 50% of them within 0° - 45°.

By selectively picking the very classic, densely executed cargo trading routes as shown in the below Figure 5, the conclusion seems the same: For long-haul voyages at service speeds around 12-16 knots, the apparent wind shows a propensity to concentrate closer to the 0° - 90° directional zone most of the voyage time (over 70%), and 50% of the time at 0° - 45° close haul range, despite if some routes are more windy than others.

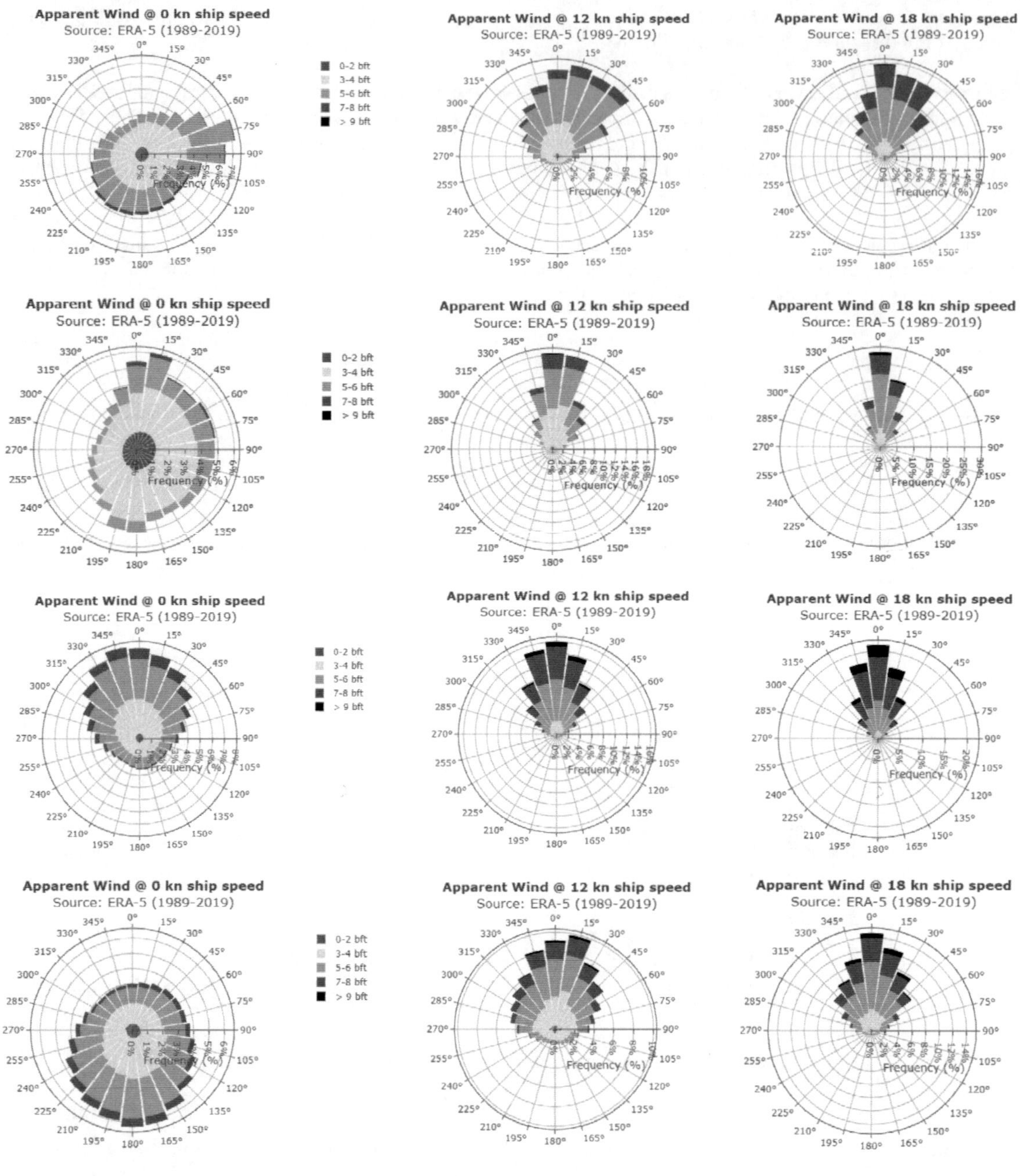

Figure 5: Apparent wind density probabilities [10] for shortest Routes (TW=AW@0kn, AW@12kn, AW@18kn), from top to bottom: **Sao Paolo** (Brazil) to **Singapore / Port Hedland** (Australia) to **Qingdao** (China) / **Rotterdam** (Netherlands) to **New Jersey** (US) / **Tokyo Bay** (Japan) to **Vancouver** (Canada).

Heuristic No 3:

For 70-80% of the sailing time, the useable winds will be coming from 0° - 90°, with close-haul to headwinds (0° - 45°) occurring by abt 50-60% of the time in major long-haul cargo ship trade voyages.

6. CRITICAL WASP FUNCTIONAL PROPERTIES

When over 50% of the time the apparent wind has to become usable through a narrow 0° - 45° close-haul window, then the debate around any specific wind propulsion system performance on this particular area becomes very relevant.

Within the Zone B, and in this narrow close-haul window, Wing Sails and Suction wings have the ability to generate certain forward thrust (even of small magnitude), with a time-wise high harvesting potential– for such highly probable occurrences, whereas Rotor Sails have a 'no-positive-net-thrust' blind zone which usually appears within 0° -20/25° AWA, depending on the AWS. For low AWS, the Rotor Sails need to spin faster so that they elevate their VR to deliver meaningful thrust through a higher C_L (Figure 1), in expenditure of electrical spin power, whereas within Zone C, very high wind speeds will lower the VR (despite the high RPM) and the C_L capability of the Rotor Sail but will improve the C_L/C_D ratio.

The Wing Sail needs to have a very good C_L/C_D ratio to exploit this frequently encountered zone, especially by reducing their C_D well below 0.5, thus being of a very good aerodynamic design, but which needs a very high accuracy of the optimal AoA alignment with the incoming apparent wind, hence controlled by a sophisticated local sensor system, while ship motions will affect this optimal AoA alignment through time (esp. in Zone C).

In such conditions, an interesting application is the Airborne Kite: by having the ability to fly above 200-400m over sea level, the TWS is magnified by altitude, so that the AWA to the Kite is becoming locally more beamy to broad-reach, favouring even the Zone C area supporting the Kite to harvest energy more than any deck-mounted wind propulsion system. Thus, the higher the Kite flies (even well above 400m) the better for T_F energy harvesting. The issue remains how useful such T_F conditions will be through a stormy weather environment for the ship.

By focusing on the most useable and most available wind for ship propulsion, the T_F generation from close-haul and beamy winds will be examined.

In the below analysis it will also be accepted - for reasons of simplicity - that since realistically feasible wind propulsion system arrangements on the benchmark oceangoing cargo ships do not provide more than a 20% propulsion power benefit, and since the operation will mostly take place within zones A & B, the developed side forces in close-haul wind conditions may not impair the manoeuvrability and may not increase significantly the added hydrodynamic resistance of the ship. As it can be shown in the below formula analysis, the T_F - when analysed into the Lift and Drag components - is strongly depended on the C_L/C_D ratio.

$$T_F = L \cdot \sin(AWA) - D \cdot \cos(AWA)$$

$$L = \tfrac{1}{2} \rho U^2 S C_L \qquad D = \tfrac{1}{2} \rho U^2 S C_D$$

For the most probable - to be encountered during voyage time - close-haul AWA, the higher this ratio is in the Zone B, the better for wind energy harvesting.

By referring to the previously shown Figure 1 numerical examples of the hypothetical/theoretical Rotor Sail and Wing Sail, it can be observed from the C_L/C_D variance in operational parameters (RPM-Rotor Sail and AoA-Wing Sail) that the Rotor Sails present the highest C_L/C_D ratio in low VR (close to 1-1.5), while their C_L/C_D ratio is relatively stable in the other VR regions.

The Wing Sail presents an optimal C_L/C_D at a specific AoA (the optimal one) and in the rest AoA regions the C_L/C_D falls dramatically away from the optimal. This suggests that for the Wing Sails, always maintaining an optimal C_L/C_D ratio depends on the accuracy with which the optimal AoA is measured, which cannot be secured without a very accurate local incoming wind sensing system (i.e., the bridge anemometer is quite far to provide for this accurate measurement).

In the Zone B encountered close-haul AWA, and away from their blind spot, the optimal VR of the Rotor Sail will be around 2-3, hence also operating at a C_L/C_D close to the Wing Sails optimal setting. These C_L/C_D figures can also reveal that the Wing Sails designed with a very good C_L/C_D ratio (and especially with quite low C_D value at optimal AoA) are less depended by means of wind energy harvesting from continuous wind capturing weather routing than Rotor Sails, since the latter need to stay away from their blind spot zone of 0°-25°, necessitating larger course deviations. However, weather routing executed with good deviation margins and the ability of the ship to become route agile (in changing course headings) will favour the flexibility of Rotor Sails to RPM-adjust their optimal C_L/C_D setting for wider close-haul and beamy apparent winds, and especially in Zone B such benefit is maximized when high VRs can be used.

The question remains what the deviation from any theoretical weather routing simulations is compared to encountering real seas traffic, and a very good example is the very dense traffic seen across the Vietnam-China coastlines with ships passing both East and West of Taiwan (Figure 6).

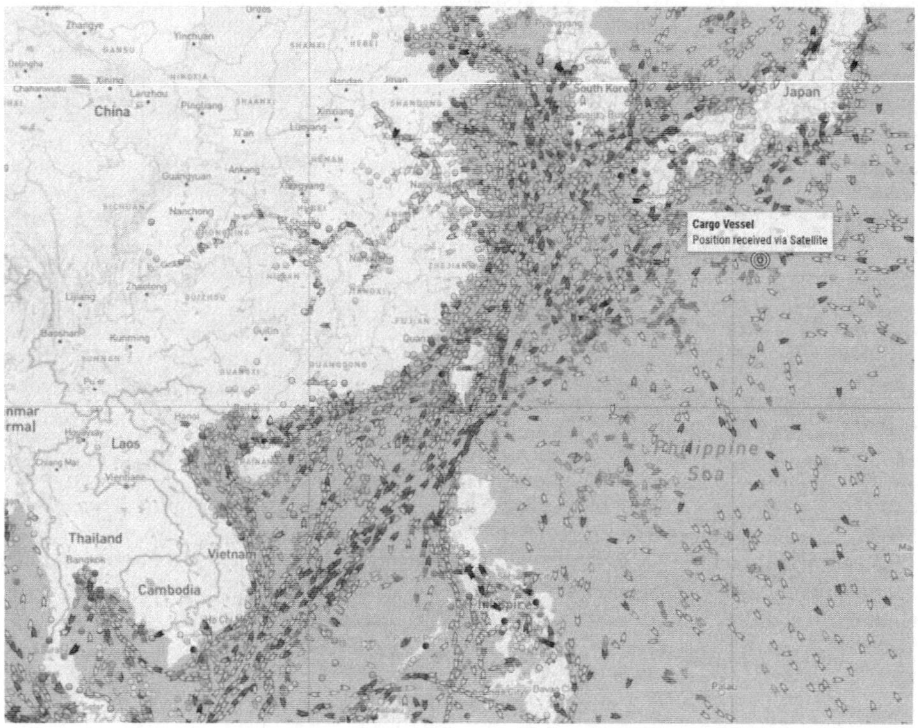

Figure 6: Traffic density across the coast of Vietnam, China, Korea and Japan (Source: MarineTraffic)

Furthermore, the commercial cargo ships are not sailing at same constant speed over the total voyage time, while the 4th IMO GHG study [11] revealed that for all oceangoing cargo vessels a considerable amount of time is spent in slow speed transiting – indicatively 40-50% for bulk carriers/oil tankers, 65% for container vessels and 20-30% for other ship types.

The analysis done above indicate that it is meaningful for wind propulsion system developers and engineers to focus on optimizing their aerodynamic device to be able to demonstrate superior performance within this narrow 0° - 45° close-haul window, by striving to elevate as high as possible the C_L and reduce as much as possible the C_D, and for the Operators to find means to increase wind energy harvesting in this particular operational window. The above example also shows that every different wind propulsion system type needs different operational optimization techniques, so apples-to-apples comparison is meaningless to be done on stand-alone wind propulsion unit performance comparison basis, but on ship profile basis.

Heuristic No 4:

The most important functional properties of a wind propulsion system are: 1) the C_L/C_D ratio achieved in close haul/beamy wind ranges, and 2) the accuracy with which a wind sensing system can provide every time the optimal C_L/C_D setting.

7. MECHANICAL SAILS CAPACITY FACTOR

Every mechanical sail has such properties and is designed with the scope of maximizing the developed L on its surface while minimizing D when operated through an incoming wind.
Eventually, for a given surface area and a set of aerodynamic properties (C_L, C_D), the mechanical sail will develop a maximum Thrust for a maximum wind energy generating condition (in this case wind speed), for a maximum wind power generation to take place.

But what is the maximum operational delivered power definition for a given mechanical sail?

That depends on the max to be encountered wind speed and the structural strength of the mechanical sail to withstand any developed mechanical aerodynamic forces. The structural strength of mechanical sails is determined by excess/extreme wind forces of over 50m/sec wind speeds so this is not reasonable as a value for defining the maximum delivered power of a mechanical sail. Hence, a lower wind speed than the extreme must be considered, but which one?
By observing how the long-established Wind Turbine industry approaches this definition, we can receive useful feedback.

The power curve of a wind turbine looks like the one shown in the Figure 7 [6]:

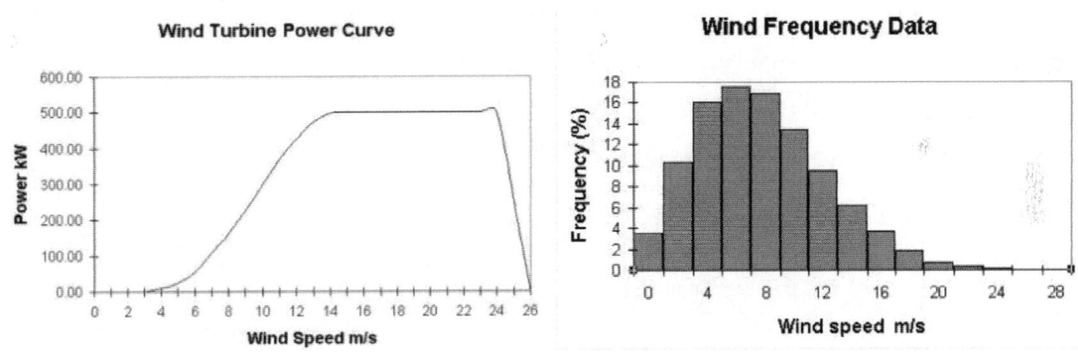

Figure 7: A typical power curve of a Land Wind Turbine and associated wind speeds frequency data at the site of installation (example) [6].

The maximum wind speed considered is around 24 m/sec (abt 46.65 knots), and it is defined as the higher wind speed that the Wind Turbine can operate continuously so that power is generated uninterruptedly. Statistically, in a given area, very high wind speeds may occur well over 50 knots, but these are considered rare events (i.e. extreme storms, hurricanes, etc) which can even be threatening to the Wind Turbine structural safety.

It can also be observed in the same Figure 7, that the wind frequency data presents a similar Weibull distribution with a higher probability of wind speeds occurring at 4-12 m/sec, but these are of the low power harvesting area for a wind-turbine: the ultimate power capacity of 500kW is used for less than 15% of the time.
Such Wind turbine applications have received a commonly accepted and adopted definition, the **Capacity Factor**.

The Capacity Factor is defined as the average annual power output generated by the Wind Turbine, divided by the maximum power capability (rated peak power). For example, if a 5MW wind turbine located in a specific wind available area is producing an average yearly power of 2MW (again on basis of long-term wind statistics), the capacity factor is then 40% (2000kW/5000kW).

In ship-fitted WASP systems such a capacity factor can be calculated equivalently, once a rated peak power definition can be used demonstrating the ability of a given WASP system to deliver continuously a maximum possible T_F for a given high wind force magnitude, while the average power generated per year for certain routes, weather conditions and ship utilization will be much smaller.

An effort is made hereby to calculate the capacity factor for certain market introduced mechanical sail technologies. For the sake of simplicity, which is not far from reality, it can be accepted that the maximum T_F coming from a mechanical sail is occurring when the mechanical sail is aligned with a full beamy apparent wind, so that the total generated vertical lift on its surface equals the total T_F (trigonometry actually shows that the max T_F will occur in a slightly close-hauled beamy AWA of abt 75°-80°, but such depends on the ship speed, so for the sake of simplicity a full 90° beamy AWA is hereby selected). It is also assumed that the maximum theoretical incoming wind speed marginally tolerable for WASP is of 50 knots, which represents a BF 10 condition (stormy weather with up to 12m wave heights), and which marginally may not affect the ship's stability, course keeping and overall structural safety.

Under such assumptions, the capacity factor (CF_{WASP}) for a WASP system arrangement will be described by the formula:

$$CF_{WASP} = \frac{P_{Yearly\ Average}}{P_{90AWA/50AWS}}$$

Supposing we have a ship that sails with an average speed of 12 knots and requires 7000 kW propulsion power in such condition, and by using simple mathematical formulas and the theoretical aerodynamic properties of our exemplary Rotor Sail and Wing Sail, below observations can be made:

Table 1: Capacity factor (CF_{WASP}) calculation

	Rotor Sail	Wing Sail
Dimensions: Height x Diameter/Span (m)	30 x 5	40 x 12
Max C_L (1)	10 (VR=5)	1.61 (AoA 12°)
Min C_D	1.5 (VR=1.5)	0.55 (AoA 12°)
Max Forward Thrust (2) (kN)	339.72 (VR=1.828)	271.18
Max Wind-assisted Net* propulsion power (3) $P_{90AWA/50AWS}$ (kW) (% of total propulsion power)	2 875.76 (41.08%)	2 391.35 (34.16%)
Average Delivered Net Propulsion power ($P_{D\ AVERAGE}$) (kW), 12 knots ship speed (4) MEPC77 wind statistics.	238.50	143.09
Capacity Factor per unit	**8.29%**	**5.98%**

(1) Optimal AoA value (Wing Sail), Max RPM 180 (Rotor Sail)
(2) The VR's relevant CL is used.
(3) Propulsion efficiency etaD 0.7, Rotor Sail max e-consumption of 120kW at max RPM of 180 – the electrical power is deduced.
(4) At 10m freeboard installation, considering Helmann exponent.

In the Table 1 the MEPC77 global wind statistics data Wi,j are used to multiply with the polar diagram $F(V_S)_{i,j}$ of these specific wind propulsion units' performance figures for different TWS and for a ship speed of 12 knots, where:

$$P_{D\,AVERAGE} = \left(V_S \frac{\sum_{i=1}^{m}\sum_{j=1}^{n} F(V_S)_{i,j}\, W_{i,j}}{etaD} \right) - \left(\sum_{i=1}^{m}\sum_{j=1}^{n} P(V_S)_{i,j}\, W_{i,j} \right)$$

It can be observed that in such marginal for the ship - but favourable for the wind propulsion system - performance conditions, the propulsive power contribution by one single unit can be up to 41%, provided there is no main engine propulsion limitation (i.e. high EPL, barred speed limit, etc) for allowing such high propulsion power replacement by the winds.

On the wonder how the ship's speed may affect the CF_{WASP}, various speed calculations can be run for the exemplary Rotor Sail and Wing Sail accordingly:

Table 2: Variation of $P_{D\,AVERAGE}$ per 1 knot difference.

	Rotor Sail (30x5)		Wing Sail (40x12)	
Vs (kn)	kW	δ kW	kW	δ kW
16	302.58	5.43%	170.06	3.85%
15	287.00	5.84%	163.75	4.19%
14	271.17	6.35%	157.17	4.58%
13	254.97	6.91%	150.28	5.02%
12	238.50	7.62%	143.09	5.57%
11	221.61	8.49%	135.54	6.21%
10	204.27	9.58%	127.61	7.04%
9	186.41	10.99%	119.22	8.13%
8	167.95		110.26	

The first observation is that the lower the ship's speed the lower the wind power harvesting, which is counter-intuitive considering that we would expect when the ship is sailing slower, better exploitability of Zones A & B would provide for higher probabilities of positive net T_F generation. The broader probability density exploitation is indeed happening, but since the T_F depends on an AWS square component, the lower wind power harvesting is due to the reduction of the incoming AWS to the wind propulsion units – hence the slower ship speed, the worse the CF_{WASP}. While the higher the ship speed, the smaller the wind power harvesting benefit, which above 16 knots seem to become less than 5%. However, the benefits of slower steaming are better reflected to the propulsion power replacement which is much higher (as the M/E propulsion power is dependent on the 3rd order of speed) and in the daily fuel consumption which is much improved. By using the simplified example of the Rotor Sail and Wing Sail similar CF_{WASP} indicators can be derived for the Figure 5 trade routes, as shown in Table 3.

Table 3: $P_{D\ AVERAGE}$ and CF_{WASP} per Figure 5 route.

Average Delivered Net Propulsion power (kW)*/ Capacity Factor	Rotor Sail 30 x 5	Wing Sail 40 x 12
Sao Paolo – Singapore	287.70 / (10%)	175.37 / (7.33%)
Port Hedland – Qingdao	142.00 / (4.94%)	80.59 / (3.37%)
Rotterdam – New Jersey	303.40 / (10.55%)	177.93 / (7.44%)
Tokyo Bay – Vancouver	357.01 / (12.42%)	226.73 / (9.48%)
La Plata – Le Havre	205.65 / (7.15%)	117.92 / (4.93%)

*At 10m freeboard installation for both, taking account Helmann exponent, for a ship speed of 12 knots and propulsion efficiency etaD 0.7

As it can be seen from the Table 3 route-specific analysis the CF_{WASP} can range from 3% - 12.5% depending on the system type, the wind intensity and direction probability, with the windiest routes utilizing a 10-12% of the full system's capacity. The analysis done on the availability and utilization potential of the ocean winds in the previous paragraphs justify the relatively low CF_{WASP} compared to land-based Wind Turbine applications.

The above example is simplistic, and better performance figures have been observed in full scale and due to continuous fine-tuned, improved designs (i.e. Rotor Sails max C_L values of even 12-13 have been recorded in full scale trials, and Wing Sails max C_L performance of even up to 2.5), which could help to elevate the CF_{WASP} up to 15%-20%, but the purpose here is to provide a handful heuristic of the expected capacity factor figure through real operations. It should also be borne in mind that these performance figures reflect passive use of the systems, i.e., without operational optimization and power integration to improve overall ship efficiency: such can even double the capacity factors. And this is probably the Best Take-Away from this heuristic: No matter how improved and performance efficient the stand-alone wind propulsion systems can become, their optimal utilization on board will determine their true benefits.

Heuristic No 5:

Any given WASP system is expected to operate within a passive capacity factor range of 5%-10% on basis of global trade routes wind statistics and for typical cargo ship speeds of 11-16 knots. Systematic route optimization and integrated power management can increase the CF_{WASP}.

8. EXPECTED WASP MAIN PROPULSION POWER REPLACEMENT

Any ship owner and operator need to have a constant empirical figure to understand the power replacement benefit coming from mechanical sails, as part of a hybrid propulsion system (being diesel-wind, electric-wind, diesel-electric-wind or other). So far wind propulsion system providers come up with either generic marketing messages about fuel savings which vary from ship type and size, or with route dependent performance evaluations, which mostly tend to confuse ship owners who need clarity for Investment decisions but also for practical operations (i.e., declare to the charterer a guaranteed (penalty-free) fuel consumption of a WASP fitted ship for a future 10-days single voyage charter).
Basis on the above analysis of a CF_{WASP} definition, ship owners are encouraged to use below empirical approach.

The maximum operationally potential delivered power from a mechanical sail, for a full beamy 90° apparent wind, is based on the below formula:

$$P_{WASP\ max} = \frac{(T_F = L)\ V_S}{etaD}$$

For air density of 1.25 kg/m^3, AWS = 25.722 m/sec (50 knots), propulsion efficiency etaD = 0.7 and converting to kN, the max delivered power from a Wing Sail (P_{WSmax}) can be determined by the below formula:

$$P_{WSmax} = 0.3\, CL_{max}\, A\, V_S$$

Where A is the projected surface area of the wing sail in m^2 (height x span of the sail), V_S is the ship speed in knots and CL_{Max} is the max Lift coefficient at the optimum AoA. It is considered hereby that the Wing Sail does not consume electrical power during operating in a fixed optimal AoA position.

In the case of Rotor Sail, a similar formula can be used, with the CL_{max} value being more complicated to calculate, as it depends on the VR of the Rotor Sail, which in turn depends on the Rotor Sail design (max RPM and diameter). So far Rotor Sails with aspect ratio of 6 or 7 and 5m diameter have a max RPM of 180, and those of 4m diameter a max RPM of 225, both providing for an approximate rounded VR = 1.8 at 50 knots AWS, hence such value can be empirically used:

$$P_{RSmax} = 0.3\, CL_{VR=1.8}\, A\, V_S - P_{EmaxRPM}$$

In the P_{RSmax} formula the power absorbed by the Rotor Sail at max RPM ($P_{EmaxRPM}$) must be deduced.

To derive a generic average P_{WASP} benefit estimate for global route trading on basis of MEPC 77 global wind statistics, which can also include a certain level of effortful operational optimization, a CF_{WASP} of 10% can be chosen as representative for deck operating wind propulsion systems comprised of N mechanical sail units, expressed by the below generic formula:

$$P_{WASP\ global} \approx 0.03\, N\, b\, A\, V_S$$

The b multiplier value adjusts a statistical convergence of the $P_{WASP\ global}$ with several vendors disclosed passive use global wind statistics performance figures within 10-16 kn speed range, by further including an average 20% performance gain potentially achieved through optimization:

- Rotor Sail: b ≈ 5
- Wing Sail: b ≈ 1.3
- Suction Wing: b ≈ 4

<u>Disclaimer about the CF_{WASP}</u>: it was shown that the capacity factor for a conservative theoretical passively used mechanical sail can range from 6-13% depending on route and on the aerodynamic properties. Full scale systems, improved designs, operational measures and improved accuracy of wind measurements can provide for better values, even up to 20%. However, the capacity factor will always be route, weather time period and traffic depended, with most oceangoing cargo ships executing variant annual voyage mix (routes and speed of sail), hence an average capacity factor of 10% is deemed as best reflecting trade-offs between real seas operational downside and potential performance enhancement upside.

Heuristic No 6:

A conservative, empirical average anticipated wind-assist propulsion power contribution during global trade routes can be calculated by the simplified generic formula P≈0.03NbAVs, with values selected appropriately as per mechanical sail type.

9. CONCLUSIONS

The above analysis demonstrates that the shipping industry can adopt 6 simple but reasonable heuristics as empirical means to understand and evaluate the mechanical sails performance. Such heuristics already provide an awareness that:

- Most available ocean winds are BF2-6 occurring from all possible directions,
- Most usable winds exist within BF4-6 at abt 53% of the long-term utilization time,
- For 70-80% of the sailing time at any given route the useable winds will come from 0°-90°, of which 50-60% of the time being 0°-45° close haul winds,
- The most important functional property of a mechanical sail is the C_L/C_D ratio and the wind sensing accuracy,
- Any given WASP system is expected to operate within a capacity factor range of 5%-10% on basis of global trade routes wind statistics and for typical cargo ship speeds of 11-16 knots,
- The average anticipated WASP power contribution for global trade routes can be empirically calculated by a simplified generic formula for deck operating systems.

10. REFERENCES

1. Justifying method choice: a heuristic-instrumentalist account of scientific methodology, Till Grüne-Yanoff, link.springer.com/article/10.1007/s11229-020-02962-z.

2. Thinking, Fast and Slow, Daniel Kahneman, 2011

3. Wind Speed Distributions Used in Wind Energy Assessment: A Review, Huanyu Shi, Zhibao Dong, Nan Xiao, Qinni Huang, Frontiers in Energy Research, 22/11/2021.

4. Simulation model of a ship's energy performance and transportation costs, Fabian Tillig, PhD thesis, Chalmers University, Sweden 2020

5. IMO MEPC.1/Circ.896

6. Wind Propulsion Principles - The complete Guide to harvesting ocean winds for commercial shipping, Edition 3, June 2022, Konstantinos Fakiolas.

7. The complex network of global ship cargo movements, Pablo Kaluza, Andrea Kölzsch, Michael T. Gastner, Bernd Blasius, Institute for Chemistry and Biology of the Marine Environment, Carl von Ossietzky Universität, Oldenburg, Germany.

8. WindLab: Seakeeping and manoeuvring model tests with wind assisted ship propulsion, June 2020, MARIN

9. en.wikipedia.org/wiki/Betz%27s_law

10. blueroute.application.marin.nl

11. MEPC 75/7/15, Annex 1 – Executive summary of the 4th IMO GHG study 2020, proportion of CO2e emissions by operational phase and by ship type.

11. AUTHORS BIOGRAPHY

Konstantinos Fakiolas is a senior naval architect, founder and CEO of Finocean Ltd, developing energy efficient and low emission newbuilding and ship upgrades. He has been specializing in WASP since 2014, involved in active WASP installation projects & institutional research. He is the author of the first global shipping industry Book for WASP commercial ship applications, the 'Wind Propulsion Principles'.

NOVEL PLATFORMS FOR OPTIMISING RENEWABLE ENERGY POWERED SHIPS

K Goh, Knud E. Hansen A/S, Australia

SUMMARY

The concept of a 50,000 DWT bulk carrier that uses wind and solar propulsion as the main source of propulsion is presented. The concept uses a proa form with main hull and outrigger. The unique form allows the vessel to maximise wind and solar energy harvesting. Due to the unconventional form, much of the proa design must follow first-principles engineering methods. The development of the calculation model for sailing economics is presented and trialled with several Windship variants to show the benefits over a conventional motor ship.

1. INTRODUCTION

This is normally the first section in the main body of the text. This section and all subsequent sections and sub-sections should be numbered manually. Automatic numbering systems MUST NOT be used. All main headings should be in bold capitals

Reducing CO2 for shipping is a high priority but the small efficiency gains in current internal combustion technology and using wind assisted systems is unlikely to make the significant changes to greenhouse gases that are being sought. KNUD E. HANSEN has been involved with numerous studies and commercial projects involving wind propulsion since 1996. It has been a persistent goal to utilise wind as a primary propulsion system, as opposed to only wind-assisted concepts.

One of the main technical challenges of wind propulsion is that the large sail rigs get in the way of cargo stowage and loading operations. KNUD E. HANSEN is investigating an unconventional proa hull platform to overcome these cargo issues and further increase the available area for solar energy. The paper examines some of the main engineering aspects and challenges of the proa platform in view of future development and studies to verify the benefits and practicality of this radical new vessel.

The sailing economic calculation model has also been upgraded to include more complex solar power calculations and several conventional Windship variants have been tested on four global trade routes. The fleet economics for a grain trade is then presented to clearly show the benefits of the Windship over a conventional bulker in terms of cost savings and environmental benefit.

2. PROJECT BACKGROUND

Around 1999, KNUD E HANSEN had developed an impressive concept design for a fully wind powered vessel – The Windship (Figure 1). Comprehensive research into the economics and sail technology showed that a wind powered vessel for bulk and liquid cargo was feasible on certain trading routes, but in the year 2000 the additional capital expenditure was not able to be recovered due to low fuel prices.

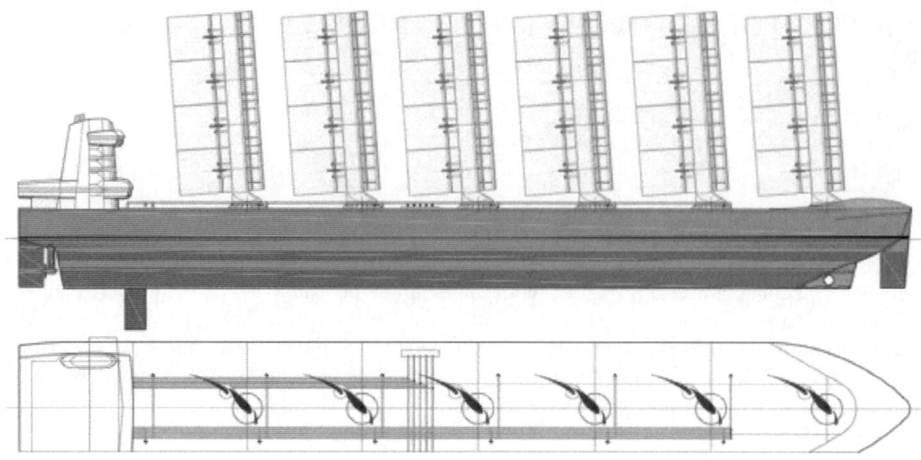

Figure 1: Knud E. Hansen Windship (circa 1999)

The current economic and environmental situation make the Windship much more attractive. In 2017 an updated Windship 2025 concept was proposed that combines the performance of the earlier wing sail rig, a hybrid power system and autonomous operation. A new calculation model was developed to investigate the capital and operational costs and was benchmarked against the previous studies to ensure accuracy. The study demonstrated that large savings in emissions the cost of operation are now possible for the Windship within technology that is available today.

From the previous Windship studies, two aspects of efficiency were particularly notable. The first was the relative inefficiency of wind power ships when sailing into the wind. Not only must the Windship tack, meaning a longer sailing distance, but the efficiency of the sailing rig itself is much lower than other wind directions since less of the lift force contributes to the propulsive effort and the drift forces are much higher. Furthermore, the effectiveness of the abaft sails is reduced considerably due to wind shadowing of the ahead sails.

The other notable aspect was how effective the solar panels were at reducing power requirements. The amount of power that photo voltaic solar panels can generate is improving. Would it be possible to install even more solar panels - enough power for significant propulsion such that heading upwind is possible without needing to tack with sail power?

3. PROA WINDSHIP CONCEPT

3.1 PROA PLATFORM

A catamaran configuration with the two hulls connected by a wide truss structure supporting the solar panels and sails was studied. However, the cost of constructing two hulls and supporting cross structure would likely double the cost of the vessel. The cross structure would need to cope with the loads from the other heavy hull as well as the sailing rig. Also, port operations would be greatly complicated by needing to unload and load both hulls.

Enter the proa (Figure 2) – a vessel with a main hull and a smaller side stabilising hull called an amah. A common sailboat type throughout the Pacific for centuries, this highly asymmetrical vessel is known for its speed and stability. This concept is gaining considerable attention, particularly from smaller eco-friendly driven shipping companies targeting high-value bulk cargo such as coffee beans that could benefit from a vessel able to load at small ports. A proa configuration is more expensive to construct than a conventional monohull, but not as much as a catamaran while still having similar advantages.

On a proa, cargo is loaded entirely into the leeward main hull, which is also close to the centre of buoyancy (Figure 3). There will be very little change in righting moment regardless of the cargo load. With a proa, righting moment is provided by the weighted amah and outrigger structure to windward, not buoyancy to leeward. The righting moment remains consistent, meaning stress, scantlings and sail area will also remain consistent between laden and unladen compared to a catamaran. This is conducive to an efficient structure through a widely varying displacement compared to other multi-hulls.

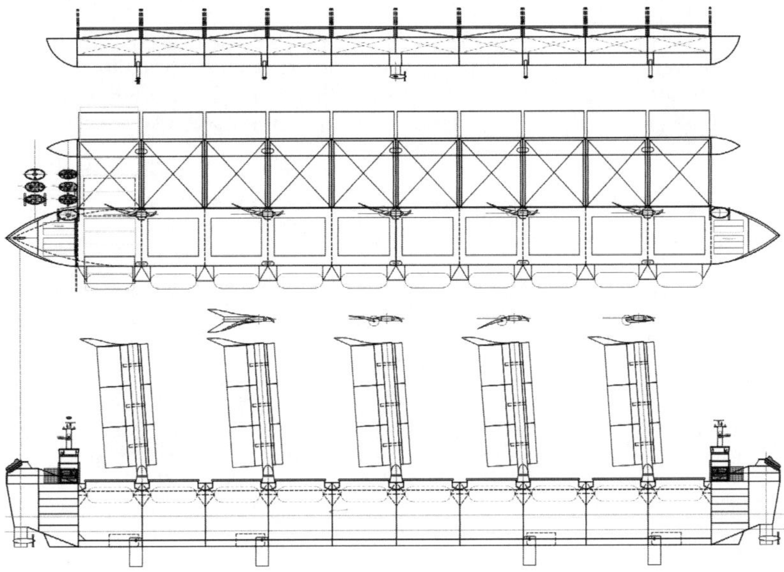

Figure 2: Proa Windship general arrangement

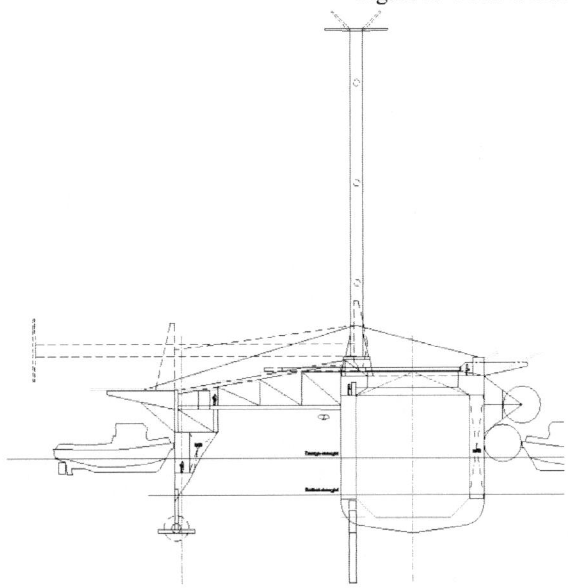

Figure 3: Proa midship section

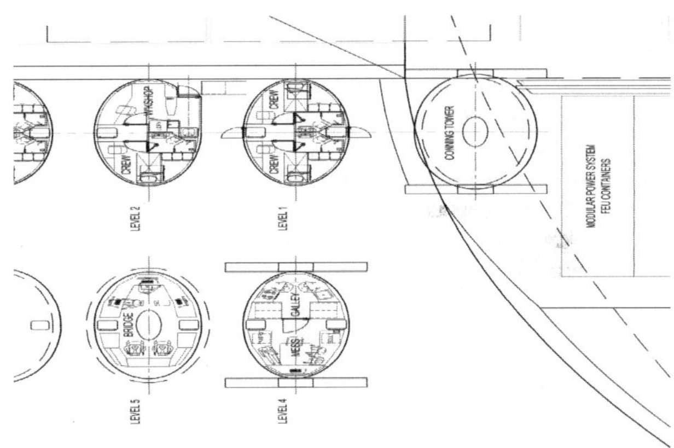

Figure 4: Proa accommodation tower

The large space between the main hull and the amah is ideal for locating solar panels and the amah can limit the vessel rolling motions and provide some protection for the solar panels. The proa is designed to always sail with the amah hull to windward. This allows the sailing rig to be located on the windward side of the main hull, moving them out of the way of cargo and loading operations. The interference of the sailing rig with the cargo is one of the main reasons limiting the development of modern ship wind propulsion.

Although the goal is to ultimately have the Proa Windship unmanned, it is expected that a crew will be necessary for longer oceanic voyages, at the time of year 2030, the goal in-service date for the Proa Windship. A crew of ten should be adequate if the vessel is designed for periodic instead of continual maintenance. As per the previous Windship 2025 study, the use of containerised generators, fuel tanks and batteries will significantly reduce the on-board maintenance requirements. So will the use of electric systems instead of hydraulics where possible.

The crew is accommodated in two conning towers, one at each end of the vessel (Figure 4). The vessel can be conned from either tower which solves the problem of masts and sails obstructing visibility. A long internal corridor on the windward side of the main hull connects the towers and also provides access to the various electrical rooms required for the solar power harvesting and controlling the sailing rig.

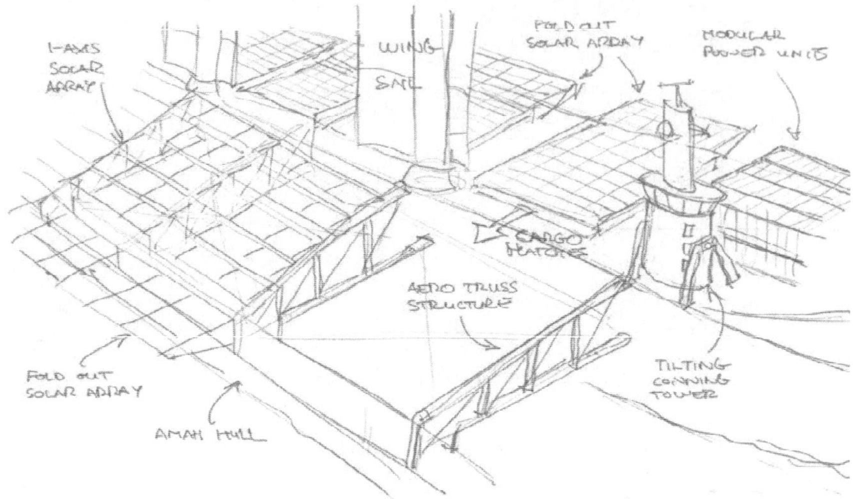

Figure 5: Proa Windship – Novel design aspects

The additional stability afforded by the proa configuration enables the main hull length-to-beam to be increased beyond the normal range of 5-7. The longer hull, 300 m length compared to 225 m, would be less cost effective in terms of steel weight per cargo deadweight, but this is offset by being able to carry more cargo volume. The longer hull can also support a larger solar array to harvest more energy. Another advantage of a longer proa configuration is the greater space available for locating the sailing rig. Masts can be placed at greater spacing to reduce the wind shadowing effect of one mast on another, or more or larger sailing rigs could be installed.

The narrower main hull also means that the cargo is less susceptible to moving when the vessel rolls, which has led to numerous bulker capsizing casualties and loss of life, due to the catastrophic nature and rapid speed of such events.

3.2 PROA SAILING

The proa configuration has some interesting sailing characteristics. Because a proa is designed to sail with the amah always to windward, the vessel does not change tack the same way as a normal sailing craft. Instead, a proa needs to stop with the sails perpendicular to the wind and then change direction like a double-ender ferry. This sailing action is called shunting, similar to the way trains can climb steep slopes with the aid of switch-back tracks (Figure 7).

A shunting vessel needs to have a symmetrical hull fore and aft. Typically, a double-ender type hull has a lower hydrodynamic efficiency than a similar capacity single ender ferry, however much of this difference is due to a lower propulsion efficiency of the forward propeller wake-field. It is expected that this will be less significant, since a proa will have longer than normal length-to-beam ratio (13 compared to about 7) with finer lines in the bow and stern.

Figure 6: Proa Windship with amah 'flying'

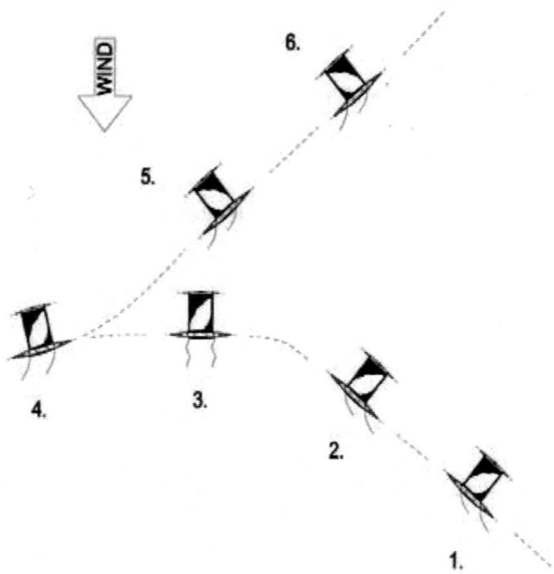

Figure 7: Proa shunting manoeuvre

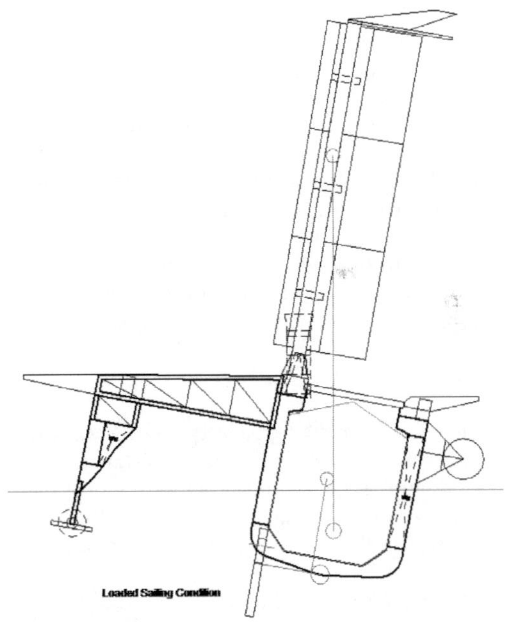

Figure 8: Proa heeled-sailing concept

To balance the proa, ballast tanks are provided on the leeward side of the main hull and in the amah (Figure 8). When the vessel is in the unloaded condition, the leeward ballast tanks are filled to offset the weight of the amah. When loading cargo, the vessel can discharge the leeward ballast water since the amah buoyancy will balance the vessel. The vessel will sail heeled to leeward by up to 10° to fly the amah free of the water to reduce resistance (Figure 6).

When sailing, ballast water can be used in either the leeward or the amah tanks to balance the wind forces on the sailing rig when flying the amah free of the water. The transfer of ballast water between amah and main hull is not fast enough to cope with wind gusts and resulting variations in the sailing rig over-turning moment.

Therefore, T-type control foils are fitted to the amah to provide active control response to changing wind forces.

The heeled sailing condition raises some interesting questions. The waterplane area of the heeled main hull will be asymmetrical and will cause the vessel to yaw towards the windward side. If the Proa Windship sails heeled the majority of time, especially at higher speeds, it is logical to shape the hull lines such that the main hull bow and stern create a symmetrical waterplane when heeled. An asymmetrical main hull will be investigated in a future study. Sailing at a heeled angle will be uncomfortable for the crew, therefore the conning towers are designed to tilt to keep the accommodations level.

3.3 PROA STRUCTURE

The main hull is designed as a full double hull to increase the longitudinal strength and provide for the leeward ballast tanks necessary to balance the amah hull. The windward side of the main hull needs to cope with additional forces from the sailing rig, the amah hull and the retractable keels. Further FEM analysis may show that the double hull could be removed on the windward side of the main hull, but until then the conservative design is implemented.

The high length-to-beam ratio of the main hull falls outside the normal ratio for such bulker vessels as defined by Class. And therefore, many Class structural rules are not applicable for this proa concept. In order to check the structural weight of the main hull, it is necessary to use first principles engineering calculation to check the Still Water Bending Moment (SWBM) for hogging and sagging. This work is on-going and will be reported in a future study.

The structure joining the amah hull to the main hull is designed as a series of trusses connected to the main hull bulkheads to reduce weight. The only purpose of this structure is to support the amah hull and the solar panels. The truss members have an aerodynamic fairing to reduce air resistance over the large, exposed airflow frontal area.

The truss structure needs to cope with both buoyancy and mass loads from the amah since the amah maybe in the water or free from the surface when it is 'flying'. The wave loads will also be significant, given the heavy weather a wind powered ship is likely to sail in. A basic truss calculation was undertaken to determine the adequacy of the structure and confirm the weight estimate.

The hydro-resistance on the amah is relatively low by comparison to the vertical loads. Therefore, horizontal cross-bracing between the trusses with steel cables is sufficient with the trusses taking the additional compression forces from the pre-stressed cables.

3.4 PROA STABILITY

The stability of the proa vessel is certainly more complex than a normal mono-hull vessel. As with most stabilised concepts, the amah is designed to have minimum contact with the water. However, the outrigger configuration creates an enormous stability margin from capsizing due to both the buoyancy and weight of the amah acting at a large moment arm. The NAPA stability model with maximum forces on the sail rig, the vessel's heel was only affected by less than 5°.

The amah is designed to have minimum contact with the water to reduce the resistance created by amah hull. The complication for a merchant vessel is the large draught variation between loaded and unloaded conditions. This is especially the case for bulkers or tankers which have high deadweight to lightweight ratios. Therefore, the Proa Windship is designed to sail heeled leeward to raise the amah out of the water when the vessel is heavily loaded.

Stability of the heeled vessel to windward is well provided by the amah buoyancy should the vessel roll towards the amah. The NAPA stability model showed that the vessel has low stability margin to leeward when sailing heeled. Protection from leeward capsizing can be provided by large pneumatic fenders supported off the leeward hull by truss structures. The righting moment of 850,000 kNm would be more than twice of the maximum sail forces. The fenders will also help balance the rolling moment created by the amah hull during wave encounters. Furthermore, the fenders could also be lowered to protect the hull in berthing.

ID: LC24 : Cargo: Wind 10 Degrees Heel

Floating position:

Draught mean moulded	12.991 m	Dead weight	56462 t	KMt	27.994 m
Thickness of keelplate	0.010 m	Light weight	15682 t	KG Solid	10.905 m
Draught mean (Below keel)	13.001 m	Displacement	72144 t	GM Solid	17.090 m
Draught FP (Below keel)	13.101 m	Seawater S.G.	1.025 t/m3	FS. cor.	0.004 m
Draught AP (Below keel)	12.902 m			KG Fluid	10.909 m
Trim	0.199 m	Heel angle	-10.2 deg	GM Fluid	17.086 m

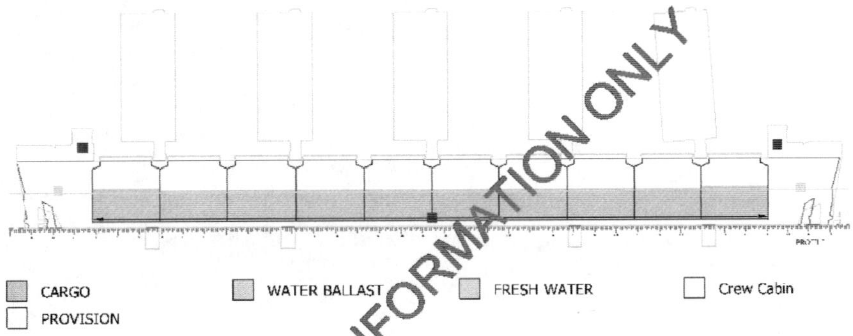

Figure 9: Proa heeled stability

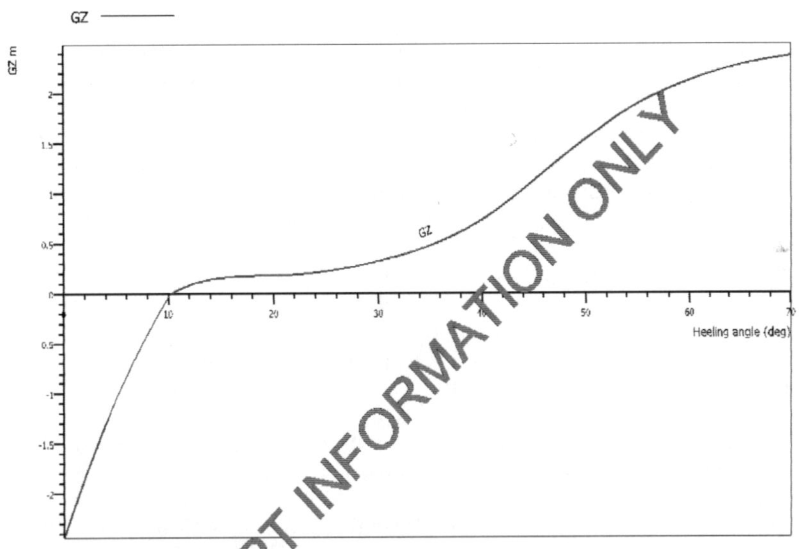

No downflooding within the calculated stability heeling range.

Criteria	TEXT	REQ	ATTV	MINGM	SIDE	STAT
AREA30	Area under GZ curve up to 30 deg.	0.055 mrad	0.063 mrad	17.016	SB	OK
AREA40	Area under GZ curve up to 40 deg.	0.090 mrad	0.149 mrad	16.814	SB	OK
AREA3040	Area under GZ curve between 30 and 40 deg	0.030 mrad	0.086 mrad	16.530	SB	OK
GZ0.2	Max GZ > 0.2	0.200 m	2.367 m	14.779	SB	OK
MAXGZ25	Max. GZ at an angle > 25 deg.	25.000 deg	70.000 deg	13.970	SB	OK
GM0.15	GM > 0.15 m	0.150 m	17.086 m	0.150	SB	OK

Figure 10: Proa heeled condition GZ curve

3.5 PROA HYDRODYNAMICS

The longer main hull of the proa configuration will have a higher frictional resistance due to the lower Wetted Surface Area (WSA) to displacement ratio compared to the shorter and wider hulls of similar capacity bulkers.

However, the hulls residual resistance comprising of wave-making and eddy currents is expected to be considerably less due to the lower Froude number, about 0.15, of the longer hull and smaller beam (Figure 11).

Due to the unusual hull dimensions of the Proa Windship, semi-empirical calculations and regression formula could not be used for resistance and powering estimations with any certainty of accuracy. Therefore, CFD has been used to compare the Proa Windship with the previous more conventional Windship hull, which is closer in main dimensions to a standard bulk carrier of similar deadweight.

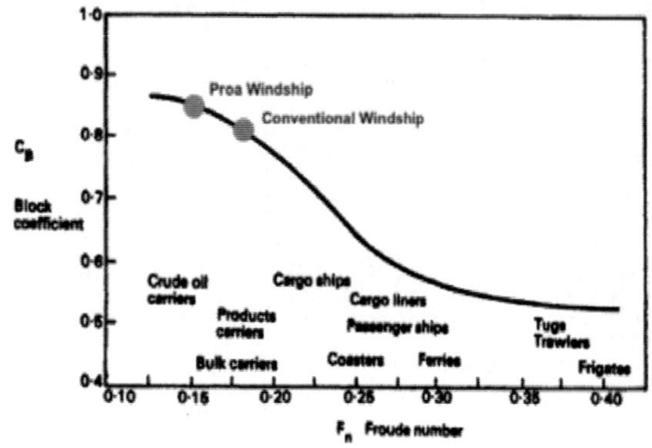

Figure 11: Proa Froude number of main hull

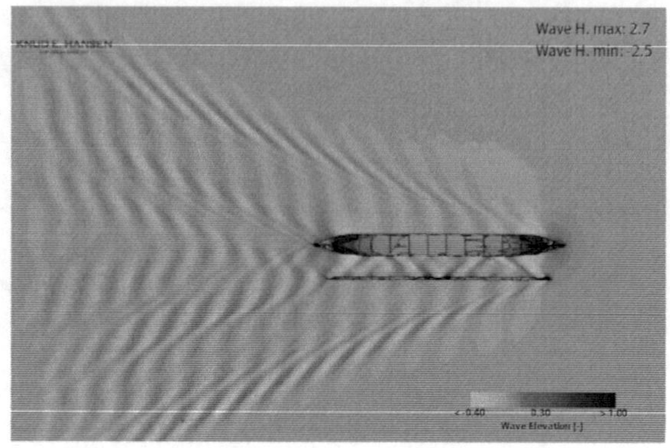

Figure 12: CFD multi-hull wave interference

The CFD calculations have been done at ballast and maximum deadweight. The ballast condition was sufficient for propeller immersion. The increased lightweight of the Proa vessel over the conventional Windship was about 3,000 tons. This is caused by the relative inefficiency of the proa structural configuration compared to a normal configuration. Furthermore, the WSA of the fully loaded proa was higher by about 48% due to the amah hull and high length-to-beam of the main hull. This increased frictional resistance considerably, but at 15 knots the overall resistance between the proa and conventional vessel was very similar due to the lower residual resistance of the longer and narrower proa hull. The effect of wave interference between the main and amah hulls still needs to be investigated (Figure 12)

With a heeled hull the CFD analysis has shown a great potential for reduced powering. At 15 knots and a 5° heel the overall resistance is reduced by 25%. At 10° the amah hull is out of the water, and the resistance is reduced by 60%. Note that these CFD calculations are only preliminary. Although the accuracy of the CFD solver is typically within 2-3% of model tank testing methods, the hull forms have not had any optimisations and trim was also not investigated.

Added resistance in waves is also expected to be reduced considerably with a narrower hull having less frontal area and a longer hull pitching less in heavy seas where a wind powered ship will spend much of its time. Although, this has not been calculated or tested at this time, a reduction of 30% for the added resistance in waves component has been included in the economic calculations for the Proa Windship based on the difference in frontal area. This will be verified in a future seakeeping study.

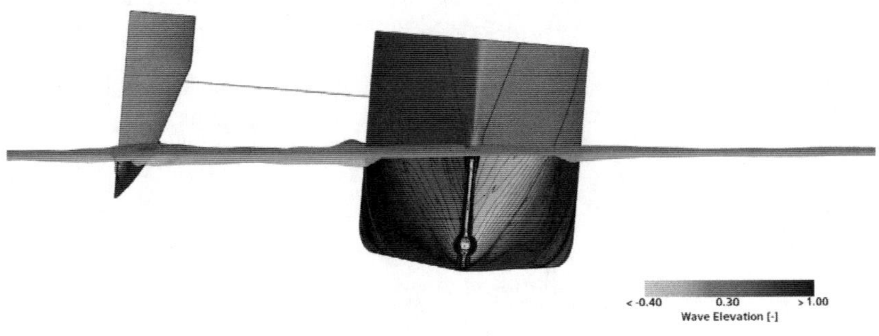

Figure 13: Hull pressure at 15 knots & 5 degree heel

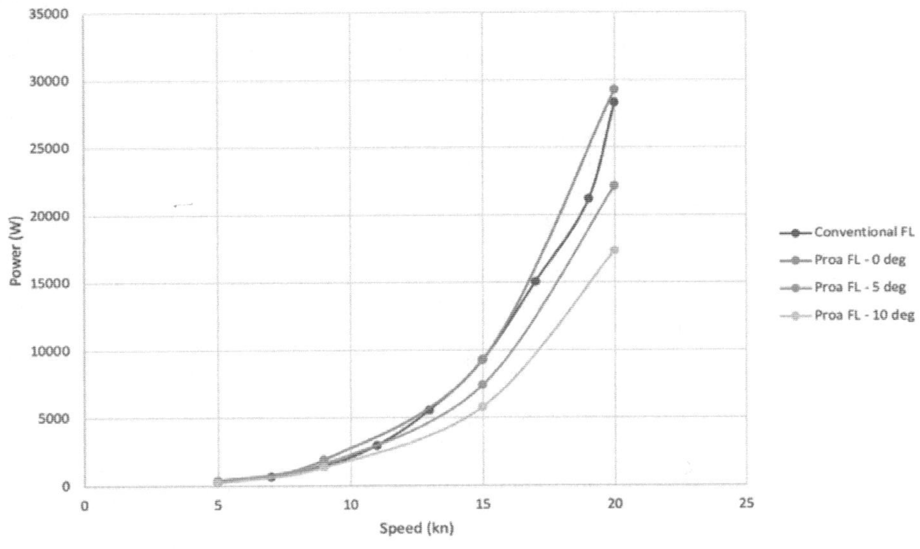

Figure 14: Powering at full deadweight

3.6 PROA HYDRODYNAMICS

As in the 2017 study, the sailing rig comprises of five wing masts. The wing mast is like an airliner wing during landing. With slats and flaps the sail has a high lift coefficient, of about 3.2. This enables the sail to be much smaller, with only a third of the area compared to a normal sail with a lift coefficient of about 1.0.

The high-lift wing mast provides lift at a wide range of apparent wind directions compared to other high-lift sail types such as Flettner rotors. The disadvantage of this type of high-lift wing sail is the high drag-to-lift ratio making them less efficient than a simple wing in upwind courses. The wing could be designed like an airliner wing to have retractable slats and flaps to reduce drag and improve performance upwind, however this would involve increased mechanisation, cost and maintenance.

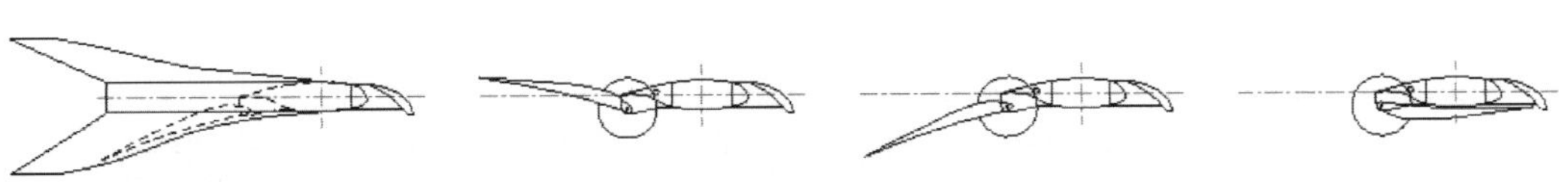

Figure 15: Wing sail section showing winglet, flap position for low drag, high-lift and reefing

The wing is comprised of several sections that can rotate around the central mast to change the direction of the wing profile, which is necessary when the vessel changes tack. The mechanism can also be used to enable the wing sections to track the sun when wind conditions are too low for effective propulsion. Each wing section has a slotted slat and a movable flap that can adjust to provide wing twist and gust alleviation. The flap can also fold forward for reefing in case of wind speeds exceeding the design limit of 50 knots.

The mast is a steel box structure that also forms the central section of the wing profile. The mast can azimuth to point the wing into the apparent wind and rotates 180° to enable the proa to swap bow to stern when shunting. The entire mast with wing sections can fold down onto the amah hull to:

- Reef the sail rig fully
- Clear traffic & rail bridges
- Prevent damage from cargo loading
- Provide better access for maintenance

A large winglet is installed at the top of each mast to reduce the induced drag from vortex shedding at the end of the wing sail. The winglet comprises of two moveable fins that can fold up or down, depending on the direction the wing profile is facing. One fin is folded onto the wing flap to provide an aerodynamic fence, while the other fin is angled up to reduce the strong vortices at the tip of the wing where low- and high-pressure streams meet.

The proa configuration also allows the sail masts to have greater separation from each other, reducing the loss of lift when the sails are in wind shadow of an upwind mast.

3.7 PROA HYDROPROPULSION SYSTEM

The hydro-propulsion system is provided by three electric pod thrusters, integrated as a steering rudder. One unit at each end of the main hull will provide about 3,000 kW of propulsive power. The third unit is a smaller 1,000 kW unit located amidship of the amah to counter the resistance which otherwise would need to be less efficiently countered by the main hull rudder thrusters. The three thrusters, each being able to azimuth 360° should give the Proa Windship the ability to berth and manoeuvre in and out of ports without tugs, reducing operating costs.

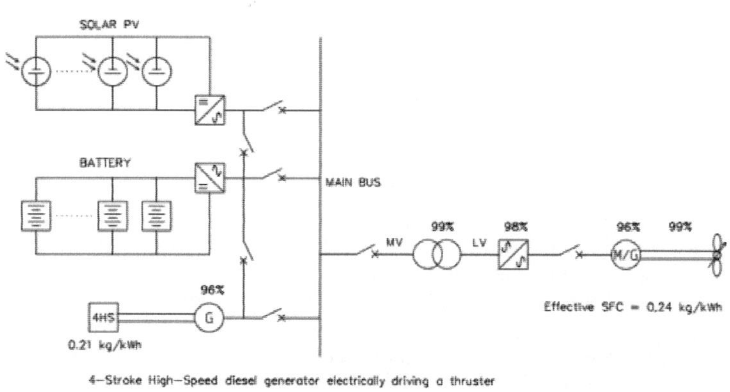

Figure 16: Hybrid propulsion concept

Since there are two main propellers, the propeller loading can be reduced to increase efficiency. With the power distributed between two propellers, a smaller propeller size is possible. In the ballast condition, this will allow the Proa Windship to run at a reduced draft or without aft trim to keep the propeller submerged, which should also reduce the propulsion power needed. The propellers will need to be of a Controllable Pitch (CP) type to work efficiently at different speeds and resistance, since the sailing rig will load and unload the propellers depending on the wind speed and direction.

The Proa Windship will have a symmetrical hull that can sail efficiently in either direction. Studies of the propulsive efficiency of similar double-ender ferries show that the overall propulsive efficiency is much better with only the astern propeller loaded because any accelerated flow created by the ahead propeller creates increased resistance on the forward hull. Therefore, the ahead propeller should be feathered if possible or turned to minimise its resistance.

No optimisation of the hydro powering system has been performed in this study. For example, it may be better to have each of the main thrusters capable of 4-5 MW. This would enable an increased loading of the aft propeller and reduce the required thrust from the ahead propeller to increase efficiency.

As proposed in the previous Windship study, the thrusters should also operate as hydro generators when the vessel speed is favourable. Model testing of the Windship with a conventional hull demonstrates that in a favourable reach (wind from the beam), the bulker vessel could sail at nearly 20 knots in a wind of 15 m/s.

The efficiency of harvesting will be somewhat lower than the propulsive efficiency since the propeller shall be optimised for propulsion and not as a turbine. A possible arrangement to improve efficiencies is to have the propeller and turbine on opposite ends of an azimuthing pod with the motor/generator in between. The non-active impeller would be declutched and feathered or folded.

3.8 SOLAR POWER SYSTEM

The amah hull is positioned about 25 meters from the main hull to limit the overall beam of the vessel to 50 meters. With a maximum draught of about 12 meters, the proa vessel should be able to access most trading ports. With an amah length of 250 meters, the area between amah hull and the main hull is 6,850 m2. Allowing a conservative 95% solar panel coverage and a panel output of 235 W/m2, the peak power output is 1,530 kWe.

A simple folding trapeze structure from the amah will expand the solar collection area by another 3,000 m2, adding a further 670 kWe. Solar panels mounted on the large cargo hatches add 700 kWe and a similar trapeze from the hatch extending to leeward, another 600 kWe. The trapeze is mounted to the sliding hatch instead of the main hull to enable the delicate panels to move out of the way during cargo operations. The rigid panel solar peak power totals to about 3,500 kWe.

The rigid structure of the wing sail also provides a perfect host for solar panels and the vertical orientation of the sails will complement the horizontal panels to even out the solar harvesting throughout the daylight hours. Both side of the wing would be covered by solar panels to maximise harvesting throughout the day. Although the sun cannot be shining on both sides of the sail at once, the side in shade can still generate about 10-25% energy due to light reflection off waves and diffuse lighting.

The curved surfaces of the wing use flexible solar panels with a lower power capacity of 160 W/m2 compared to 235 W/m2 for the rigid flat panels. Each of the five wing sails has an area of 1,000 m2, resulting in 1,500 kWe with 90% coverage and peak power of about 850 kWe, considering one side will be in shade. During periods of low wind, the wing sail is also able to act as a double-axis solar tracking system.

In addition to the type of solar panels, the harvesting capability of the Proa Windship was divided into different tracking types. The rigid panels mounted to the cargo hatches are fixed in the horizontal plane. The rigid panels in the truss structures have single-axis tracking, with the axis aligned to the vessel's length. When under sail propulsion, the flexible panels on the sail are fixed vertically at the mast azimuth determined by the apparent wind direction. When utilising the sail rig for solar harvesting, the mast and wing sections act like a 2-axis tracker.

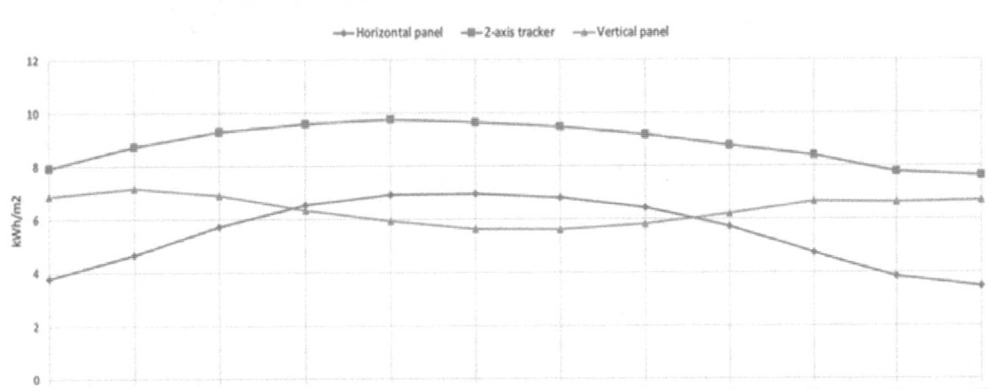

Figure 17: Daily solar radiation for various PV panel types for latitude 30 degrees north

The solar calculator showed that on equatorial routes, the average solar power that could be harvested over a day in ideal conditions was about 30,000 kWh. This would amount to an average of about 1,250 kWe across the entire day and night, allowing for about 5% loss due to converters and some battery charging. This will obviously be much lower at the higher and lower latitudes and with effects of the cloud cover. The time of year will also vary the harvesting potential.

4. TECHNO-ECONOMIC MODELLING

4.1 PREVIOUS STUDIES

The calculation model developed for the 2017 study was able calculate sailing performance by wind and auxiliary propulsion systems for any chosen sailing route and thereby also inform the fuel consumption based on generally accepted efficiencies and performance metrics. The 2017 model correlated well with the more detailed initial Windship studies from 1999 and could therefore be used to quickly verify the basic economics of the WindShip and compare them to a conventional motorship.

This model was tested with a trans-Atlantic route from Rotterdam to New York City and ballast return with an average wind speed of over 8 m/s and a long Trans-Pacific grain trade route from Vancouver to Shanghai with ballast return with a lower wind of under 7 m/s. The model indicated that the Windship was cost effective on the higher wind trans-Atlantic route at an average sailing speed of 13 knots, but only cost-effective on the trans-Pacific route at a lower speed of 11 knots.

The 2017 study also explored an autonomous Windship vessel with a modular auxiliary power system that was shown to provide some savings in building-cost and manning that would offset the additional cost of the sailing rig. The 2022 study explored an unconventional proa hull configuration that could maximise solar power harvesting and reduce propulsion requirements at sailing speeds greater than 10 knots.

4.2 REVISED CALCULATION MODEL

For the present study the calculation model from the 2017 study was upgraded to more accurately calculate the solar power harvesting based on vessel latitude and what tracking apparatus was active. For example, the solar panels on the deck and cargo hatches are fixed, while the solar panels on the extendable truss structures are able to track the sun on one axis. The solar panels on the sails are able to track the sun with along two axis, but only if they are not being used for sail propulsion which is when the wind speed is below 2.5m/s. A factor for self-shading was also included to allow for shadowing effects of the sail rig on the other solar panels. Cloud cover from the weather maps was also modelled over the sailing routes.

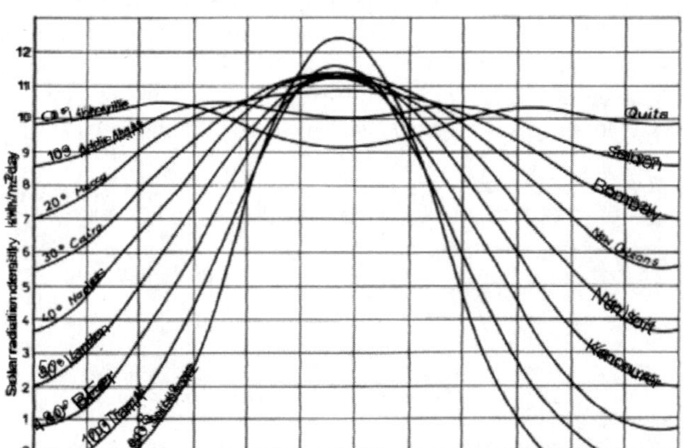

Figure 18: Global solar radiation density

In addition to the Atlantic and Pacific routes, two additional routes were investigated for the present study. An east-west route has been added between Singapore and Kuwait and a north-south route between Australia and China. Both these routes have a higher equatorial component than the trans-Atlantic and trans-Pacific routes and feature lower wind speeds but more sunlight. Both routes have an average wind speed of less than 7 m/s.

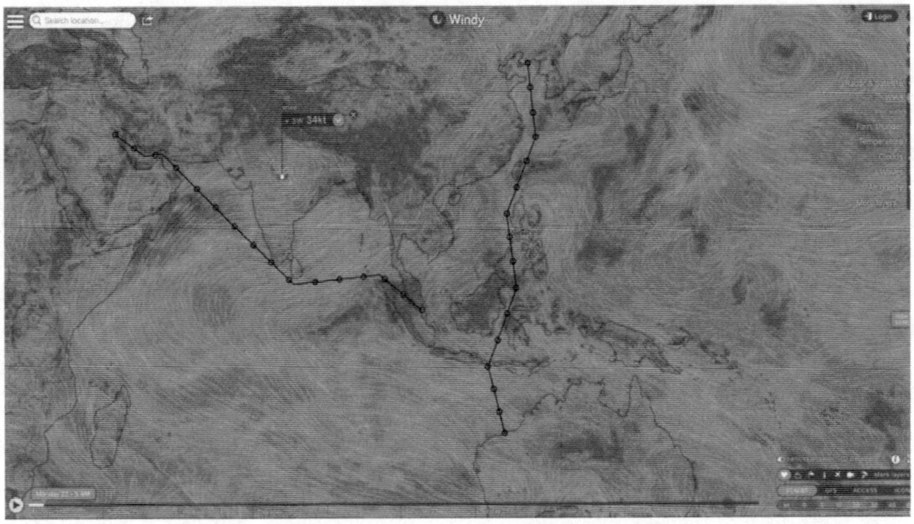

Figure 19: Kuwait-Singapore & Australia-China trade routes

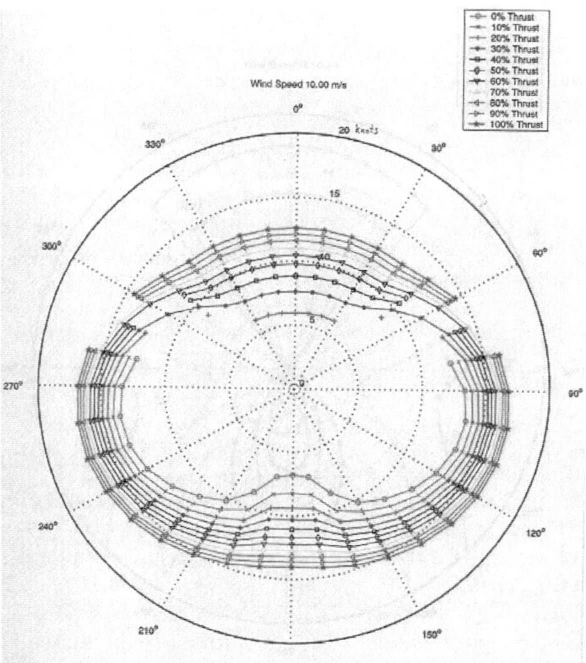

Figure 20: Windship performance at 10m/s true wind

As per the 2017 study, the calculation spreadsheet took surface wind speed and direction information from the European Centre for Medium Range Weather Forecast (ECMWF) model, which is generally considered the most accurate global model, using the Windy.com graphical interface. The spreadsheet also inputs cloud cover, to aid the solar energy calculations and wave data to estimate the added-resistance-in-waves.

The vessel course was plotted directly and divided into 16 legs. Three trials for each route were calculated and averaged for three different speeds, 8, 11 and 13 knots. The course was dead-reckoned and weather routing was not applied. A sea margin was also added to the required propulsion power based on the wind speed.

Polar plots of vessel speed from the MatLab Velocity Prediction Program (VPP) were previously calculated from tank testing of speed and resistance from wind tunnel testing of the sail rig and hull drift forces. The vessel speed relative to wind speed and direction were translated from the polar plots into the calculation spreadsheet. The effective power and theoretical shaft power was then estimated based on a propulsion efficiency of 60% to enable any additional propulsive power to be calculated.

4.3 NAVIGATION MODELLING

From the vessel course, the relative direction of the wind was calculated for each leg. From the wind speed and direction, the vessel speed was known from the VPP. If the vessel was pointing higher than 45 degrees into the wind, the vessel would tack and the additional distance from the tacked course was calculated for that leg. Additional electrical load was also allowed for turning the wing sections when changing tack.

For the sailing economic calculations, several strategies were tried. The first was allowing the WindShip to run as fast as the wind would allow. The second strategy was hydro-generation when the speed-by-wind was more than required average speed required for the route. The power generated charged the batteries and the energy saved used when auxiliary propulsion was needed if the sail power was insufficient to make the required average speed. The efficiency of the hydro-generation mode was set to 50%, which is 20% less than the propulsive efficiency, since the efficiently the CP propeller operating as a turbine will be lower.

4.4 COST MODELLING

A simple regression cost model was used to calculate new vessel building costs based on steel and equipment weights, block coefficient, installed power and machinery type. The building cost of the total sail rig was estimated at 5.0 MUSD for year 2000, which resulted in a 23% increase over a conventional motorship. Operating costs were taken from the previous study, which was based on year 2000 costs for benchmarking. Cost inflation was set to 2% per year.

Manning, maintenance, stores and administration costs where the same as the motorship. Insurance costs reflected the increased cost of the WindShip and port dues were reduced since tugs would not be needed for docking. Vessel finance was set at 5% per year over a ten-year term. All costs were calculated for the year 2025.

The WindShip fuel consumption was calculated based on the auxiliary propulsion requirement and the service electrical load. The return voyage at ballast condition was calculated at 85% of the deadweight fuel consumption. The price of fuel in the calculation was 750 USD/ton for LSHFO (2-stroke main engine fuel) and 1,000 USD/ton for LSMDO (4-stroke auxiliary engine fuel). These are cost estimates for the year 2025.

The cost of CO2 was included at 100 USD/ton. This is a cost estimate for the year 2025, although some predictions are for up to 200 USD/ton. The amount of CO2 emitted by the engines was calculated and included in the operational costs.

4.5 FLEET ECONOMICS

The results of the vessel sailing economics and cost model were then aggregated in a fleet economic study to look at overall ownership and operational costs and environmental impact. The study is based on transporting five million tons of grain per annum between the ports one-way with ballast return.

Three vessel speeds of 8, 11 and 13 knots were modelled. More vessels were needed for the slower speeds to transport the same tonnage of cargo per annum and this increased impacted negatively on the Required-Freight-Rate.

The CO2 emitted for shipbuilding was calculated based on 5.0 tons CO2 per vessel lightweight tons. (Typically, a value of 3.3 tons CO2 is used, but this only allows for the hull and superstructure steel.) The shipbuilding CO2 was then prorated over a vessel life of 25 years. The shipbuilding CO2 was not costed since it is a part of the shipbuilding cost but was included in the fleet study to give a total CO2 amount. The CO2 generated during maintenance and for ship recycling was also considered, with studies indicating that these add another 30% and 15% respectively to the vessel CO2 emissions.

Cost savings for fuel and total operation and CO2 saving were compared between the conventional vessel and the Windship relative to each speed.

4.6 VESSELS MODELLED

As per the previous studies, the 50,000 DWT Handymax bulk carrier with conventional 2-stroke propulsion would serve as the basis of the present study since the performance was already known from extensive model testing. The second vessel adds the sailing rig and solar panels.

The third vessel in addition to the second vessel, adds large extendable truss structures to the parallel mid-body to maximise the solar harvesting of a conventional-hulled bulk carrier. The truss structures add about 6,000 square meters of solar panels which have a peak power of 1,300 kWe.

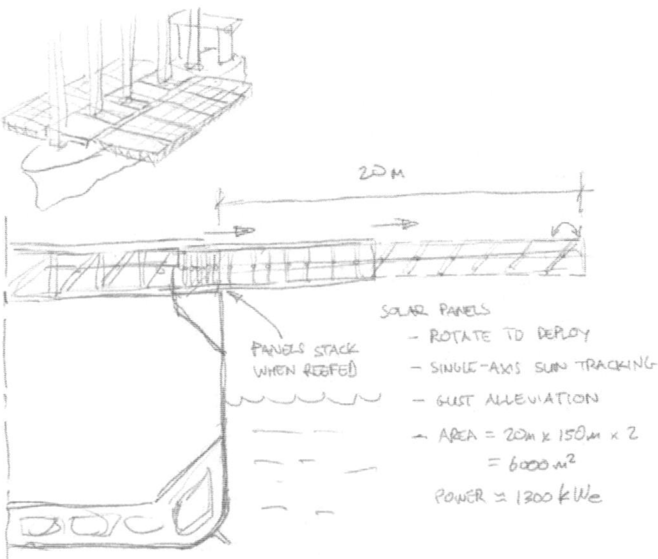

Figure 21: Retractable solar truss concept

The fourth vessel is unmanned and has 4-stroke modularised generators and an electric propulsion system. It was assumed for the vessels with solar power, that sufficient battery capacity was available to store the power generated by either solar or hydro harvesting.

The proa vessel configuration is not modelled in the economic calculations, since the maturity of the hull resistance and sailing performance is not currently at a comparable level compared to the conventional Windship hull and sailing rig which has been model tested in a wind tunnel and towing tank.

Although further performance optimisations to the sail rig were possible from the knowledge gained from the previous studies, it was decided to use the initial wind-tunnel data set as closely as possible to ensure accurate benchmarking of the current calculation model.

Figure 22: Unmanned Windship

Figure 23: Modular power system & solar harvesting

Table 1: Main particulars of conventional vessel & Windship variants

50,000DWT Bulk Carrier	**2-stroke Conventional**	**2-stroke Windship**	**2-stroke Windship + solar truss**	**Unmanned Windship + solar truss**
Length BPP	210 m	210 m	210 m	210 m
Beam	32 m	32 m	32 m	32 m
Draft	11.9 m	11.9 m	11.9 m	11.9 m
Block	0.77	0.77	0.77	0.75
Lightweight	11,500 t	12,500 t	13,500 t	13,000 t
Cargo Capacity	69,000 m^3	69,000 m^3	69,000 m^3	72,000 m^3
Propulsion efficiency	0.6	0.6	0.6	0.6
Installed prop'n power	8,000 kW	8,000 kW	8,000 kW	6,000 kW
Prop'n power SFC	0.17 kg/kWh	0.17 kg/kWh	0.17 kg/kWh	0.24 kg/kWh
Design speed	14.0 kn	14.0 kn	14.0 kn	13.0 kn
Typical elec. load	465 kWe	500 kWe	500 kWe	120 kWe
Aux. power SFC	0.20 kg/kWh	0.20 kg/kWh	0.20 kg/kWh	0.24 kg/kWh
Hydro-gen. efficiency	0.0	0.5	0.5	0.5
Solar panels on sails	-	980 kWe	980 kWe	980 kWe
Solar panels on deck	-	630 kWe	630 kWe	630 kWe
Solar panels on truss	-	-	1,250 kWe	1,250 kWe

5. MODELLING RESULTS

5.1 FUEL CONSUMPTION

Daily Fuel Consumption (DFC) is a popular metric and shows the overall fuel consumption per vessel per day. The key points to note from the graphs (Figures 24-27):

- Typical increase of fuel consumption as speed increases.
- At lower speeds the hydro-generation capacity is so high that the average daily fuel consumption is negative.
- Unmanned Windship saves considerable fuel with lower hotel and ship service loads.
- Unmanned Windship with modular high-speed gensets need more fuel than the 2-stroke Windships at 13 knots.
- Only minor difference to fuel consumption by adding the extendable truss structures for additional solar panels.

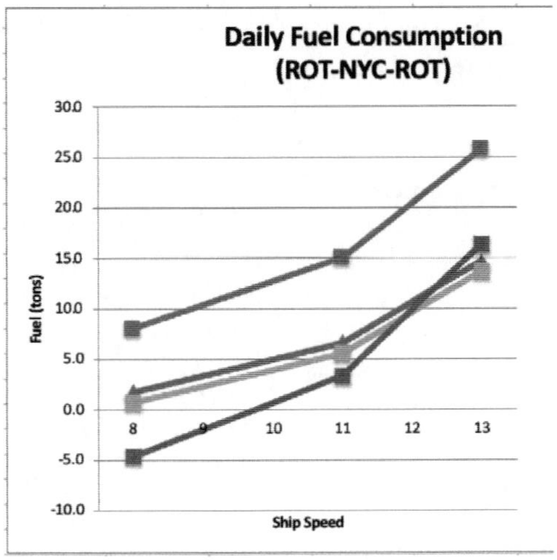

Figure 24: DFC Rotterdam – New York City

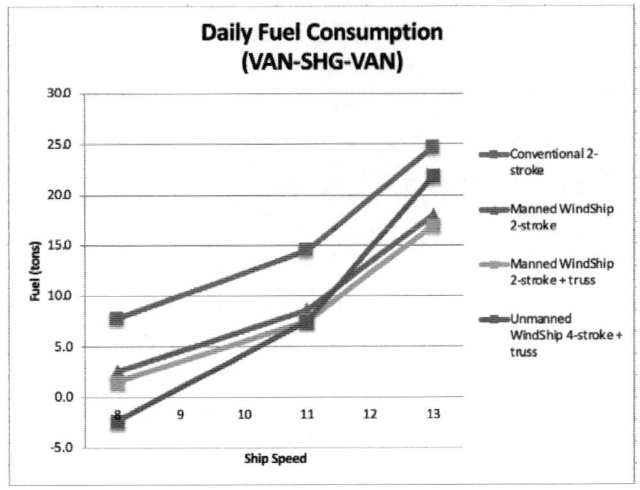

Figure 25: DFC Vancouver - Shanghai

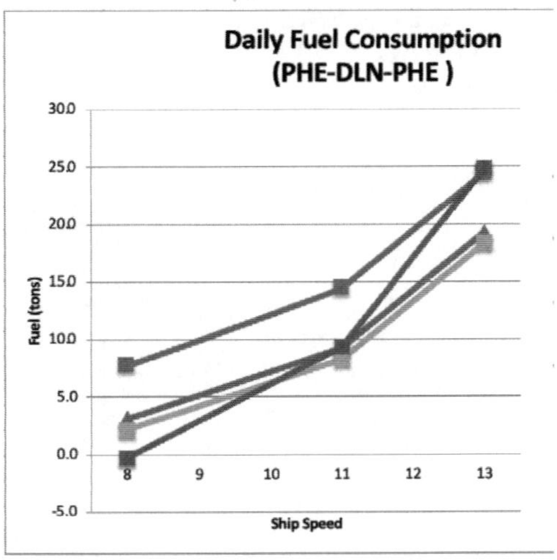

Figure 26: DFC Port Hedland – Dalian

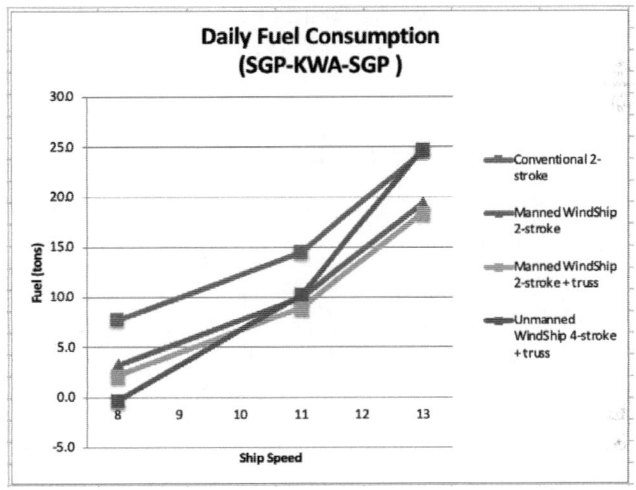

Figure 27: DFC Singapore - Kuwait

5.1 REQUIRED FREIGHT RATE

The Required-Freight-Rate (RFR) is the minimum cost per ton of freight to cover all capital and operation costs as discussed in the cost model above. No profit margin for the operator has been added. The RFR is a good indicator of overall transport efficiency. Key points to note from the graphs (Figures 28-31):

- The higher RFR cost at 8 knots is due to lower transport capacity at lower speeds.
- The higher RFR cost at 13 knots is due to the increased fuel cost at higher speed.
- The optimum speed for lowest RFR cost is 11 knots in nearly all routes and vessel variants.
- The unmanned variant has a lower RFR cost at 8 knots is, due to lower manning and fuel costs.

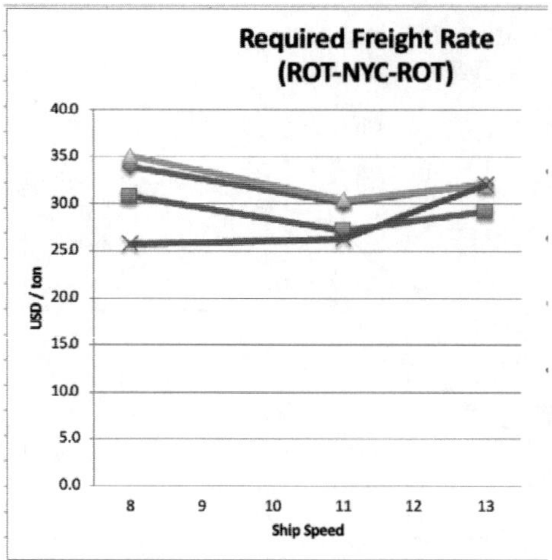

Figure 28: RFR Rotterdam – New York City

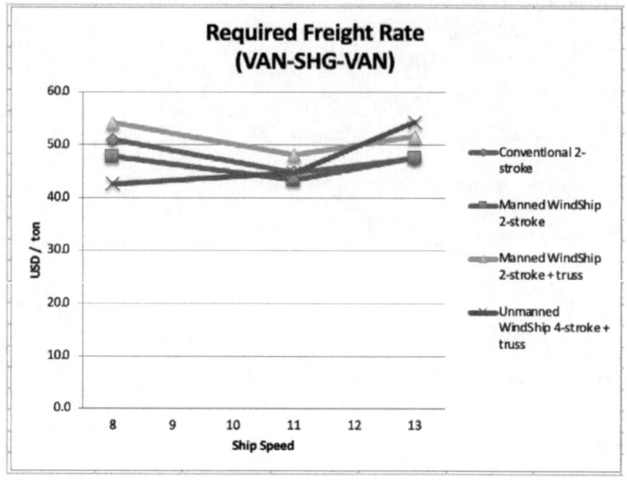

Figure 29: RFR Vancouver - Shanghai

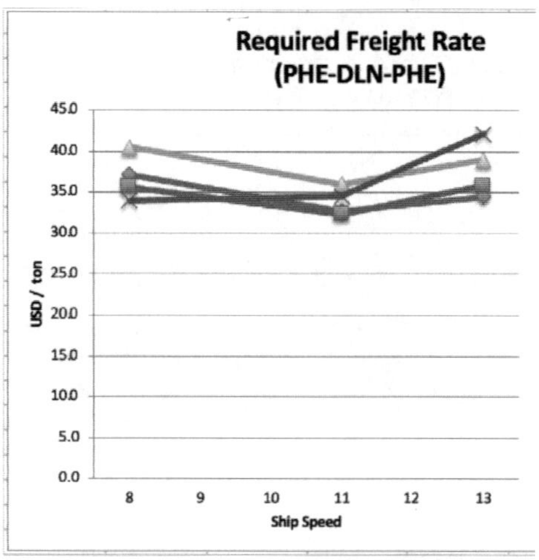

Figure 30: RFR Port Hedland – Dalian

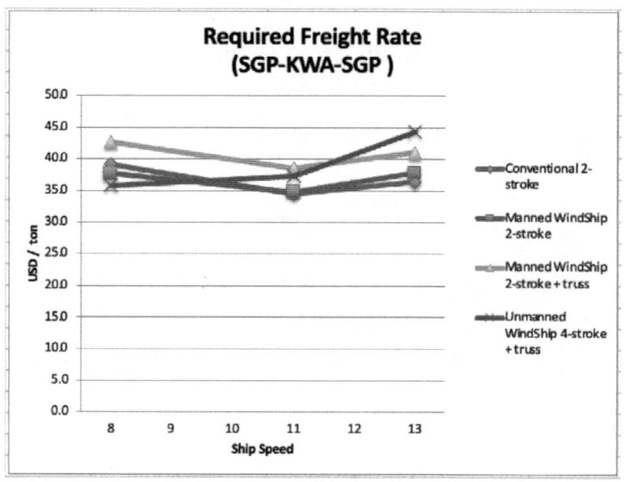

Figure 31: RFR Singapore - Kuwait

5.3 ANNUAL CO2 EMISSIONS

The annual CO2 emissions is only from fuel consumed annually by a single vessel. Since the CO2 emitted is directly related to the fuel consumed, the graphs of annual CO2 and daily fuel consumed are very similar. The CO2 from shipbuilding, vessel maintenance and end-of-life recycling is considered in the fleet study results further below.

Key points to note from the graphs:

- Lower speed always results in lower CO2
- The reduced fuel and CO2 emitted offsets the increased shipbuilding associated CO2.

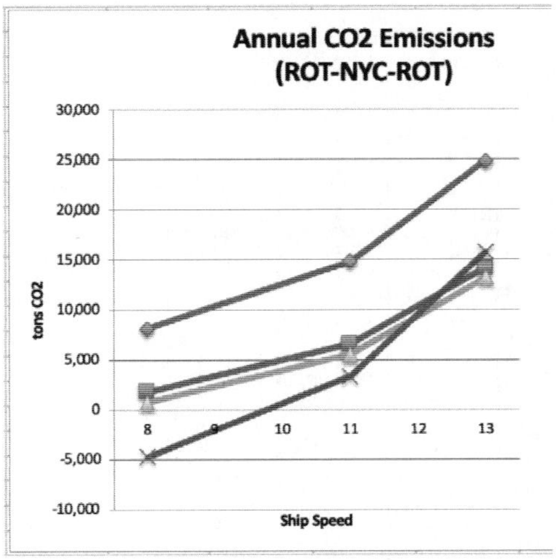

Figure 32: CO2 Rotterdam – New York City

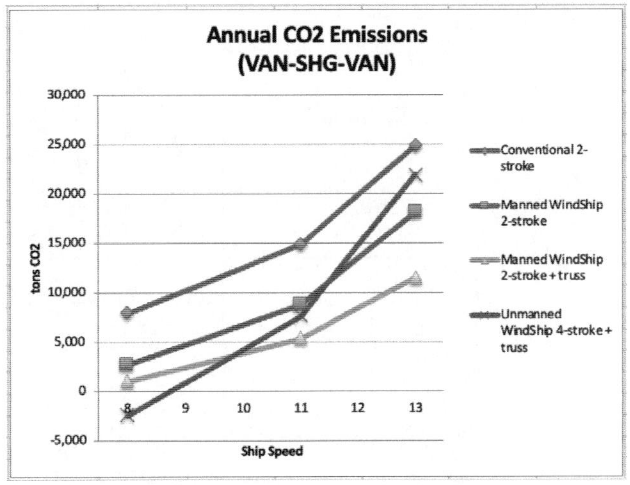

Figure 33: CO2 Vancouver - Shanghai

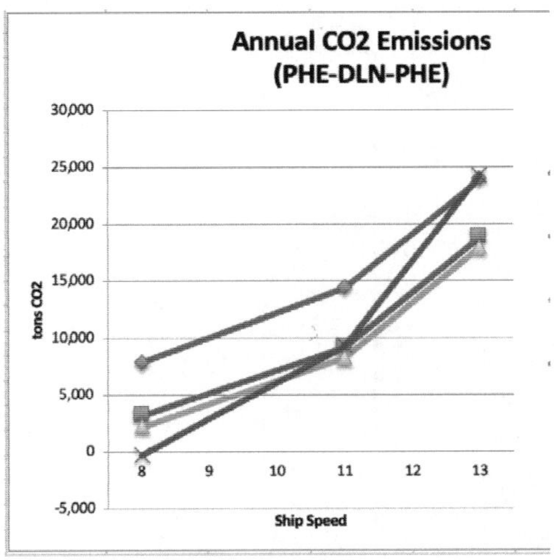

Figure 34: CO2 Port Hedland – Dalian

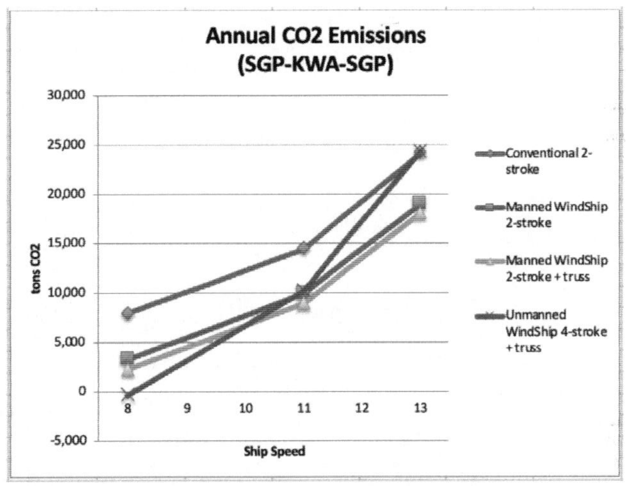

Figure 35: CO2 Singapore - Kuwait

5.4 FLEET STUDY

The results of the fleet economics for shipping five million tons of grain annually are shown below. They compare the conventional bulker and the simplest Windship variant with an average sailing speed of 8, 11 and 13 knots on the various sailing routes and the number of vessels needed for the trade. Windship fuel consumed, operating cost and CO2 savings are compared as percentage to the conventional vessel for the fleet. Key points to note from the tables (Tables 2-5):

- The cost saving increases on the windier routes.
- Sailing at 11 knots gives the lowest overall cost, being the best combination of reduced fuel cost and not needing too many vessels.
- Overall CO2 saving which includes fuel and lifecycle is improved on the windier routes due to the fuel saving.
- The CO2 saving is less than half compared to the fuel saving due to the increased shipbuilding CO2 of the more complicated Windship.

Table 2: Fleet economics Rotterdam - NY City route

ROT-NYC-ROT	Convent'l @ 8 kn	Convent'l @ 11 kn	Convent'l @ 13 kn	WindShip @ 8 kn	WindShip @ 11 kn	WindShip @ 13 kn
Grain per annum (tons)	5,000,000	5,000,000	5,000,000	5,000,000	5,000,000	5,000,000
Grain/year/ship (tons)	400,615	534,750	619,901	400,615	534,750	619,901
Ships required	13	10	9	13	10	9
Daily fuel consumption (tons)	8.0	15.1	25.8	0.7	5.6	13.7
Fuel saving	**0**	**0**	**0**	**91%**	**63%**	**47%**
Ship operating cost per year (kUSD)	13,446	15,807	19,361	12,384	14,729	18,511
Fleet annual cost (MUSD)	175	158	174	161	147	167
Cost saving	**0**	**0**	**0**	**8%**	**7%**	**4%**
Required Freight Rate (USD/ton)	**34.0**	**30.2**	**32.1**	**30.8**	**27.2**	**29.2**
Shipbuilding CO2 (tons)	29,900	23,000	20,700	39,000	30,000	27,000
Maintenance CO2 (tons)	16,445	12,650	11,385	21,450	16,500	14,850
Recycling CO2 (tons)	8,073	6,210	5,589	10,530	8,100	7,290
Fuel CO2 (tons)	105,658	148,195	223,995	23,171	65,124	127,263
Total annual CO2 (tons)	160,076	190,055	261,669	94,151	119,724	176,403
CO2 saving	**0**	**0**	**0**	**41%**	**37%**	**33%**

Table 3: Fleet economics Vancouver - Shanghai route

VAN-SHG-VAN	Convent'l @ 8 kn	Convent'l @ 11 kn	Convent'l @ 13 kn	WindShip @ 8 kn	WindShip @ 11 kn	WindShip @ 13 kn
Grain per annum (tons)	5,000,000	5,000,000	5,000,000	5,000,000	5,000,000	5,000,000
Grain/year/ship (tons)	266,780	359,614	419,502	266,780	359,614	419,502
Ships required	19	14	12	19	14	12
Daily fuel consumption (tons)	7.8	14.5	24.7	2.6	8.6	18.0
Fuel saving	**0**	**0**	**0**	**67%**	**41%**	**27%**
Ship operating cost per year (kUSD)	13,436	15,811	19,366	12,829	15,856	20,484
Fleet annual cost (MUSD)	255	221	232	244	222	246
Cost saving	**0**	**0**	**0**	**5%**	**0%**	**-6%**
Required Freight Rate (USD/ton)	**51.0**	**44.8**	**47.5**	**47.8**	**43.4**	**47.6**
Shipbuilding CO2 (tons)	43,700	32,200	27,600	57,000	42,000	36,000
Maintenance CO2 (tons)	24,035	17,710	15,180	31,350	23,100	19,800
Recycling CO2 (tons)	11,799	8,694	7,452	15,390	11,340	9,720
Fuel CO2 (tons)	150,386	207,615	298,833	50,917	122,975	217,424
Total annual CO2 (tons)	229,920	266,219	349,065	154,657	199,415	282,944
CO2 saving	**0**	**0**	**0**	**33%**	**25%**	**19%**

Table 4: Fleet economics Port Hedland – Dalian route

PHE-DLN-PHE	Convent'l @ 8 kn	Convent'l @ 11 kn	Convent'l @ 13 kn	WindShip @ 8 kn	WindShip @ 11 kn	WindShip @ 13 kn
Grain per annum (tons)	5,000,000	5,000,000	5,000,000	5,000,000	5,000,000	5,000,000
Grain/year/ship (tons)	364,085	487,287	565,840	364,085	487,287	565,840
Ships required	14	11	9	14	11	9
Daily fuel consumption (tons)	7.7	14.4	24.5	3.1	9.2	19.3
Fuel saving	**0**	**0**	**0**	**60%**	**36%**	**21%**
Ship operating cost per year (kUSD)	13,357	15,643	19,005	13,074	16,047	20,821
Fleet annual cost (MUSD)	187	172	171	183	177	187
Cost saving	**0**	**0**	**0**	**2%**	**-3%**	**-10%**
Required Freight Rate (USD/ton)	**37.1**	**32.7**	**34.5**	**35.7**	**32.4**	**35.8**
Shipbuilding CO2 (tons)	32,200	25,300	20,700	42,000	33,000	27,000
Maintenance CO2 (tons)	17,710	13,915	11,385	23,100	18,150	14,850
Recycling C02 (tons)	8,694	6,831	5,589	11,340	8,910	7,290
Fuel CO2 (tons)	110,276	157,899	214,915	44,434	100,863	169,192
Total annual CO2 (tons)	168,880	203,945	252,589	120,874	160,923	218,332
CO2 saving	**0**	**0**	**0**	**28%**	**21%**	**14%**

Table 5: Fleet economics Singapore – Kuwait route

SGP-KWA-SGP	Convent'l @ 8 kn	Convent'l @ 11 kn	Convent'l @ 13 kn	WindShip @ 8 kn	WindShip @ 11 kn	WindShip @ 13 kn
Grain per annum (tons)	5,000,000	5,000,000	5,000,000	5,000,000	5,000,000	5,000,000
Grain/year/ship (tons)	345,697	463,300	538,448	345,697	463,300	538,448
Ships required	15	11	10	15	11	10
Daily fuel consumption (tons)	7.7	14.5	24.6	3.2	10.0	19.4
Fuel saving	**0**	**0**	**0**	**59%**	**31%**	**21%**
Ship operating cost per year (kUSD)	13,376	15,677	19,105	13,122	16,427	20,921
Fleet annual cost (MUSD)	201	172	191	197	181	209
Cost saving	**0**	**0**	**0**	**2%**	**-5%**	**-10%**
Required Freight Rate (USD/ton)	**39.2**	**34.5**	**36.5**	**37.7**	**34.8**	**37.8**
Shipbuilding CO2 (tons)	34,500	25,300	23,000	45,000	33,000	30,000
Maintenance CO2 (tons)	18,975	13,915	12,650	24,750	18,150	16,500
Recycling C02 (tons)	9,315	6,831	6,210	12,150	8,910	8,100
Fuel CO2 (tons)	118,948	158,959	241,635	49,057	109,309	190,000
Total annual CO2 (tons)	181,738	205,005	283,495	130,957	169,369	244,600
CO2 saving	**0**	**0**	**0**	**28%**	**17%**	**14%**

6. CONCLUSIONS

6.1 SUMMARY

The results from the revised calculation model are well aligned with the previous studies. An average speed of 11 knots is optimum for the lowest operating cost for a Handymax bulker operating on a range of global trade routes. Windier routes provide greater savings. The total operating cost of a Windship is almost always lower than a conventional motorship except on highly equatorial routes with lower average wind speeds, in which case the Windship cost penalty is very low.

Regarding sailing strategy, it was found that hydro-generation was very effective at reducing the overall fuel consumption. The vessel should always use hydro-generation when the vessel speed was above the average speed required for the route. Tacking should be avoided if possible since the additional distance added to the navigation requires a higher average speed for comparisons with conventional motor vessels.

The proa configuration was investigated and offers the benefits of greater solar harvesting and promises reduced powering requirement at speeds above 10 knots. This is expected to make a wind powered vessel more viable by having a second source of renewable energy for more efficient sailing, particularly on upwind courses. Although initial calculations and modelling have been undertaken, the performance predictions are still too immature to be included in the economic study presented herein.

First principles engineering studies have confirmed the basic viability of the proa configuration for a large merchant vessel, but many technical challenges remain and will need more thorough modelling and investigation. These include the longitudinal structural strength of the main hull and the stability of the vessel particularly when sailing at a heel angle to minimise the amah resistance.

The unusual configuration and physical characteristics of the proa means that many Class rules are not applicable and will need special consideration with regard to engineering integrity, vessel handling and safety.

6.2 FUTURE STUDIES

Technical studies and modelling for the proa concept to a maturity that the platform can be assessed with the economic and sailing calculation model will be undertaken.

Further investigation into propellers capable of performing as a turbine is necessary to prove the viability of the hydro-generation operational mode.

Further refinement of the economic and sailing calculation model to improve accuracy and ease of use. Several more trade routes will be investigated for the performance of the Windship to access the potential cost benefits.

Calculations will be performed for different sailing and harvesting strategies. For example, always tacking into the wind and not tacking to sail the shortest distance. Solar and hydro harvesting strategies will be investigated to determine when the wings should be used as sails or solar trackers.

7. REFERENCES

1. MN (1996), Modern Wind Ships – Phase 1, Knud E. Hansen A/S, https://www.osti.gov/etdeweb/servlets/purl/515478

2. ROSANDER, M.; BLOCH, J.O.V. (2000), Modern Wind Ships – Phase 2, Knud E Hansen A/S, https://www2.mst.dk/udgiv/publications/2000/87-7944-019-3/pdf/87-7944-020-7.pdf

3. S.D. Chatzinikolaou & N.P. Ventikos (2014), Assessing Environmental Impacts of Ships from a Life Cycle Perspective

4. K.GOH (2017), Wind Powered Vessels with Hybrid and Autonomous Technologies, Knud E Hansen A/S

8. AUTHORS BIOGRAPHY

Ken Goh holds the position of General Manager, at Knud E. Hansen Australia and specialises in mechanical engineering and advanced concepts in ship design, energy efficiency, alternative fuels and advocating virtual reality. His project experience includes ferries, merchant, offshore, research, icebreakers, high-speed, yachts and naval vessels. He is also active in the development of autonomous and wind powered vessels.

"WIND HUNTER"- THE ZERO EMISSION CARGO SHIP POWERED BY WIND AND HYDROGEN ENERGY

K Ouchi, Ouchi Ocean Consultant Inc., Japan: ouchi@athena.ocn.ne.jp
K Shima and **K Kimura,** Mitsui O.S.K. Lines Ltd, Japan

SUMMARY

When a sailing ship with large rigid sails navigates in a sufficiently windy area, the thrusts from sails are strong enough to not only drive the ship at the proper speed but also rotate a large underwater turbine. The turbine generates electricity for the water electrolysis to generate hydrogen onboard. The hydrogen is attached to toluene and changed in the form of methyl-cyclo-hexane (MCH), which is in liquid form under ambient temperature and pressure. In the case of weak winds when the sails cannot generate sufficient thrust, the MCH is led to the dehydrogenation device. Using the hydrogen generated, the fuel cell supplies electricity to the electric motor propeller. Thus, the ship can navigate at a constant speed regardless of wind speed and direction. The concept of this ship is one of the best candidates for a Zero CO_2 emission ship, because it is operated by only wind energy.

NOMENCLATURE

L	Length (m)
B	Breadth (m)
D	Depth (m)
d	Draft (m)
Disp	Displacement (tf)
DW	Dead Weight (tf)
W	Wet Surface Area (m^2)
S	Sail Area Total (m^2)
Cx	Thrust Coefficient of Sail
Vw	Apparent Wind Speed (m/s)
T	Total Thrust of Sails (tf)
R	Total Resistance (tf)
Rh	Hull Resistance (tf)
Rtp	Turbine Resistance or Propeller Thrust (tf)
Fn	Froude Number
Vs	Ship Speed (kt)
Ct	Hull Resistance Coefficient
Dia	Diameter of Turbine/Propeller (m)
A	Circle Area of Turbine/Propeller (m^2)
Ptp	Power Output on Turbine or Propeller (kW)
Pgm	Power Output on Generator or Motor (kW)
H2	Hydrogen Generation or Consumption (m^3/day)
MCH	MCH Generation or Consumption (m^3/day)
ENDU	Days for Filling or Consuming MCH at Tank (day)
MCHT	MCH Storage Tank Capacity (m^3)
TOLT	Toluene Storage Tank Capacity (m^3)

1. INTRODUCTION

In order to realize the sustainable society, CO_2 emission from the activities of mankind needs to be reduced drastically.
Following the Paris Agreement in 2015, IMO (International Maritime Organization) adopted the resolution in April 2018 that all maritime voyages to emit zero CO_2 within this century, and to halve CO_2 emission by the year of 2050. This resolution will strongly accelerate the change of ship's propulsion energy source from the fossil oil to zero emission energy source. The innovation of zero emission ship propulsion system needs to be researched and realized urgently.

2. CANDIDATES FOR ZERO EMISSION SHIP

Table 1 shows features of conventional ships and candidates for zero emission ships. And it also indicates their energy source, power generation device, propulsor device and remarks.
As for the nuclear ships, the practical technology has already been established and vastly applied in the naval ships and submarines. It is not so technically difficult to build nuclear cargo ships. However, the nuclear merchant ships are difficult to be introduced in the international shipping industry because of the negative opinions from the general public about the safety and nuclear disaster.
Regarding the CO2 recovery ships, the main part of ship propulsion system is the same as that of the conventional ships. However, the problems lie not only on the size of additional CO_2 recovery device, but also on the cost of transporting intruding discharged CO_2 to the storage site.
As for the battery-operated ships, even in the case of the most advanced lithium battery, the battery weight is quite heavy and it may not be suitable as the propulsion system for high-powered long-distance cargo ship.
Both fuel-cell ships and hydrogen engine ships are under development in maritime community. However, their technology is not yet so popular, because of the high cost of hydrogen related equipment.
Among the above candidate ships, it seems that the ships powered by wind as the main basis are one of the most suitable methods for the zero emission ships. Ocean winds have been utilized since the ancient times. Such sailing ships used to enjoy their golden era in the 19th century. However, because the wind speed and direction on the sea were not constant, it was impossible to keep the ships on time schedules. Since the industrial revolution, especially after the invention of steam turbine and diesel engine, steam turbine and diesel engine ships, which ran without wind, replaced wind powered sailing ships.
In this paper, we would like to propose "WIND HUNTER" which could be possible to run by both wing sails using wind energy and electric propellers using fuel cell with hydrogen energy. If the hydrogen is originated from a renewable energy source from wind energy, such ship is completely free from oil burning and CO_2 emission.

Table 1: Candidates for Zero Emission Ship

Kind of Ship	Energy Source	Power Generation	Propulsor	Remarks
Conventional	Oil, Gas	Diesel Engine	Propeller	Much CO2 Emission
Nuclear	Uranium	Steam Turbine	Propeller	Uranium Supply, No Public Acceptance
CO2 Recovery	Oil, Gas	Diesel Engine	Propeller	Needs of CO2 Recovery Plant
Battery	CO2 Free Electricity	Lithium Battery	E. Motor & Propeller	Small Capacity, Heavy Weight of Battery
Fuel Cell	CO2 Free Hydrogen	Fuel Cell	E. Motor & Propeller	CO2 Free Hydrogen Supply
Hydrogen Engine	CO2 Free Hydrogen	Hydrogen Gas Engine	Propeller	CO2 Free Hydrogen Supply, NOx Emission
Biofuel	Biofuel	Diesel Engine	Propeller	High-Cost Fuel, Pressure on Cultivated Land for Food
Wind powered	Ocean Wind	Wing Sail	Wing Sail	Few Thrust in Weak Wind

3. WIND CHALLENGER PROJECT

We have been developing the WIND CHALLENGER Project which is an extraordinarily large rigid wing sail for driving large merchant ship [1]. Figure 1 shows the concept of the WIND CHALLENGER sail. Its height can be changed from about 53m (Full sail) to about 23m (Reef Sail), breadth is about 15m, and maximum breadth of wing is about 3m. The shape of wing section is crescent type.
The telescopic massive steel spar which located inside the sails receives the various wind forces via enclosed rigid wing sail made of GFRP. This sail can be rotated by a pinion gear automatically to keep the proper angle which derive the maximum thrust for the ship.
In case of stormy and windy weather, the reefing is done automatically by the hydraulic cylinders inside of the spars depending on the stress of root part of the spar.
Figure 2 shows the sea trial photo of first WIND CHALLENGER Ship. Mitsui O.S.K. Line's bulk carrier "SHOFU MARU" which was delivered on Oct. 7th 2022, from Oshima Shipbuilding Co., Ltd. She is 100 kDW bulk carrier fitted with one WIND CHALLENGER Sail in the forecastle deck.

The advantage of WIND CHALLENGER sail is the big change in the height and the area between full and reef sail, compared with traditional sails. Therefore, we can install larger sail than before and get stronger thrust. Table 2 shows the principal particulars of "SHOFU MARU".

Figure 3 shows the example of the performance of energy saving ratio in various wind speed and direction at the ship speed of 14.3 knots based on the calculation at the stage of development of WIND CHALLENGER Project in 2015. In the case of TWS (True Wind Speed) 15 knots and TWA (True Wind Angle) 85 degrees, the fuel saving ratio was estimated about 9% according to our CFD calculation. Some zig-zag line in Figure 3 means the step of reefing, i.e., trigger for reefing operation is activated by the value of the moment which is sensed at the root of spar. Specifically, the energy saving ratio for "SHOFU MARU" per WIND CHALLENGER sail is estimated to be about 5% on Japan-Australia east coast round-trip and about 8% on Japan-North America west coast round-trip at present.

Figure 1: WIND CHALLENGER Sail

Figure 2: "SHOFU MARU" at Sea Trial

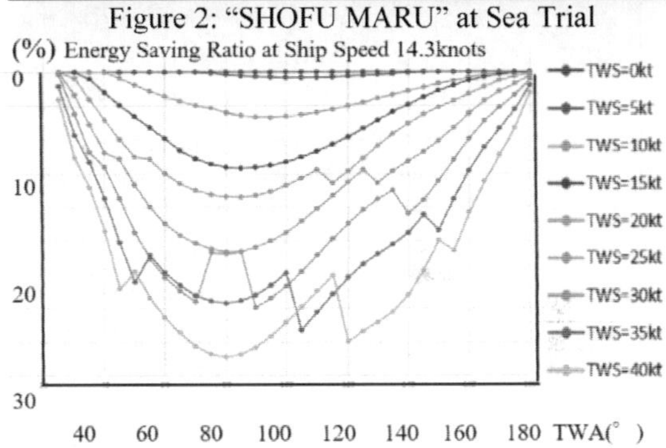

Figure 3: Example of Energy Saving Ratio Calculated in 2015

Table 2: Main Particulars of "SHOFU MARU"

Item	Value	Remarks
Length	235.00 m	Over All
Breadth	43.00 m	
Depth	20.05 m	
Design draft	12.80 m	Moulded
Gross Tonnage	58,209	
Dead Weight	100,422 ton	Scantling
Main Engine	MITSUI MAN B&W 6S60ME-C10.5-EGRBP	
Ship Speed	14.3 knot	85%MCO, 15%SM, Design draft
Sail Area	abt. 700m2	Telescopic Rigid Sail

4. WIND HUNTER PROJECT

Figure 4 shows schematic drawing for the various forces acting on zero emission sailing ship, which is WIND HUNTER PROJECT. Difference between WIND HUNTER PROJECT and WIND CHALLENGER PROJECT is while wind assists main engine propulsion for fuel saving, propulsion and energy only comes from wind and achieve zero emission for WIND HUNTER.

To keep the constant ship speed, the resistance of hull (Rh) has some constant value, however at the weak wind, the ship thrust (T) is smaller than (Rh), so that the propeller thrust (Rtp) is needed. In case of strong wind, the turbine resistance (Rtp) generates excessive power [2].

Figure 5 shows the schematic drawing of main components and energy flow for WIND HUNTER which has mainly two major remarkable modes. One is the production and stowing of the hydrogen, which is done by the water turbine generator, the water electrolyser and the hydrogeneration device in the case of strong wind. The other is consumption of the hydrogen by activating the dehydrogenation device, the fuel cell and the electric motor driven propeller in the case of weak or no wind.

To achieve zero emission for WIND HUNTER, an acquisition of large amount of wind energy is important. The size and performance of WIND CHALLENGER Sail has the potential to run the ship at the cruising speed and generate proper amount of hydrogen which is made from kinetic energy of the wind via electricity. The volume of hydrogen gas is too large to store onboard the ship, therefore, it is necessary to change hydrogen gas into other hydrogen carriers such as liquid hydrogen, compressed gas hydrogen, Methyl-Cyclo-Hexane (MCH), ammonia, metal hydride, etc. For this research, a chemical hydride MCH/Toluene system shown in Figure 6 is chosen from the viewpoint of safety, economy and handling. MCH which is made from Toluene and hydrogen with the catalyst is one of the excellent hydrogen carriers, because its shrinkage rate from hydrogen gas to MCH liquid is approximately 1/500 in the condition of ambient temperature and pressure.

As for the electric generator and motor propeller, we propose the common usage to simplify both components together in the system.

Figure 5 shows a summary for the concept and energy flow of the WIND HUNTER which has two modes for propulsive operation:

1) In the case of strong wind: Red line. The excessive energy, which is produced more than the energy needed for ship's cruising speed, generates electricity, then activates the electrolysis device and produces hydrogen gas. The hydrogeneration device enables hydrogen gas to be attached to toluene and makes MCH to act as the hydrogen carrier and stores it in the MCH tank
2) In the case of a weak or no wind: Blue-dot line. The energy is less than the energy needed for ship's cruising speed and electric driven propeller rotates with electrics via fuel cell, i.e., fuel cell is driven by the hydrogen gas which is separated from MCH stowed in the MCH tank using the dehydrogenation device.

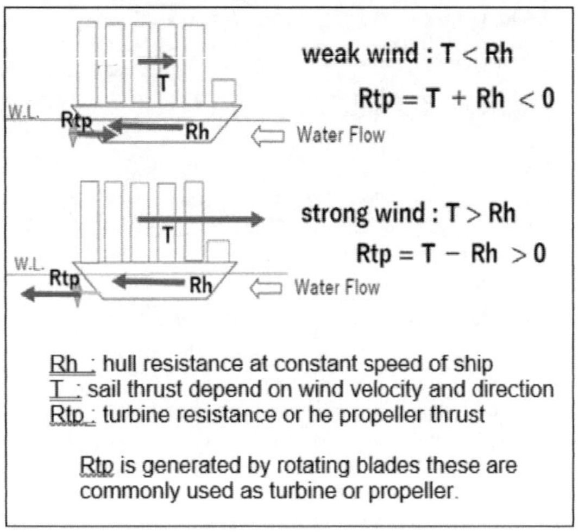

Figure 4: Forces Acting on Zero Emission Sailing Ship

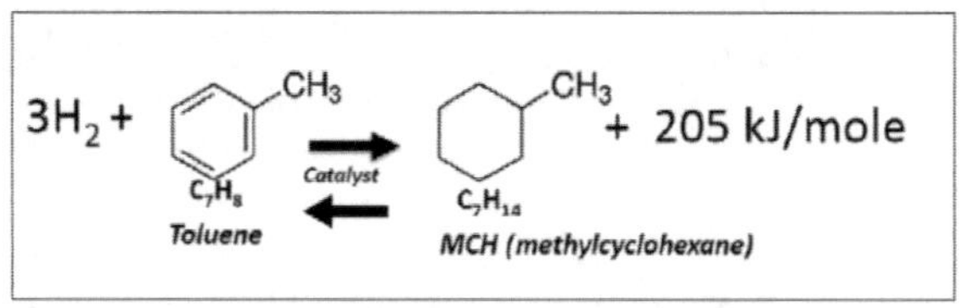

Figure 6: Chemical Hydride System

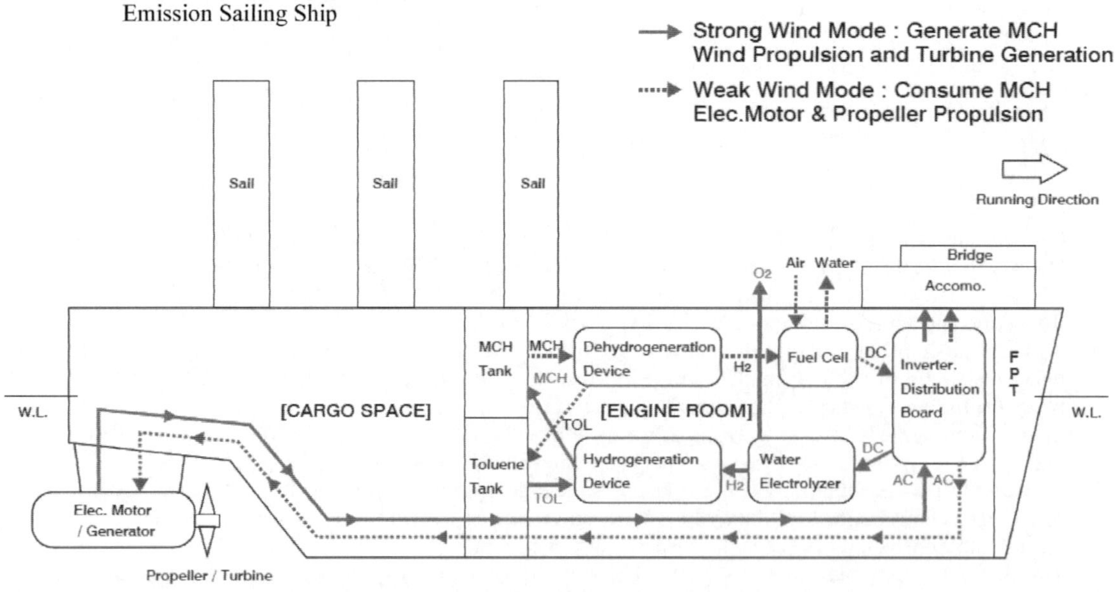

Figure 5: Main Components and Energy Flow of the WIND HUNTER

5. ARRANGEMENT AND PARTICULARS

Figure 7 shows and is proposed the rough arrangement and main particulars for 220m length Bulk Carrier type Zero Emission Sailing ship [3]. The bridge is to be in the fore part of the ship to keep forward visibility. The accommodation and engine room are also in the fore part or near by the bridge for the convenience of operation of the ship.
The hull form is so called Buttock Flow type with large skeg in the ship's centre line. This hull form avoids the speed reduction by the hull wake at the point where the turbine/propeller are fitted, and the skeg keeps the course stability. Two pod type turbine/propellers with generator/motors are installed in parallel at the end of the hull to acquire or give the kinetic energy of water flow.
Six sails are installed on the ship centre line at every hatch end space.

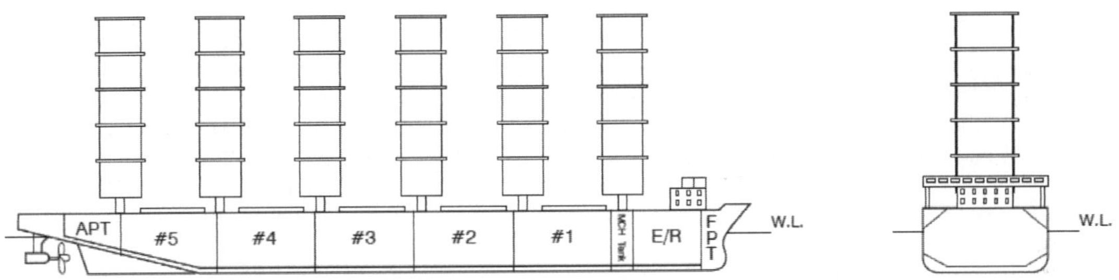

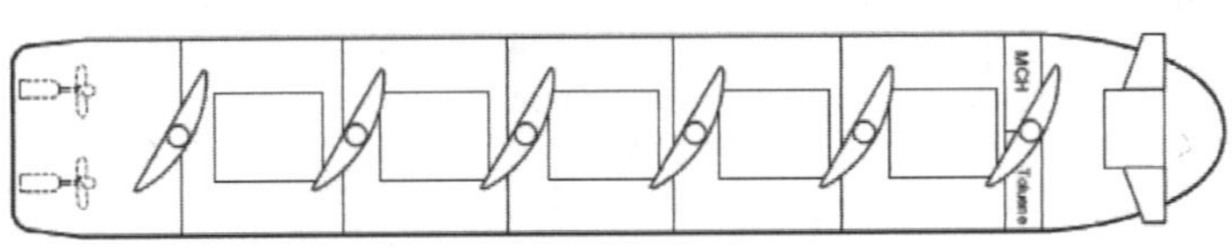

L x B x D:	220m x 36m x 20m
d:	13m
DW:	75,500tf
Sail:	(80mx22mx6) 10,500m^2
Ship Speed:	Average 12knots
Fuel Cell:	4,000kW
Turbine/Propeller (Pod):	9.6mΦ x 2
MCH/Toluene Tank:	1,000m^3 each

Figure 7: Rough General Arrangement and Main Particulars

6. CASE STUDY FOR PERFORMANCE

A case study for performance of the WIND HUNTER is carried out in various cases of wind velocity and ship speed. Table 3 shows the summary table for the various performance factor and estimation results for 75kDW Bulk Carrier mentioned in Section 5 and Figure 7 The definition and explanation of each symbol is shown in the column of Term & Remarks. Here, the total area of sails is about 10,500m^2, and the diameter of turbine/propeller (Dia) is about 9.6m

6.1 STRONG WIND CASE

Regarding the strong wind case which is explained in the red line in Figure 5, 'Strong Wind' column in table 3 shows the estimation for hydrogen generation capacity and related data. In this case, the apparent wind velocity (Vw) 15m/s drives the ship not only at the speed (Vs) 14knots but also generate the electric power (Ptp) about 10MW using the turbine generator. This electricity is converted to hydrogen gas (H2) about 50Mm^3 by the water electrolyser whose conversion ratio is assumed 5kWh/m^3. The hydrogeneration device makes the MCH about 100m^3/day using the hydrogen gas and toluene with catalyst. Here, the shrinking rate of hydrogen gas to MCH is assumed 1/500. The MCH tank capacity (MCHT) is 1,000m^3, therefore the endurance (ENDU) to full up the tank is about 10 days.

6.2 WEAK WIND CASE

Regarding the weak wind case which is explained in the blue dot line in Figure 5, 'Weak Wind' column in table 3 shows the estimation for hydrogen consumption capacity and related data. In the extreme condition of apparent wind velocity (Vw) 0m/s, the propellers are required to emit the power (Ptp) about 2.2MW and the fuel cell to emit the power (Pgm) about 2.8MW for driving ship speed of 10knots, here, the propeller efficiency is assumed 0.78. The MCH about 100 m3/day equivalent to the hydrogen (H2) about 50Mm3/day is consumed and this hydrogen is changed to the electricity by the fuel cell whose conversion ratio of 1.35kWh/m3. Therefore, the endurance of voyage using the fuel cell (ENDU) is about 10 days in case of ship speed 10knots.

6.3 MARGINALWIND CASE

The marginal wind is the special wind velocity at the condition; Rtp = 0, T = Rh, shown in Figure 4 and Table 3. The ship neither generates nor consumes hydrogen in case of above condition. 'Marginal Wind' column in table 3 shows the wind velocity 6.5m/s is the sufficient wind velocity of ship speed 12knots, and the endurance (ENDU) at this condition is unlimited.

6.4 VOYAGE PLAN

For operating such kind of ship, it is important to keep the balance of the generation and consumption for maintaining punctual service. The voyage plan should be organized taking keen consideration on the wind and climate forecasting, voyage route, MCH reserve capacity and cruising speed [4].

For example, for 14 days voyage, if the ship has two no wind (Vw: 0m/s, Vs: 10knots) days, ten marginal wind (Vw: 6.5m/s, Vs: 12knots) days and two strong wind (Vw: 15m/s, Vs: 14knots) days, average cruising speed is 12knots and MCH consumption (about 200m3) is same as MCH generation. Thus, many various patterns of voyage plan are considered to meet specific voyages.

Table 3: Performance Estimation of WIND HUNTER

Symbol	Unit	No Wind	Marginal Wind	Strong Wind	Term & Remarks
L	m	220.0	220.0	220.0	Length
B	m	36.0	36.0	36.0	Breadth
D	m	19.5	19.5	19.5	Depth
d	m	13.0	13.0	13.0	Draft
Disp	tf	85,457	85,457	85,457	Cb:0..83, Buttock Flow Hull Form
DW	tf	75,457	75,457	75,457	Dead Weight (LW:10,000tf)
W	m2	12,276	12,276	12,276	Wet Surface Area
S	m2	10,560	10,560	10,560	Sail Area Total (H80mxB22mx6sets)
Cx		1.9	1.9	1.9	Thrust Coeffecient of Sail
Vw	m/s	0.0	6.5	15.0	Apparent Wind Speed (Cross Wind)
T	tf	0	54	282	Total Thrust of Sails: T=R
R	tf	0	54	282	Total Resistance: R=Rh+Rtp
Rh	tf	37	54	73	Hull Resistance
Rtp	tf	-37	0	208	Turbine Resisance or PropellerThrust
Vs	kt	10.0	12.0	14.0	Ship Speed
Fn		0.116	0.140	0.163	Froude Number : Vs/(gL)^0.5
Ct		0.0022	0.0022	0.0022	Hull Resistance Coeffecient
Dia	m	9.6	0.0	9.6	Diameter of Turbine/Propeller
A	m2	145	0	145	Sircle Area of 2 Turbine/Propeller
Ptp	kW	2,179	0	11,081	Power Output on Turbine or Propeller
Pgm	kW	2,794	0	10,306	Power Output on Generator or Motor
H2	m3/d	49,663	0	49,467	Hydrogen Generation or Consumption
MCH	m3/d	99	0.0	99	MCH Geneneration or Consumption
ENDU	day	10.1	∞	10.1	Days for Filling or Consuming MCH at Tank
MCHT	m3	1,000	1,000	1,000	MCH Storerage Tank Capacity
TOLT	m3	1,000	1,000	1,000	Toluene Storage Tank Capacity
Remarks		Thrust by Propeller	Thrust by Sail	Thrust by Sail-Turbine	

7. SUBJECT FOR FUTURE DISCUSSION

Following is the agenda for the future discussion towards the realization of the WIND HUNTER as a large size cargo
ship.

- Selection of a hydrogen carrier

Chemical hydride (Toluene/MCH) is being used here, but other carriers such as liquid hydrogen, pressurized hydrogen, ammonia hydrogen absorbing alloy can also be considered. A discussion to compare in terms of safety, cost, handling and business opportunities is required

- Weight reduction of WIND CHALLENGER Sails
  To enable increasing the area and height of WIND CHALLENGER Sail, further development of lighter and stronger material such as CFRP for moving spars inside of the wing sail will be needed instead of steel spars.
- Common usage of Generator/Motor and Turbine/Propeller
  The ship uses MCH production mode in the strong wind area, and MCH consumption mode in the weak wind area. Therefore, the turbine/propeller device is used as energy collector in strong wind, and the energy emitter in the weak wind. The research on the optimum blade shape of the device will be required.
- Customizing the fuel-cell specification
  Development of the fuel-cell to meet the power demand of the two-mode operation will be needed.
- Integral usage of heat energy
  Heat energy which is absorbed in or emitted from components such as fuel cell, water electrolyser, hydrogeneration and de-hydrogeneration should be researched and be used properly.
- Voyage plans and selection of voyage routes
  Development of the voyage programs is necessary considering the punctuality of voyage schedules, the weather forecast, ship cruising speed, and safety [4].

8. CONCLUSION

We propose WIND HUNTER as a Zero CO_2 Emission ship abiding by the resolution of 2018 IMO meeting. The total energy for the WIND HUNTER is supplied by the ocean wind and WIND HUNTER has two modes of operations: the energy collection mode in the strong wind; and the energy consumption mode in the weak wind. The generator turbine (commonly used with electric propellers), water electrolyser, and hydrogeneration device will be used to collect energy. The de-hydrogeneration device, fuel cell, and electric motor (commonly used with generator turbines) will be used to emit energy.
The performance study on a 75kDW Bulk Carrier is carried out at the various assumption of wind velocity, ship speed, hydrogen and MCH storage capacity. And we propose that this concept of ship is suiting to the forthcoming hydrogen society as one of the solutions for zero emission society.
Technologies of the above components has basically already been established either in land or marine fields. The WIND HUNTER will be one of the most promising technologies for the next generation merchant ships, and the acceleration of its research and development will be expected.
In addition to zero emission cargo carrier, it is also now under investigation to create hydrogen production vessel, i.e., the vessel sails out from port with full of Toluene. Then, the vessel goes to windy area, produce hydrogen and store as MCH onboard. Once tanks are filled up enough, the vessel comes back to port for discharging MCH for shore usage.

9. REFERENCES

1. OUCHI, K., UZAWA, K., KANAI, A. AND KATORI, M. 2013. "WIND CHALLENGER" the Next Generation Hybrid Sailing Vessel. Proceedings of The Third International Symposium on Marine Propulsors (smp'13) Launceston, Tasmania, Australia.

2. OUCHI, K., AND HENZIE, J. 2017. Hydrogen Generation Sailing Ship. The proceedings of OCEANS2017 Aberdeen Scotland, IEEE & Marine Technology Society.

3. OUCHI, K., OMIYA, T. 2019. Zero Emission Sailing Ship - Concept and Design - RINA Wind Propulsion Conference, Proceedings of Royal Institution of Naval Architects, London UK,

4. DAVIES, G., WASEDA, T. 2019. Optimal Routing of a Wind-Powered Vessel Using Ensemble Feather Forecast Data, RINA Wind Propulsion Conference, Proceedings of Royal Institution of Naval Architects, London UK,

10. AUTHORS BIOGRAPHY

Dr. Kazuyuki Ouchi:
CEO, Ouchi Ocean Consultant Inc. He is responsible for R&D Adviser for Mitsui O.S.K. Lines, especially on WIND CHALLENGER Project and Zero Emission Sailing Ship. His previous experience includes Professor of Graduate school of Frontier Sciences The University of Tokyo.

Mr. Kentaro Shima:
He is responsible for General Manager, Technology Research Center, Mitsui O.S.K. Lines Ltd.

Mr. Keisuke Kimura:
He is responsible for Research Engineer, Technology Research Center, Mitsui O.S.K. Lines Ltd.

ABOUT THE BENEFIT OF OPTIMIZING BOTH SPEED AND ROUTE IN WIND-ASSISTED SHIPS MULTI-OBJECTIVE WEATHER ROUTING

M Dupuy [1 2], D Dudka [2], L Letournel [2], F Rongère [2], G Vincke [2]

[1] : LIX CNRS, École Polytechnique, Institut Polytechnique de Paris, 91128 Palaiseau, France
[2] : D-ICE Engineering, 44321 Nantes, France

SUMMARY

Optimizing ship navigation with regards to meteorological and oceanographic forecasts is a crucial issue for the marine industry. Different approaches have been proposed, but a few are general enough to solve all applications and therefore compute route optimization for sailing, motor, and hybrid-propelled ships. This is a topic of interest as wind propulsion is being pushed on the shipping industry by the IMO regulations, to reduce its carbon footprint. Various algorithms are available in the state-of-the-art and an important characteristic is the number of control variables considered for the optimization. In this study, we compare and analyze different optimization strategies, amounts of wind assistance, and service speeds. We quantify the added value in terms of fuel savings of the simultaneous optimization of route and speed and how it is enhanced by wind propulsion.

1. INTRODUCTION

Optimizing ship navigation with regards to meteorological and oceanographic forecasts is a crucial issue for the marine industry. Depending on the application, some variations of this problem occur: for racing sailboats, the objective is to minimize travel time, respecting safety constraints while for merchant ships we aim at reducing the voyage consumption, respecting trip duration and maximum ship motions.

For each variation of the problem, different approaches have been developed, but a few are general enough to solve all applications and therefore compute route optimization on sailing, motor, and hybrid-propelled ships. This is a topic of interest as wind propulsion is being pushed on the shipping industry by the IMO regulations, to reduce its carbon footprint. In this context, D-ICE is often acting as a third party for shipowners to compare the performances of wind propulsion systems with statistical weather routing. Through these studies, we noticed that weather routing can enhance the expected fuel savings up to a factor of 2.

Various algorithms are available in the state of the art and an important characteristic is the number of control variables taken into account for the optimization. In this study, we aim at quantifying the benefits in terms of savings of the simultaneous optimization of route and speed for different kinds of ships. Previous studies showed the benefits of considering two control variables for motor ships [1, 2, 3], and here we investigate the impact of wind propulsion.

2. SHIP WEATHER ROUTING

2.1. PROBLEM STATEMENT

Considering a vessel moving on the ocean with a set of control variables modeled as scalars (for example heading angle h and speed through water v), we can introduce the following vectors: the ship position vector $x(t) = [\varphi(t), \lambda(t)]$ where φ, λ are respectively ship latitude and longitude at time t, the control vector $u(x, t) = [h(x, t), v(x, t)]$, and the environmental conditions vector $e(x, t)$ composed of wind, waves, and (ocean) current at ship position x and time t. Other environmental variables can be considered if needed. These vectors may have admissible values because of operational constraints (navigation area, propulsion setup, maximum weather conditions, etc...).
The ship engine power p can be determined using the ship position x, the control vector u, and the environmental condition vector e as mentioned in section 3.2.
Considering a route from a starting position and time (x_s, t_s) to an arrival position and time (x_f, x_f), with $C(u, x)$ being a k-dimensional vector-valued cost function, the route cost vector can be expressed as:

$$J = \int_{t_s}^{t_f} C(u, x)dt$$

The control problem is to find the optimal controls $u^*(t)$ that minimizes J. The route $x^*(t)$ determined by $u^*(t)$ is called the 'optimal route'. When $k > 1$ (multi-objective optimization), we may have conflicting objectives to optimize, so determining optimality may be impossible: instead, we aim at finding a set of *Pareto-optimal solutions*.

2.2. STATE OF THE ART

Even if the general problem is continuous due to the nature of the ocean, methodologies for finding paths are generally discrete. More precisely, the control variables considered in the problem often have predefined discrete values. We can distinguish three kinds of methodologies. First, fixed-grid approaches that are based on a predefined discretization of the search space, such as dynamic programming or graph-based shortest paths algorithms. Second, dynamic-grid approaches that discover on-the-fly the search space during the algorithm, such as the isochrone and isopone methods. Finally, evolutionary approaches that start from an initial population of solutions and then, from mutations and crossover operations, explore other candidates. Nonetheless, some continuous approaches like calculus of variation have been proposed for single-objective optimization but are not used operationally. For a more complete survey of ship route optimization, see the work of Dupuy *et al.* [4]. The choice between these methods depends on the objective(s) to optimize, the constraints to satisfy, and the number of control variables.

In this work, we are studying multi-objective weather routing for wind-assisted ships and therefore we face some challenges because each of the previously introduced methods has its area of application. For example, dynamic-grid methods are more adapted to sailing ships, whose optimal routes are strongly impacted by the weather conditions, while fixed-grid methods are more adapted to merchant ships, which need to optimize multiple objectives and whose optimal routes stay around the direct one.

More precisely, for time-optimal weather routing with the heading as a single control variable (often the case for sailing ship's weather routing), the isochrone method is the reference. Initially developed by Hanssen and James [5, 6], various versions of the algorithm have been proposed that differ by the way isochrones are modeled, expanded, and simplified [7]. Other approaches based on graph shortest path algorithms (Dijkstra or A* adaptations for ship routing [8, 9, 10, 11], or on dynamic programming have also been proposed [12].

For the fuel-optimal weather routing problem with the heading as a single control variable, Hagiwara used the previously mentioned dynamic grid isochrone method [7], but fixed grid methods are mainly used. As an example, Takashima et al. adapted the Dijkstra algorithm for fuel optimization [13], and De Wit [14] and Calvert *et al.* [15] proposed the dynamic programming algorithm 2DDP. Some work also includes the optimization of two control variables such as the adaptation of the isochrone method by Klompstra *et al.* called isopone [16]. Aligne *et al.* [1] and Shao and Zhou [17], proposed an extension of the 2DDP to include a second control variable, called 3DDP, and Zaccone *et al.* proposed some speed-up techniques [18]. Wang *et al.* adapted the Dijkstra algorithm with the 3DDA [19].

For multi-objective optimization applied to ship weather routing, a lot of different approaches have been proposed. Some graph-based methods, such as Veneti *et al.* [20], tackle bi-objective optimization with two control variables. Other works such as Chauveau *et al.* [21] and Park and Kim [22] proposed two-phase approaches to handle each control variable one after the other.
A lot of methods based on evolutionary algorithms have also been developed for multi-objective ship routing with either one or two control variables (such as Veneti *et al.* [23] and Lee *et al.* [2] that are based on NSGA-II genetic algorithm, Szlapczynska and Szlapczynski [24] and Vettor and Guedes Soares [25] based on SPEA algorithm, etc).

The main differences between these methods remain in the number of control variables considered, and also in the geographical area covered by the search. This is a matter of importance as depending on the wind assistance, hybrid ships may have a behavior similar to a sailing ship or to a motor ship, and often in between.

2.3 OUR APPROACH

At D-ICE we developed a graph-based approach to compute optimal paths for mono and multi-objective time-dependent shortest path problems. A future paper will be dedicated to an exhaustive presentation of the algorithm, with some benchmarks with other state-of-the-art approaches but we will roughly describe the principle here.

As the ocean is a continuous space there is no straightforward way to mesh it, and obviously, the method impacts the accuracy of the optimization. Various methods can be used for ocean meshing depending on the application (such as unstructured meshing for sailing ships) but in this work, we will use a distribution of positions perpendicularly away from the great circle route. Once the space is discretized, we expand the graph with parallel edges labeled with each discrete value of the second control value, the ship speed.

More precisely, we consider a graph $G = (N, A)$ with N the set of nodes and A the set of arcs. Arcs are directed, which means that crossing the arc (i, j) may have a different cost than crossing the arc (j, i) (supposing both arcs are in A, which means that G can be a directed acyclic graph or not). From section 2.1, let us consider a starting node n_s and an arrival node n_f that can be identified in G from (x_s, t_s) and (x_f, t_f). Let us consider X_{pq} a path of length m from node p to q in G. This path can be described with a sequence of arcs $((p, n_2, v_2), (n_2, n_3, v_3), \ldots, (n_{m-1}, q, v_{m-1}))$ containing nodes and speed through water values. Considering the cost function mentioned in Section 2.1, we can attribute to each arc a cost vector c. More precisely, $c_{ijv}(t)$ is a vector of dimension k of the approximated costs to cross the arc (i, j, v) starting from node i at time t with speed through water v. Moreover, in our model, we consider the environmental conditions as constant for crossing the arc (frozen arc model).

The total cost vector of a path X_{pq} is calculated from each arc as follows:

$$C(X_{pq}) = \sum_{(i,j,v) \in X_{pq}} c_{ijp}$$

In this work, we consider a bi-dimensional cost function representing the travel time and the energy consumption in kWh.

3. STUDY

3.1 OBJECTIVE AND PROTOCOL

As mentioned in the introduction, the objective of this study is to quantify the benefits of optimizing both route and speed for ship navigation. This means considering two control variables in the optimization algorithm, which is not always the case as mentioned in the state-of-the-art section. We aim at evaluating this impact on a conventional ship, and on wind-assisted ships.
As explained in the formulation of the problem, the two control variables that can be considered in the optimization are the ship course (impacting ship route) and speed through water.

We will consider 4 scenarios of optimization depending on whether the route and/or the speed is optimized:

- GCR - CS : Great Circle Route and Constant Speed
- GCR - OS : Great Circle Route and Optimized Speed
- OR - CS : Optimized Route and Constant Speed
- OR - OS : Optimized Route and Optimized Speed

This study covers a range of departure dates from 01.01.2016 to 01.01.2020, with a weekly frequency. All the computations will be done without an estimated time of arrival (ETA) constraint, to get the full Pareto front.
From all these results, we will evaluate and compare the performances of each scenario, in terms of voyage consumption reduction. We will consider 3 voyage durations for the analysis, based on 3 service speeds chosen according to the ship speed discretization: 12, 14, and 16 knots.
Besides, we will analyze each scenario's global results, to understand the behavior of the optimization process that explains performances.

3.2 INPUTS DATA

3.2 (a) Ship model

We consider a generic Ro-Ro vessel, whose main characteristics are listed in the following table, with only one loading condition, but two sails configurations. A twin-screw propeller arrangement was preferred.

Table 1: vessel main particulars

Parameters	Value	Units
Lpp	120	m
B	20	m
T	4.5	m
displacement	8 000	tons
propellers diameter	3.4 x 2	m
rudders lateral area	11 x 2	m²

The vessel setup has been done using empirical relations: Holtrop and Menen [26] for the calm water resistance and propulsion prediction, with Wageningen B-Series for the propeller Open Water curves and Fujii simplified rudder model [27], DTU relations [28] for the wave added resistance, Yoshimura's relations [29] for the drift added resistance and Fujiwara's relations [30] for the wind loads on hull and superstructures.

3.2 (b) Generic suction wing model

A generic suction wing model has been developed based on the data provided in Malavard [31]. A 30m height and 6m chord wing has been considered, with different levels of suction to limit the active power to 165kW. A visualization of the thrust along the vessel speed is plotted in the following figure, for vessel velocities ranging from 8kt to 20kt, and a True Wind Speed of TWS=20kt.

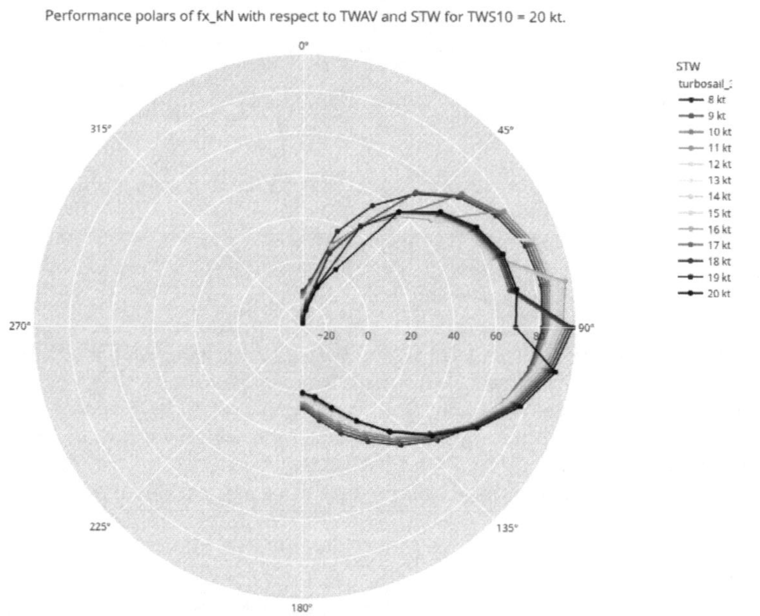

Figure 1: Thrust of the generic suction wing for TWS=20kt

A 4 sails configuration and a 6 sails configuration have been studied. However, no interactions between sails or between sails and hull have been considered.

3.2 (c) Performance polar generation

The performance of the vessel, with and without the sails, has been computed using an inhouse Performance Prediction Program (PPP), called ESPERADO. Unlike a VPP (velocity prediction program), the PPP solves the brake power required to sail at a required vessel velocity. To do so, static kinematic equilibriums are solved for different wind, waves, and vessel conditions, along the surge, sway, and yaw degrees of freedom. An optimization problem under operational constraints is superimposed on the solver to determine the best solution for all conditions.

It results in 5 dimensions (STW, TWS, TWA, HS, WA) polar tables, required by the weather routing solver, containing all solved quantities and load components. While the sea state wave height (HS) and angle (WA) are considered as inputs, the wave period is indexed on the wave height.

Based on the model decomposition and superposition principle, loads applying on the vessel are supposed independent. Special dedicated interaction models are considered when interactions between models are required.

The main models featured in the PPP are:

- hydrodynamic models: calm water resistance, wave-added resistance, drift-added resistance
- aerodynamic models: wind loads on hull and superstructures
- actuator models: propeller, rudder, sails

The hull/propeller/rudder interactions have been considered using the classic MMG [Yasukawa 2015] model.

The sail loads are modeled using 3 dimensions tables (STW, TWS, and TWA), in which the sail is calibrated to provide the maximum thrust, with respect to the maximum active power. A simple depowering coefficient is then applied to the sail loads, with the help of the optimizer, to reduce the sail loads to the minimum required to sail at a given vessel speed without trailing propellers.
An alternate solution would be to incorporate the generic suction sail model in the PPP and trim the wing using the optimizer. We chose to work with 3 dimensions sail tables, to be able to consider data from sail providers who do not wish to share intellectual properties regarding their system model and trimming algorithms.

A polar visualization of the brake power performance of the vessel, in the three configurations (no sails, 4 sails, 6 sails), is plotted in the following figures.

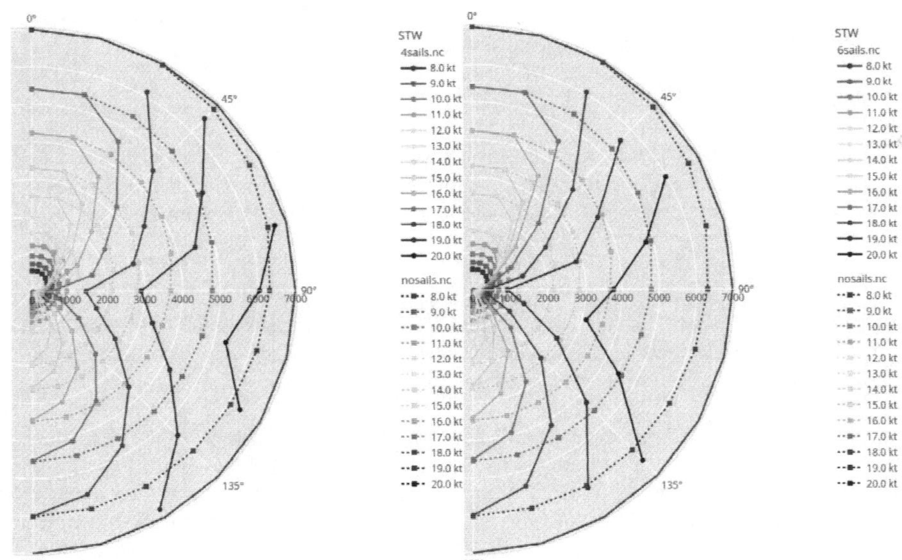

Figure 2: Brake power performance of 4 sails (left) and 6 sails (right) compared to no sails configurations for TWS=20kt.

3.2 (d) Environmental fields

In this study, we use the reanalysis dataset ERA5 from the European Center for Medium-Range Weather Forecast. From this dataset we get 10m components of wind, and significant height, period, and direction for combined wind waves and swell.
The advantage of taking these weather data is that it is a merging between models and observations and therefore is close to the weather that really occurred.

We decided not to include ocean currents, which are more area specific rather than wind and waves. As the purpose of the study is to understand the behavior of weather routing optimization steps, we thought it would be clearer to only consider wind and waves.

Table 2: Weather models

Environmental field	Model	Provider	Time resolution	Spatial resolution
Wind (10m zonal and meridional components)	ERA 5	ECMWF	6h	0.25°
Waves (combined wind waves and swell)	ERA 5	ECMWF	6h	0.25°

3.2 (e) Route and operational constraints

The chosen route is a transatlantic, from Europe to the Caribbean, as many wind-assisted projects will be opened on this line or an equivalent one (e.g. Neoline, Canopée, etc).

More precisely, the route starts at position (48.2, -5.3) and ends at position (19.19, -67.3). The longitudinal resolution (distance between two successive waypoints on the route) is 90 nautical miles, which represents a delay of around 6.5 hours between two control changes at 14 knots. We computed both the eastbound and westbound directions but, in this analysis, we will focus on the westbound direction for an easier understanding.

We decided to take large maximum weather conditions (maximum true wind speed of 60 knots and significant waves height of 8m) to be not too restrictive on the optimization and only ensure ship safety. However, as the added resistances from waves and wind are considered in the ship model, the over-consumption in hard weather is well considered in the optimization.
As mentioned before, we do not include an ETA constraint to get the full Pareto front for each computation, and then export the optimal paths for each service speed ETA.

4. RESULTS AND ANALYSIS

In the following, we compare the performances of each step of the optimization. For each ship and service speed, we compute the benefits by comparing the energy consumption of the optimal path compared to the GCR-CS one. We also propose some analysis to understand which parameters mainly drive the optimization performances.

4.1 RESULTS

In Figure 3 are presented the benefits of each optimization scenario, depending on the service speed and the kind of ship. First of all, we see that decreasing the service speed and increasing the amount of wind propulsion led to better savings, for each ship and optimization scenario.

As an example, the benefits of speed optimization increase from 0.3% to 1.2% for the motor ship when decreasing the service speed from 16kt to 12kt while it increases from 1.7% to 7.8% for the 4-sails hybrid ship, and from 3.4% to 14.4% for the 6-sails hybrid ship. In the same way, the benefits of route optimization increase from 4.5% to 13.1% for the motor ship while it increases from 13.4% to 32.9% for the 4-sails hybrid ship, and from 20.4% to 42% for the 6-sails hybrid ship. And finally, the benefits of route optimization increase from 5.8% to 15.2% for the motor ship while it increases from 15.7% to 39.2% for the 4-sails hybrid ship, and from 24.2% to 50.9% for the 6-sails hybrid ship.

It is important to note how the added value of considering two control variables is enhanced by the wind propulsion. Indeed, for a motor ship at a 14kt of service speed, the relative savings associated with speed, route, and combined speed and route optimization are respectively 0.4%, 8.1% and 9.6%. For the 4-sails hybrid ship, these benefits are respectively 3.8%, 22.2% and 26.1%, and for the 6-sails hybrid ship, 8.1%, 31.6% and 38.6%.

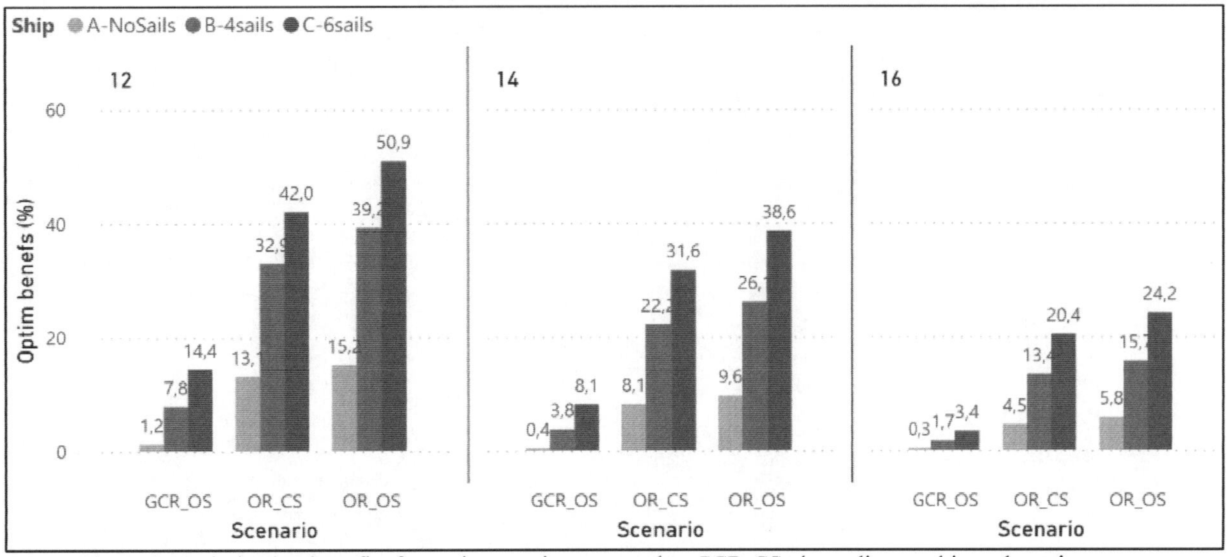

Figure 3: Optimization benefits for each scenario compared to GCR-CS, depending on ship and service speed

4.2 OPTIMIZATION ANALYSIS

4.2 (a) Speed optimization: GCR-OS scenario

As presented previously, important savings can be achieved with speed optimization, especially for wind-assisted vessels. In Figure 4 which represents the evolution of ship speed occurrences for the 4-sails hybrid ship, we see that the more we decrease the service speed, the more the algorithm has the freedom to optimize vessel speed.

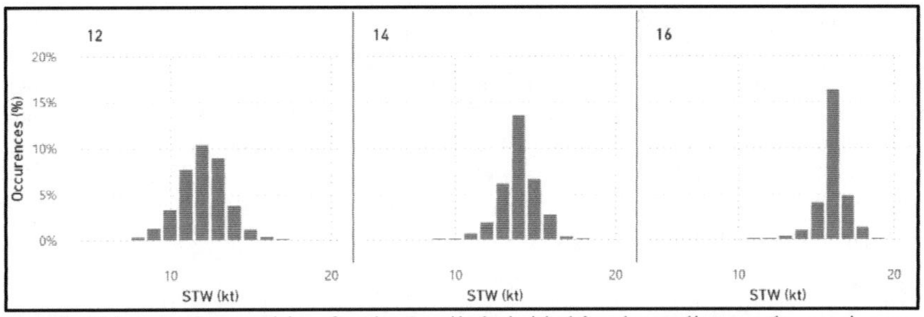

Figure 4: Ship speed repartition for the 4-sails hybrid ship, depending on the service speed

In Figure 5 is presented the evolution of the delta between ship speed and service speed, which gives an idea of the optimization strategy, when the algorithm "accelerates" or "decelerates". The set of visuals represents in lines the different ships and in columns various environmental conditions (wind speed and angle, waves significant height and angle). What we see is mainly that the savings are achieved because the optimization accelerates in good weather conditions and decelerates in bad weather conditions.

Another interesting result is that for the 6-sails hybrid ship, we also see a deceleration when facing low wind speed and downwind conditions. The notion of "bad" weather conditions is indeed specific to the ship. For a motor ship, this is facing high waves and winds, from the bow. For a hybrid ship, bad weather conditions are also facing high waves and winds from the bow, but additionally low wind speeds and downwind conditions.

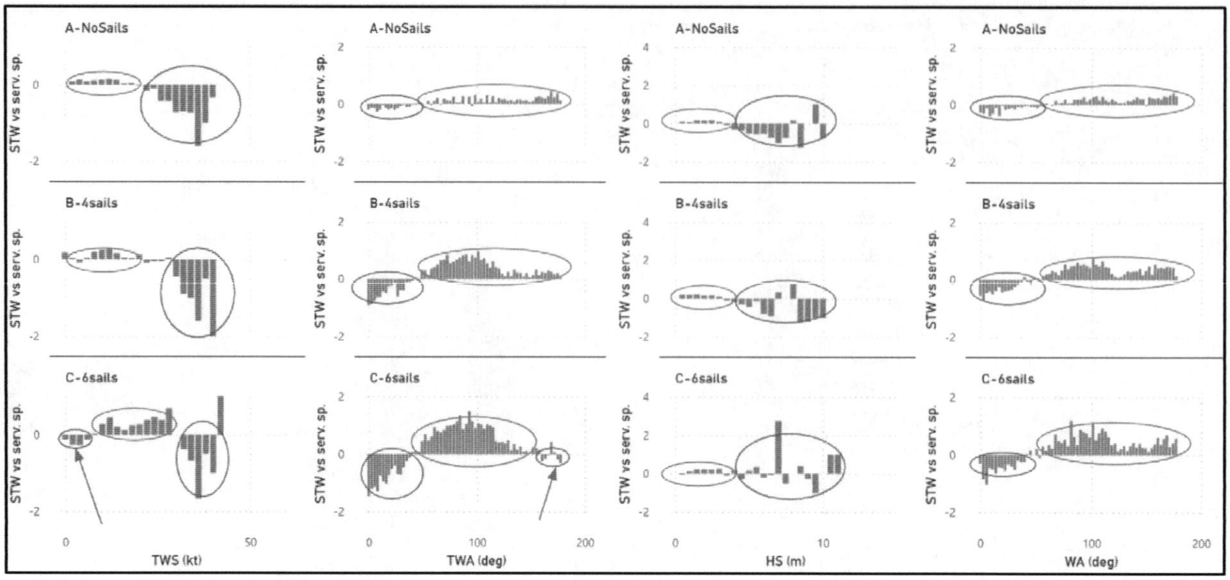

Figure 5: Evolution of ship speed difference with service speed (red/green circles show deceleration/acceleration trends)

4.2 (b) Route optimization: OR-CS scenario

The important savings associated with route optimization are related to the freedom of the algorithm to find a route with favorable weather conditions.
When we compare weather conditions faced in scenario GCR-CS with the one of scenario OR-CS for the 4-sails hybrid ship with a service speed of 14kt, we see that the route optimization helps to meet favorable weather, especially in terms of wind angles. This is related to the wind propulsion system, which is more effective with reaching winds as presented in Section 3.2 (b).

In Figure 6 we can see the percentage of time (color evolving from blue to red) sailed in specific weather conditions and in Figure 7 are compared the TWA and AWA repartition. First, we see that wind angles of the OR-CS are better for wind propulsion: the mean value of TWA increases from 77.07° to 97.4°, and the mean value of AWA increases from 33.09° to 40.20° compared to GCR-CS. Secondly, we see that the standard deviations decrease also (from 53.48° to 40.13° for TWA and from 32.05° to 25.99° for AWA). This means that the algorithm focuses more on favorable ship operating points.

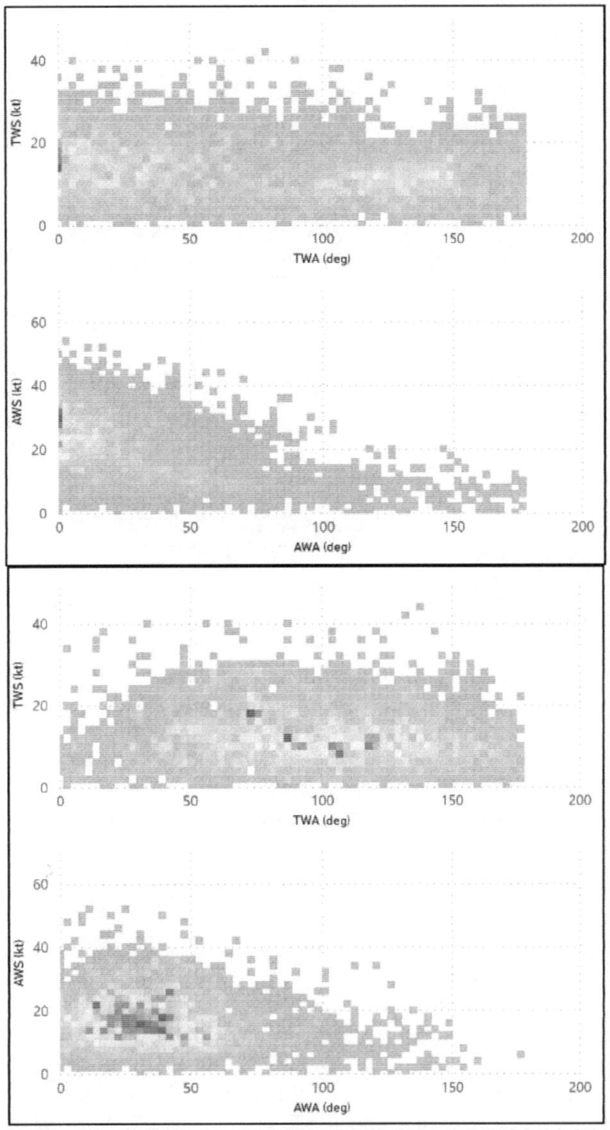

Figure 6: Weather conditions repartition for the GCR-CS (left) and OR-CS (right). Color represents the percentage of time for specific weather conditions (true wind / apparent wind).

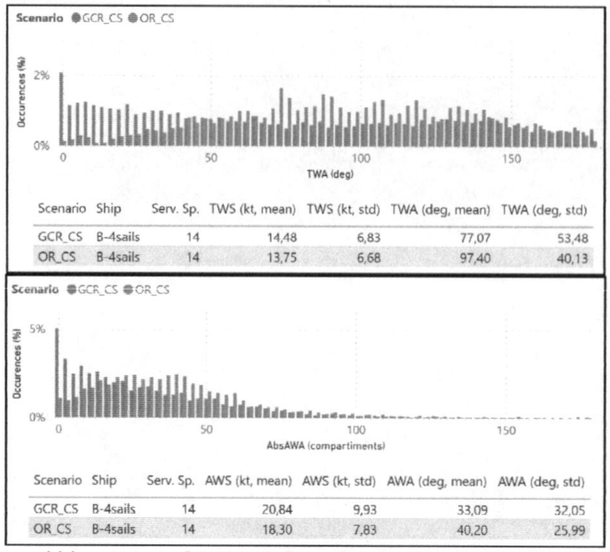

Figure 7: Main metrics and histograms of TWA (left) and AWA (right) for GCR-CS and OR-CS scenarios

4.2 (c) Route and speed optimization: OR-OS scenario

These savings are reached thanks to the combination of both previously highlighted strategies: speed monitoring depending on the bad or good weather conditions, and the ability to face favorable weather conditions thanks to route optimization.
The combination of speed and route optimization also helps to focus slightly more on optimal operating points of the ship, especially in terms of apparent wind angle.
Finally, it is important to highlight that the added value of this scenario in terms of savings becomes significant for wind-assisted ships.

4.3 LIMITS OF THE STUDY

We tried to do a representative study, first by the choice of a generic Ro-Ro vessel and of a generic wind propulsion system. Then, we chose a classic transatlantic route and computed a large set of voyages with one departure per week for 3 years.
However, as the computing time increases with the graph resolution, we had to make a compromise. The graph parameters and especially the width may have led to slightly pessimistic results in terms of route optimization. We see indeed in Figure 8 that a large part of the graph has been used during the 3 years of voyages.
Another important aspect is that these results are specific to the ship model, the wind propulsion system, and the route. Further work should be undertaken to evaluate these impacts on other representative ships such as containers, tankers, or bulk carriers. The impact of each optimization scenario should be probably different.
We also used a generic suction wing model, but a lot of actors are today developing wind propulsion systems whose performances may be rather different from this one, which may impact the conclusions.
Finally, all these computations used historical weather data, meaning that the optimization always chose the best path, knowing exactly the weather evolution. In operational routing, forecast uncertainties may lead to suboptimal choices and finally to less global savings.

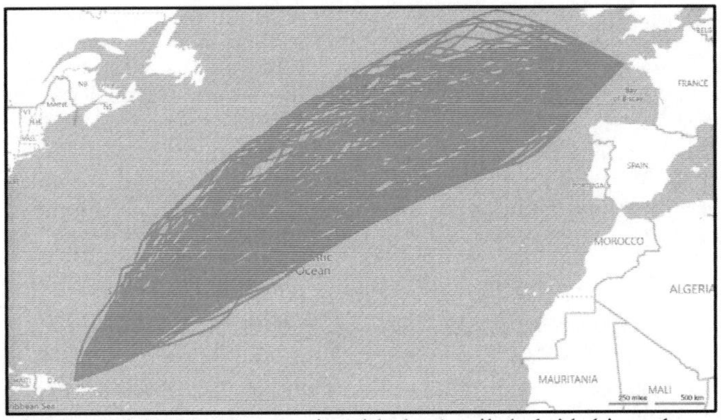

Figure 8: All optimal routes of the 0R-0S scenario with the 4-sails hybrid ship and a service speed of 14kt

5. CONCLUSIONS

Considering two control variables complicates the optimization and therefore it is important to evaluate its impact and added value in terms of savings.
In this work we wanted to study the impact of control variables on weather routing performances, depending on the ship service speed, and the amount of wind assistance.
The main result that has been highlighted is that the more we have wind assistance on the ship, the more we have an added value at considering two control variables in the optimization. This is an important result because as presented in Section 2.2, a lot of methods only consider one control variable. While the shortfall of considering one control variable is not that important for motor ships, it is for wind-assisted ships.

Another interesting result was the understanding of algorithm behavior related to the control variable considered in the optimization. First, the savings associated with speed optimization are due to ship speed monitoring: acceleration in good weather conditions and deceleration in bad weather conditions. Besides, we saw that the definition of such good and bad weather conditions evolves with the amount of wind assistance. Then, we showed that the important savings achieved with route optimization are because the algorithm finds routes that face favorable weather conditions. Finally, optimizing both route and speed leads to added savings compared to previous steps of optimization thanks to the combination of both strategies, and slightly improved apparent wind conditions.

Finally, as statistical weather routing is used for wind-assisted ships performance evaluation, we must use optimization strategies with two control variables. The added value in terms of percentages may be decisive for the viability of the project.

6. ACKNOWLEDGEMENTS

We wanted to thank the D-ICE team for their support in this study. We also wanted to thank the Royal Institution of Naval Architects and the International WindShip Association for organizing this conference and for the tremendous work of education and promotion of wind propulsion.

7. REFERENCES

1. F. ALIGNE, M. PAPAGEORGIOU, AND E. WALTER, 'Incorporating Power Variations into Weather Routeing and Why It May Lead to Better Results', *IFAC Proceedings Volumes, 31(30):269–274*, 1998.

2. S. M. LEE, M. I. ROH, K. S. KIM, H. JUNG, AND J. J. PARK, 'Method for a simultaneous determination of the path and the speed for ship route planning problems', *Ocean Engineering, 157:301–312*, 2018

3. H. WANG, W. MAO, AND L. ERIKSSON, 'Benchmark Study of Five Optimization Algorithms for Weather Routing', *In Volume 7B: Ocean Engineering,* 2017

4. M. Dupuy, C. D'Ambrosio, L. Liberti, 'Optimal Paths on the Ocean', *P.M., Prokopyev, O.A. (eds) Encyclopedia of Optimization. Springer,* 2020

5. R. W. JAMES, 'Application of wave forecasts to marine navigation', *Technical report, U.S. Naval Oceanographic Office, Washington, D.C.*, 1957

6. G. L. HANSSEN AND R. W. JAMES, 'Optimum Ship Routing', *Journal of Navigation, 13(3):253–272*, 1960

7. H. HAGIWARA, 'Weather routing of (sail-assisted) motor vessels', *PhD thesis, Delft University of Technology, Delft*, 1989

8. E. KANOULAS, Y. DU, T. XIA, AND D. ZHANG, 'Finding Fastest Paths on A Road Network with Speed Patterns', *In Proceedings - International Conference on Data Engineering*, 2006

9. Y. W. SHIN, M. ABEBE, Y. NOH, S. LEE, I. LEE, D. KIM, J. BAE, AND K. C. KIM, 'Near-Optimal Weather Routing by Using Improved A* Algorithm', *Applied Sciences, 10(17):6010*, 2020

10. G. MANNARINI, G. COPPINI, P. ODDO, AND N. PINARDI, 'A Prototype of Ship Routing Decision Support System for an Operational Oceanographic Service', *TransNav: International Journal on Marine Navigation and Safety of Sea Transportation, Vol. 7(no. 1)*, 2013

11. D. SEN AND C. P. PADHY, 'An approach for development of a ship routing algorithm for application in the North Indian Ocean region', *Applied Ocean Research, 50:173–191*, 2015

12. C. T. WANG AND P. CHRETIENNE, 'Routage des voiliers et programmation dynamique', *RAIRO - Operations Research, 27(1):61–76*, 1993

13. K. TAKASHIMA, B. MEZAOUI, AND R. SHOJI, 'On the fuel saving operation for coastal merchant ships using weather routing', *Marine Navigation and Safety of Sea Transportation*, 2009

14. C. DE WIT, 'Proposal for Low Cost Ocean Weather Routeing', *The Journal of Navigation, 43(3):428–439*, 1990

15. S. CALVERT, E. DEAKINS, AND R. MOTTE, 'A Dynamic System for Fuel Optimization Trans-Ocean', *The Journal of Navigation, 44(2):233–265*, 1991

16. M. B. KLOMPSTRA, G. J. OLSDER, AND P. K. G. M. VAN BRUNSCHOT, 'The isopone method in optimal control', *Dynamics and Control, 2(3):281–301,* 1992

17. W. SHAO AND P. ZHOU, 'Development of a 3D Dynamic Programming Method for Weather Routing', *Methods and Algorithms in Navigation, pages 181–187*, 2011

18. R. ZACCONE, E. OTTAVIANI, M. FIGARI, AND M. ALTOSOLE, 'Ship voyage optimization for safe and energy-efficient navigation: A dynamic programming approach', *Ocean Engineering, 153:215–224*,1 2018

19. H. WANG, W. MAO, AND L. ERIKSSON, 'A Three-Dimensional Dijkstra's algorithm for multi-objective ship voyage optimization', *Ocean Engineering*, 2019

20. A. VENETI, C. KONSTANTOPOULOS, AND G. PANTZIOU, 'Continuous and Discrete Time Label Setting Algorithms for the Time Dependent Bi-Criteria Shortest Path Problem', *INFORMS Computing Society Conference, pages 62–73*, 2015

21. E. CHAUVEAU, P. JEGOU, AND N. PRCOVIC, 'Weather Routing Optimization: A New Shortest Path Algorithm', *IEEE 29th International Conference on Tools with Artificial Intelligence, pages 687–694*, 2017

22. J. PARK AND N. KIM, 'Two-phase approach to optimal weather routing using geometric programming', *Journal of Marine Science and Technology, 20(4):679–688*, 2015

23. A. VENETI, C. KONSTANTOPOULOS, AND G. PANTZIOU, 'An evolutionary approach to multi-objective ship weather routing', *6th International Conference on Information, Intelligence, Systems and Applications, pages 1–6,* 2015

24. J. SZLAPCZYNSKA AND R. SZLAPCZYNSKI, 'Preference-based evolutionary multi-objective optimization in ship weather routing', *Applied Soft Computing*, 2019

25. R. VETTOR AND C. GUEDES SOARES, 'Development of a ship weather routing system', *Ocean Engineering, 123:1–14,* 2016

26. J. HOLTROP, 'A statistical re-analysis of resistance and propulsion data', *International Shipbuilding Progress, ISP, Volume 31, Number 363,* 1984

27. H. FUJI, 'Experimental researches on rudder performance', *J. Zosen Kiokai 107, 105-111,* 1960

28. M. MITTENDORF, U. D. NIELSEN, H. B. BINGHAM, S. LIU, 'Towards the uncertainty quantification of semi-empirical formulas applied to the added resistance of ships in waves of arbitrary heading', *Ocean Engineering, 251, 111040.*, 2022

29. H. YASUKAWA, Y. YOSHIMURA, 'Introduction of MMG standard method for ship maneuvering predictions', *Journal of Marine Science and Technology, 37–52*, 2015

30. T. FUJIWARA, 'A new estimation method of wind forces and moments acting on ships on the basis of physical component models', *Journal of the Japan society of naval architects and ocean engineers, 2, 243–255,* 2005

31. L. MALAVARD, 'Un nouveau propulseur éolien de navire', *La vie des Sciences*, *1*, 57-72, 1984

PRELIMINARY STUDY ON THE PROPELLER AND ENGINE PERFORMANCE VARIATION WITH WIND PROPULSION TECHNOLOGIES

M Reche-Vilanova, **H B Bingham,** and **H N Psaraftis**, Technical University of Denmark, DK
M Fluck, and **D Morris**, North Windship Technologies, UK

SUMMARY

Wind Propulsion Systems (WPS) for commercial ships are vital to achieving the IMO targets on energy efficiency and GHG emissions. Most WPS will operate in a hybrid mode alongside actual main propulsion units. This will affect the propeller and engine operating conditions and thus, their performance. The present paper discusses a preliminary assessment of commercial ship propellers and engine performance variation as a function of the wind power installed for two propeller plant types (Fixed Pitch Propeller, FPP, and Controllable Pitch Propeller, CPP) at constant speed operational mode. The contribution is based on empirical and analytical methods requiring minimal input data. It aims to provide general trends and contribute basic knowledge on this matter. A cost model is included for a cost-benefit assessment of both propeller types. This leads to advice on which systems to install as a function of WPS installation size.

NOMENCLATURE

K_T	Thrust coefficient (-)
P_B	Brake power (kW)
P_D	Delivered power (kW)
η_D	Propulsive efficiency (-)
η_O	Open water efficiency (-)
J	Advance number (-)
P/D	Pitch ratio (-)
Z	Blade number (-)
CPP	Controllable Pitch Propeller
EAR	Expanded disk area coefficient (-)
FPP	Fixed Pitch Propeller
GHG	Greenhouse Gases
IMO	International Maritime Organization
MBM	Market Based Measures
MCR	Maximum Continuous Rating
PPP	Performance Prediction Program
SFOC	Specific Fuel Oil Consumption (g kWh^{-1})
VLSFO	Very Low Sulphur Fuel Oil
WPS	Wind Propulsion System

1. INTRODUCTION

Maritime activity has a severe impact on the environment by negatively affecting marine ecosystems and emitting greenhouse gases (GHG). In fact, international shipping is responsible for about 3% of global anthropogenic emissions according to the International Maritime Organization [1] and is expected to increase up to 250% by 2050 in a business-as-usual scenario [1]. To meet the Paris Agreement [2] and the 2030 agenda for sustainable development [3], we need green and sustainable shipping. To achieve this, the IMO has set its own strategy: total annual GHG emissions should be reduced by at least 50% by 2050 compared to 2008 [4]. Driven by strategic targets on polluting emissions, ship owners and operators are re-thinking their propulsion systems. Thus, there is a growing interest in the potential of wind propulsion systems (WPS) as an alternative green propulsion. Wind is an impactful emissions reduction technology for shipping, with an abundant, widely available, and cost-free energy source waiting to be harnessed.

Nevertheless, wind propulsion ship dynamics differs from both conventional cargo ships and sailing yachts. To predict potential fuel savings and emissions reduction, Performance Prediction Programs (PPPs) have been developed. Despite significant progress on this topic in the literature, there is still a substantial lack of knowledge on key performance aspects of the performance of wind-assisted (<50% wind power) and wind powered (>50% wind power). One of these key performance aspects that concerns ship owners and operators is the influence of WPS installations on actual propeller and engine performance. WPS will generally operate together with conventional power plants in a hybrid condition. For instance, in constant speed operation mode and active WPS contributing forward thrust, the propeller and, consequently, the engine are unloaded. Thus, their operating points are moved away from their design conditions, which may significantly influence the overall performance. In most of the actual PPPs, these influences are assumed to be negligible, and propeller and engine performance are set to be constant and equal to design conditions. This is an unrealistic assumption and may greatly influence the potential savings from WPS installations, and, consequently, their business case. In addition, it is also important to assess which propeller type is most suitable for each WPS installation. Generally, tankers and bulk carriers, which are the most WPS-friendly thanks to their low speed and free deck space, have a Fixed Pitch Propeller (FPP) installed. A FPP is cheaper than a Controllable Pitch Propeller (CPP), but its operating range is smaller due to its inability to trim its blades to match thrust requirements. Therefore, we must predict which conditions result in fuel consumption losses or gains between both types of propellers. This will assist decision makers to determine if a change is required or not to their existing propeller plant.

To our knowledge, there are only two studies which investigate these phenomena. Molland and Hawksley [5] attempted to identify the influence of auxiliary wind propulsion on the operation of the propeller and propulsion machinery for two operating conditions. They also presented two test cases to numerically investigate the various influences for three different WPS thrust contributions. On the other hand, Naaijen and Koster [6] studied the performance of a cargo ship with a kite installed which accounted for the variation of propeller and engine performance.

Despite the light that these articles have shed on the subject, there are still many unresolved questions. Based on these observations, the present paper aims to contribute knowledge on this topic by showing general trends on the propeller and engine performance variation and its corresponding effects for all sizes of wind propulsion installations (from generating 0% to 100% of the required thrust) on commercial ships sailing at constant speed. In addition, cost-related predictions are included to assess the cost-effectiveness of two different propeller plants (Fixed and Controllable Pitch Propeller, FPP and CPP) as a function of the wind power installed.

The results presented are generated thanks to the development of a generic, fast, and easy tool based on empirical and analytical methods with only the ship main particulars and general specifications as the minimum input data. The tool is meant to provide a sufficient level of accuracy to be used as a preliminary performance assessment tool. The tool is briefly outlined in this paper, but the full description with several operational modes and its corresponding results will be available in a forthcoming journal paper [7].

The rest of this paper is structured as follows: Section 2 provides a general overview of the developed tool including assumptions, input data, and methods. Sections 3 presents and discusses the results generated after applying the tool to a real-life ship test case. Finally, section 4 presents the conclusions and discusses future work.

2. GENERIC PROPELLER AND ENGINE PERFORMANCE AND COST-BENEFIT TOOL

The developed tool is a mathematical model able to predict the theoretical propeller and engine performance and assess the cost-benefit of a FPP and a CPP for wind-assisted (<50% wind power) and wind powered (>50% wind power) ships. Its main novelty is that it can predict these performance variations for any WPS installation size and determine the appropriate propeller type. The WPS installation size is defined as the percentage of its net thrust contribution to the overall calm water resistance, 0% being the baseline scenario of the conventional ship with no WPS installed, and 100% the fully wind powered ship. Note that this is agnostic on which WPS could lead to a given contribution, and the influence of side forces and leeway angles are ignored at this stage. Instead, we aim to exhibit generic trends on propeller and engine performance if a WPS installation is able to generate a given percentage of the net calm water resistance. We avoid focussing on specific one-off conclusions for specific designs by using this simple model approach, which leads to general design guidelines. Thus, the tool focuses on both retrofit and new built installations, as well as wind-assist and wind-powered ships. Results generated from this tool are expected to fill a knowledge gap on this topic.

The tool is based on analytical and semi-empirical methods, and the widely used regression propeller charts. It is a fast and easy tool which only requires the ship main particulars and general propeller and engine data as minimum input data. All missing input data is then approximated by basic analytical and semi-empirical approaches. Thanks to its modularity, user specified engine data and costs can be included to replace the generic approaches.
The tool is divided into three main models operating in serial: the *propeller model*, the *engine model*, and the *cost model*. Together, they can provide generic performance prediction of the entire power plant (including propeller and engine), related costs, emissions and fuel savings, and a preliminary comparison on the cost-benefit of FPP and CPP for different WPS installation sizes. Results are numerically presented in the form of graphs as a function of the WPS net forward thrust contribution in percentage.

2.1 ASSUMPTIONS

Due to its simple approach and the nature of this preliminary study aiming at finding generic trends, the following was assumed. We kindly ask the reader to review the assumptions here described carefully to correctly interpret the applicability and limitations of the results presented in this paper.

- The ship sails under calm water condition at constant speed operation mode.
- The WPS net thrust contribution is a percentage of the calm water resistance, from 0% (zero WPS thrust) to 100% (WPS is fully driving the ship). It is generated from a "ghost" rig, independent of WPS technology and generating thrust only without side forces.
- The WPS operates in a hybrid mode alongside main propulsion units. The propeller and engine are unloaded as a function of the WPS net thrust contribution to keep constant speed.
- Resulting WPS side forces and leeway angles, which may potentially influence the resistance and the flow seen at the propeller are negligible.
- The flow seen at the propeller is uniform and curl-free. We assume that commercial ships are long enough that small leeway angles would only insignificantly impact the flow seen by the propeller.
- Minimum allowed operational engine power is 10% of MCR according to engine manufacturer MAN's recommendations.
- The engine is a low speed two stroke diesel with a direct coupling. This means that the propeller and engine rotational speed are equal.
- No shaft generator (PTO), motor (PTI) or auxiliary marine engines are considered.

2.2 PROPELLER MODEL

The newly developed model to assess propeller performance variation of wind propulsion ships works as follows: first, the calm water resistance, at the design speed, is computed based on Hollenbach's method [8]. This method estimates single and twin-screw vessel resistance based on test data from the Schiffbau Versuchsanstalt in Vienna. It has a narrower range of applicability than Holtrop and Mennen's method but with more reliable results. Next, based on the calm water resistance value, an array of WPS thrust contributions is generated where 0% is the base-line scenario without WPS, and 100% is when ship is fully propelled by the WPS.

The FPP is modelled based on the derived polynomials of the Wageningen B-series [9]. The derived polynomials express the torque and thrust coefficients, and the open water efficiency in terms of blade number, Z, the expanded disk area coefficient, EAR, the advance ratio, J, and the pitch ratio, P/D. The valid pitch ratio range is from 0.5 to 1.4. A Reynolds number correction is also included following [10]. Next, the $K_T(J)$ technique [11] is used to easily determine which value of advance ratio satisfies the different propeller thrust requirements without knowing the propeller rotational speed. Knowing the operating advance ratio allows to determine, among other things, propeller thrust, efficiency, rotational speed, and power delivered as a function of WPS net thrust contribution.
To model CPPs, the same FPP derived polynomials are used with a correction of 1% decrease in the open water efficiency due to their relatively larger hub [12]. However, the main difference is that the CPP process combines the $K_T(J)$ technique for different pitches since CPPs are not constrained to only one curve but variable within the applicable range due to their ability to trim their blades. The additional output of the CPP model is the resulting optimal pitch for each WPS net thrust contribution.

2.3 ENGINE MODEL

The engine model determines the influence of WPS on the main engine and quantifies how significant the changes are in terms of fuel consumption and corresponding emissions. The output of the propeller model is used as input. Assuming a direct coupling between the propeller and to account for engine and a shaft efficiency of 0.99, the brake power is computed from the propeller delivered power. Then, accounting for the installed power, the main engine percentage load of Maximum Continuous Rating, MCR, as function of the WPS net thrust contribution is calculated. This value is key to determine the Specific Fuel Oil Consumption, SFOC, which varies with the percentage load of MCR. It is generally known that a marine engine operates at its optimum point at around 70-80% of MCR. Here, the minimum $SFOC$ is set to 165 g/kWh and its variance as a function of WPS net thrust contribution is based on an approximation for two stroke standard MAN engines [13]. Next, the fuel consumption of the main engine for each thrust requirement is calculated based on brake power and $SFOC$. Finally, the corresponding CO_2 emissions are computed. There is a linear relationship between fuel burned and CO_2 produced, with the proportionality constant being known as the 'carbon coefficient', which in this model is set at 3.11 tons of CO_2 per ton of fuel consumed following IMO GHG study guidelines [14].

2.4 COST MODEL

The cost model developed here is coupled to the propeller and engine performance variation models previously presented. It is divided into two main parts: the cost calculation and the cost-benefit assessment. In the first, it monetarizes the fuel consumption reduction and its corresponding emissions savings for each propeller type and each WPS net thrust contribution compared to the baseline scenario. Note that the baseline scenario is the same commercial ship with no WPS installed and sailing at design conditions. To convert the fuel consumption into operational costs, the fuel is assumed to be Very Low Sulphur Fuel Oil (VLSFO) with a price per tonne of 700€. On the other hand, the corresponding emissions are transformed to operational costs by applying Market-Based Measures (MBMs). The carbon price is assumed to be equivalent to the current EU ETS with an average of 90€ per tonne of CO_2.
In the second part, the cost model assesses the cost-benefit of both CPP and FPP installations as a function of the WPS net thrust contribution based on the payback period which helps determine how long it takes to recover the initial costs associated with an investment.

To objectively assess the cost-benefit of each propeller for the given conditions, it is key to account for both their savings and price differences. Most commercial ships have a Fixed Pitch Propeller (FPP) plant installed [12]. Therefore, we must determine up to which WPS installation size, a FPP is good enough and where a replacement by a CPP is economically interesting. If a CPP may be installed, then the extra investment should be accounted for in the calculations, for both retrofits and new builds. Note that CPP are generally more expensive than FPP. Here, it is assumed that the price of a normal FPP is 1,000,000€ and that a FPP costs 70% of the CPP cost, a realistic estimate according to MAN (personal communication). Note that all costs are rough estimates and subject to material and energy price variation.

3. RESULTS AND DISCUSSION

The generic propeller-engine performance and cost-benefit tool was implemented in a use case. The aim was to apply the three models previously described to an example ship. The results are presented and discussed in this section for the two propeller types in constant-speed operation mode.

The simulated vessel was the open-source oil tanker ship model KVLCC2 with main particulars presented in Table 1. The installed propeller is a 4 bladed FPP with a diameter of 9.86 m and a pitch ratio of 0.72 at 70% of disc radius. To achieve its service speed of 15.5 Knots, a two stroke MAN B&W 7G80ME-C10.5 with an installed power of 32,970 kW was chosen as the main engine. Note that this choice was made thanks to MAN Energy Solutions cooperation in this study. The simulated ship is an existing model with a conventional ship hull which is not designed to withstand WPS side forces. We acknowledge, that it is unrealistic for such a hull to achieve fully wind-powered operation. Nonetheless, the full range of WPS net forward thrust contribution (from 0% to 100%) is studied to show trends for what would happen if the given conditions could be achieved on an arbitrary hull.

Table 1: KVLCC2 main particulars.

Parameter	Dimension
Length Over All	320.00 m
Length of the Water Line	325.50 m

Breadth	58.00 m
Designed Draft	20.80 m
Block Coefficient	0.81 -

The simulated FPP propeller is the Wageningen B-series equivalent to the given propeller parameters. The baseline operating point is defined as the FPP's open-water efficiency with no WPS installed (0%) operating at design conditions. It is found following the $K_T(J)$ technique for the required calm resistance at design speed. It can be seen in Figure 1 where the baseline open-water efficiency is 0.57 (intersection point between dashed green line and orange curve). The corresponding CPP is assumed to have the same geometry as the baseline FPP (same D, Z, and EAR). However, the CPP can trim its pitch within the model valid range, from 0.5 to 1.4 modelled with a step size of 0.01. Note that the main drawbacks of CPPs are their slightly lower efficiency, in this study 1% less than FPPs (due to their larger hub), and their higher cost. The real engine data generously provided by MAN is used.

WPS will generally operate together with thermal power plants under a hybrid condition. For constant speed operation mode, the ship's total thrust requirement is kept constant. However, the thrust requirement from the conventional marine propulsion plant is not constant but dependent on the net thrust contribution from the WPS and the propeller's efficiency at this point of operation. With increasing WPS force, the propeller, and consequently the engine, is unloaded to meet thrust requirements. To do so, the propeller rotational speed is reduced, which results in an increase of the advance ratio, J. Thus, the open water efficiency, η_O, will increase with higher WPS thrust contribution, up to a point, where the lower limit of load is reached, and the efficiency decreases again. This can be seen in Figure 1 for three different percentages of WPS thrust contribution following the $K_T(J)$ technique: 0% (baseline scenario, no WPS installed), 50% (the WPS resulting net thrust corresponds to 50% of the calm water resistance), and 80% (the WPS resulting net thrust corresponds to 80% of the calm water resistance). In addition, the reduction in propeller rotational speed leads to less risk of cavitation and less harmful noise for marine mammals.

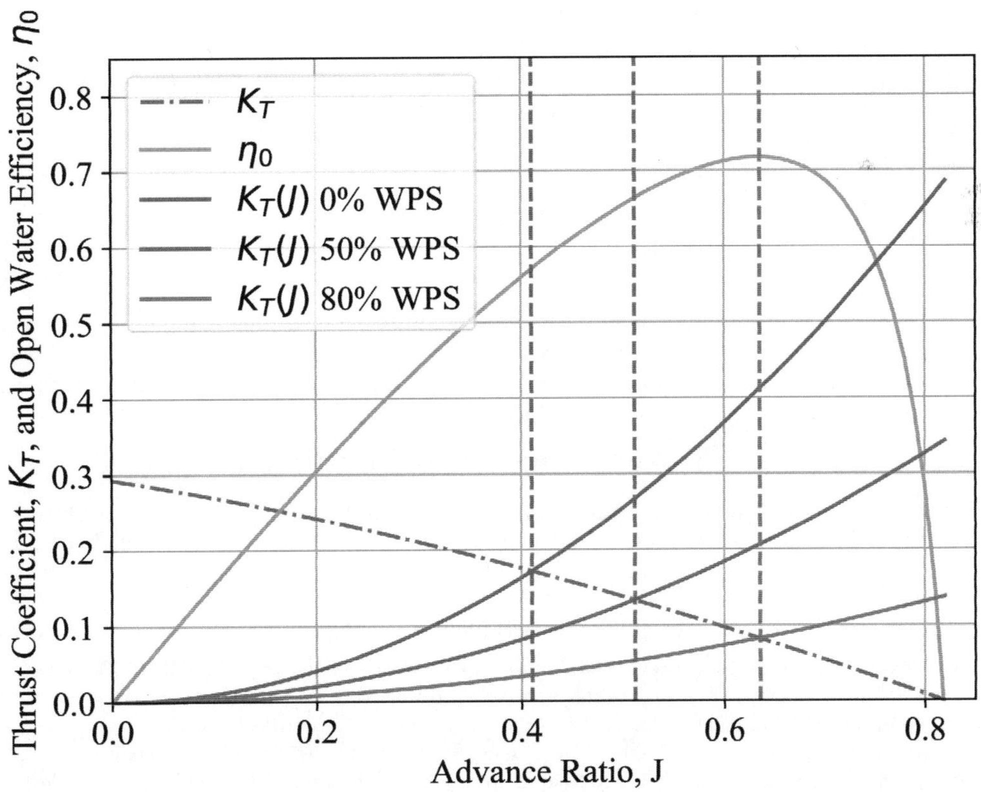

Figure 1: FPP Wageningen B-series chart for a 4-blades, 0.72 P/D, and 0.43 EAR propeller with different thrust requirement curves $K_T(J)$: with 0% WPS contribution (in green), 50% (in red), and 80% (purple). The

resulting η_0 operation points for each propeller thrust requirement are shown as the intersection between the dashed lines of each thrust requirement and the η_0 curve.

Based on these findings, Figure 2 presents the open water efficiency variation, $\Delta\eta_O$, as a function of the WPS net thrust contribution in percentage for the simulated FPP and CPP in constant speed operation mode. The results are plotted in absolute value and in percentage of variation with respect to the baseline scenario for comparison. As seen, increasing the WPS net thrust contribution results in an increase of the open water efficiency for both FPP and CPP up to a point, where the minimum load limit is reached, and the propeller starts stalling. Then, the efficiency falls dramatically. However, this happens with high WPS forces (about 80% contribution for the FPP and 95%, for the CPP) and there, the thrust requirement is low enough for the power demand not to increase dramatically, which is seen in Figure 3. The FPP outperforms the baseline scenario for most of the WPS net thrust contribution percentages. This indicates that the FPP is not only good enough for small-medium size WPS installations but outperforms design conditions of the oil tanker with no WPS. On the other hand, the CPP increases the open-water efficiency slightly less than the FPP due to the relatively larger hub but can keep increasing the efficiency for a wider range of WPS net thrust contribution thanks to its ability to trim the pitch.

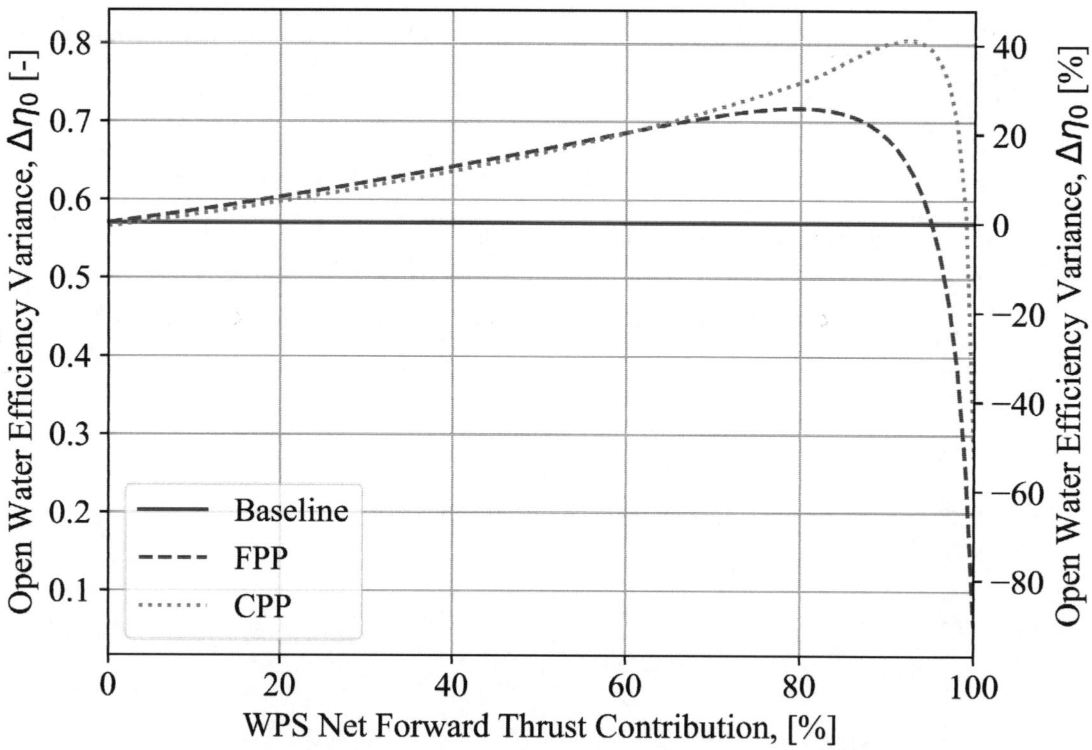

Figure 2: Open-water efficiency variance, $\Delta\eta_O$, in absolute and in percentage values as function of increasing WPS net forward thrust contribution in percentage for a FPP and a CPP, compared to the baseline scenario.

Next, the effects of the propulsive efficiency variation due to WPS on the delivered power, P_D, and the resulting engine brake power, P_B, were investigated. The propulsive efficiency, η_D, which is used to compute the delivered and brake power, is proportional to the open water efficiency with its variation following the same trend but with higher magnitude. Note that the brake power corresponds to the delivered power divided by the shaft efficiency. As seen in Figure 3, increasing WPS thrust contribution in constant speed operation mode results in a lower brake power compared to the baseline scenario. However, accounting for the propulsive efficiency variation with a WPS installation is key as it impacts with significant differences in power savings. In fact, up to 20% and 30% higher power savings are seen for most of the WPS contribution range compared to when neglecting the effect for both FPP and CPP efficiency changes, respectively. On the other hand, for high WPS percentages (higher than 80%), these higher savings are reduced due to the lower thrust requirements and the dramatic fall of their propulsive efficiency. At lower percentage of WPS contribution, the FPP can reduce brake power more than the CPP thanks to its higher propulsive efficiency. However, once the CPP outperforms the FPP in efficiency, this is also reflected in a higher reduction in engine brake power. The results are also plotted as a function of the operating engine load, % MCR, for each new thrust requirement point. As recommended by MAN Energy Solutions, marine engines cannot operate lower than 10% of installed MCR and are advised to keep it higher than 20%. Thus, the minimum operational line at 10% of MCR is also included in Figure 4. Under 10% of MCR, the lubrication and cooling systems may not perform correctly which can cause engine damage. So, technically, we can never reach higher than 80% power savings with a hybrid set up (WPS and engine) in constant speed operation mode. In such a condition, where the minimum engine load is reached, the engine would need to be turned off and reduce the ship speed or keep the engine power fixed and increase ship speed. Another solution for a wind-powered ship aiming at saving more than 50% without compromising the speed, is to install the minimum power allowed by IMO [15] or operate a diesel-electric engine. By reducing the installed power, we would be able to achieve a higher percentage of WPS contribution before reaching the minimum limit of operating %MCR. In addition, if a minimum power is installed, it may be sensible to include a CPP, which would be able to adapt to the thrust requirements and perform in heavy weather conditions. Nonetheless, in this study, any results which cause an engine load lower than 10% of MCR are considered unrealistic. Thus, the minimum reachable conditions are kept constant after surpassing the limit. With the simulated oil tanker, these limits are reached at 82% of net WPS forward thrust contribution for the CPP, and at 83% for the FPP.

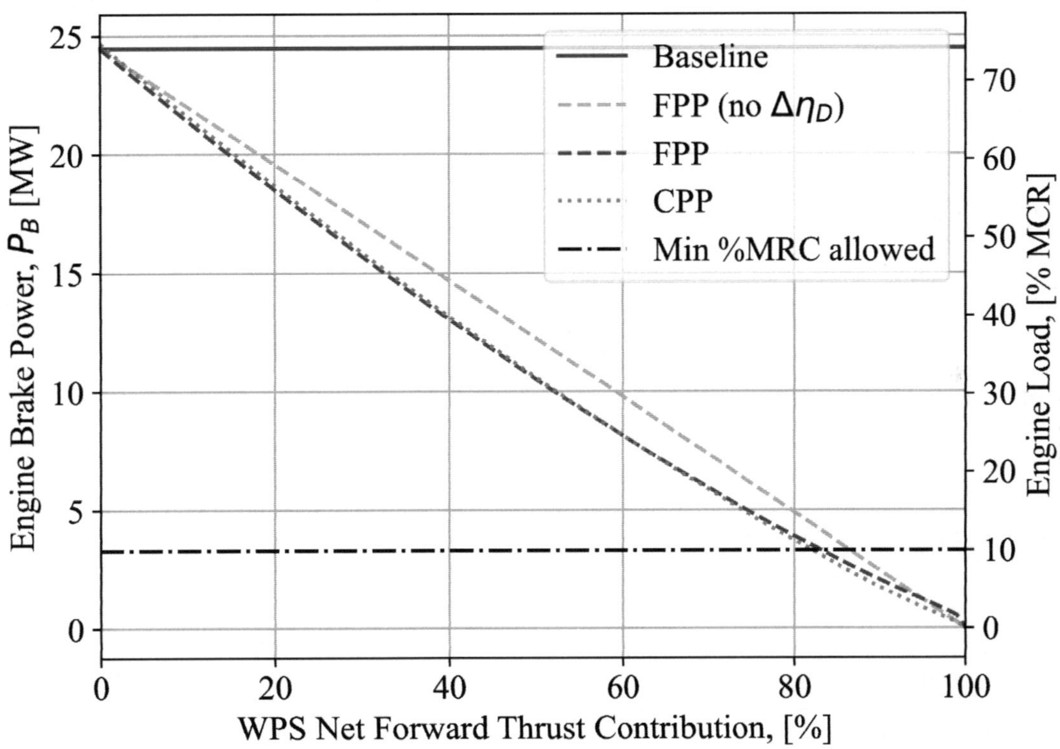

Figure 3: Engine brake power, P_B, and engine load, % MCR, as function of increasing WPS net forward thrust contribution in percentage for a FPP and a CPP, compared to the baseline scenario with no WPS and the FPP scenario where no η_D variance is considered. In black, the minimum engine load allowed by engine manufacturer, which is 10% of MCR.

With the decrease in engine load, the SFOC also varies. Its variation as a function of WPS contribution compared to the baseline scenario was computed for the actual installed power. The results are shown in Figure 4. As seen here, the CPP reaches slightly faster the maximum SFOC variation because it reduces faster the engine load. Figure 5 shows the resulting effects of the combination of the engine performance variation parameters, the ΔSFOC, on the fuel consumption. As seen, WPS can significantly reduce the fuel consumption of a ship. Due to a reduction of the main engine load, the SFOC increases. This has a penalty on the fuel savings compared to the scenario where only the good effects of the propulsive efficiency variation are accounted (the green dashed line). However, comparing with the scenario where none of these effects (neither $\Delta\,\eta_D$ nor ΔSFOC) are considered, extra savings are achieved. Thus, accounting for propulsive and engine performance variation for both FPP and CPP scenarios, results in up to 12% higher fuel savings than without considering these effects for most of the WPS range of net forward thrust contribution. The beneficial effects of the increase of propulsive efficiencies are higher than the penalty of the increased SFOC.

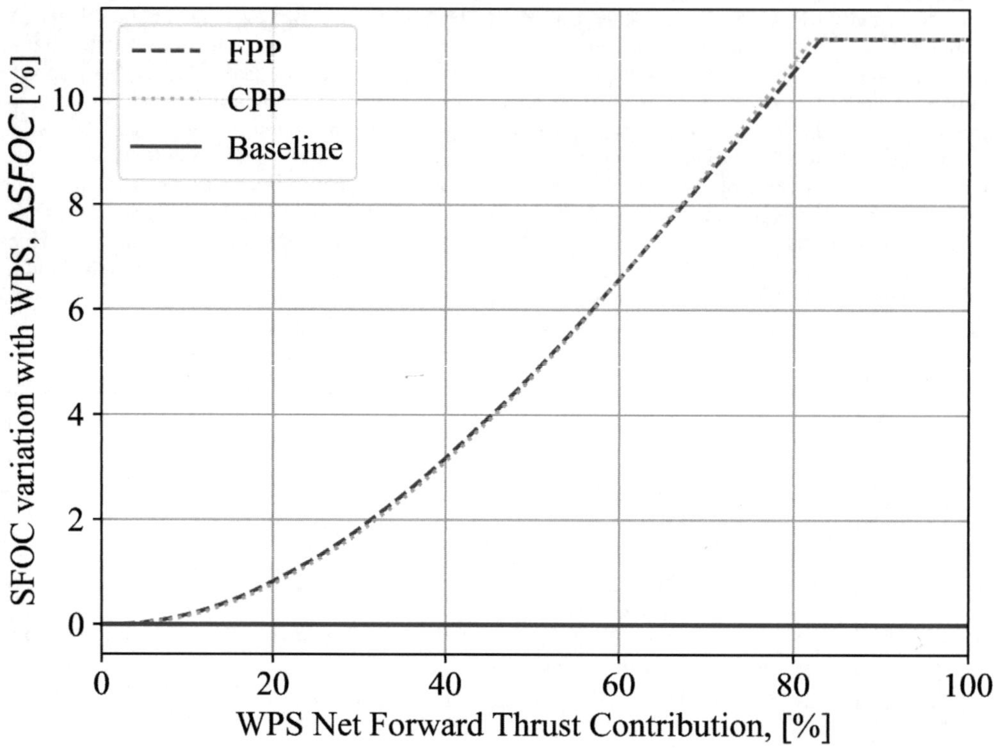

Figure 4: Percentage of SFOC variation as function of increasing WPS net forward thrust contribution in percentage for a FPP and a CPP, compared to the baseline scenario with no WPS.

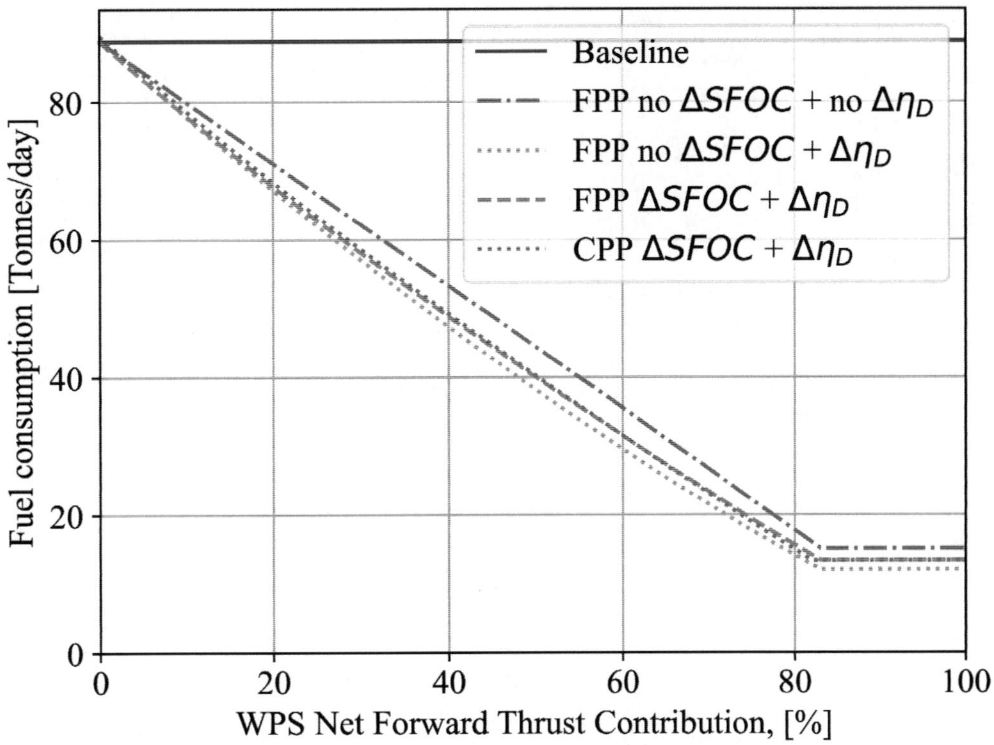

Figure 5: Fuel consumption as function of increasing WPS net forward thrust contribution in percentage for a FPP and a CPP, compared to the baseline with no WPS and the scenario where no η_D and $SFOC$ variance is considered.

Finally, the difference in savings for both propellers are translated into costs and resulting money saved. The fuel price is assumed to be 700 € per tonne and the CO_2 emission costs are set to 90 € per tonne. Note that the prices are highly dependent on the market. The results for the simulated vessel are presented in Figure 6. It seems clear that it is not economical to install a CPP for a small contribution of WPS for the studied vessel. In fact, we would lose money due to the higher power and fuel savings achieved by the FPP, compared to the CPP, at up to 60% WPS contribution. However, this starts changing for bigger WPS installations, where the ship is mainly wind-powered. CPP become more relevant as they save more fuel and corresponding pollutant emissions. In figure 6 only realistic values, where we are operating higher than 10% of MCR, are plotted. If the minimum power allowed was installed, since the 10% of MCR minimum would be reached later, the cost benefits seen by including a CPP would be significantly increased in the wind-powered vessel.

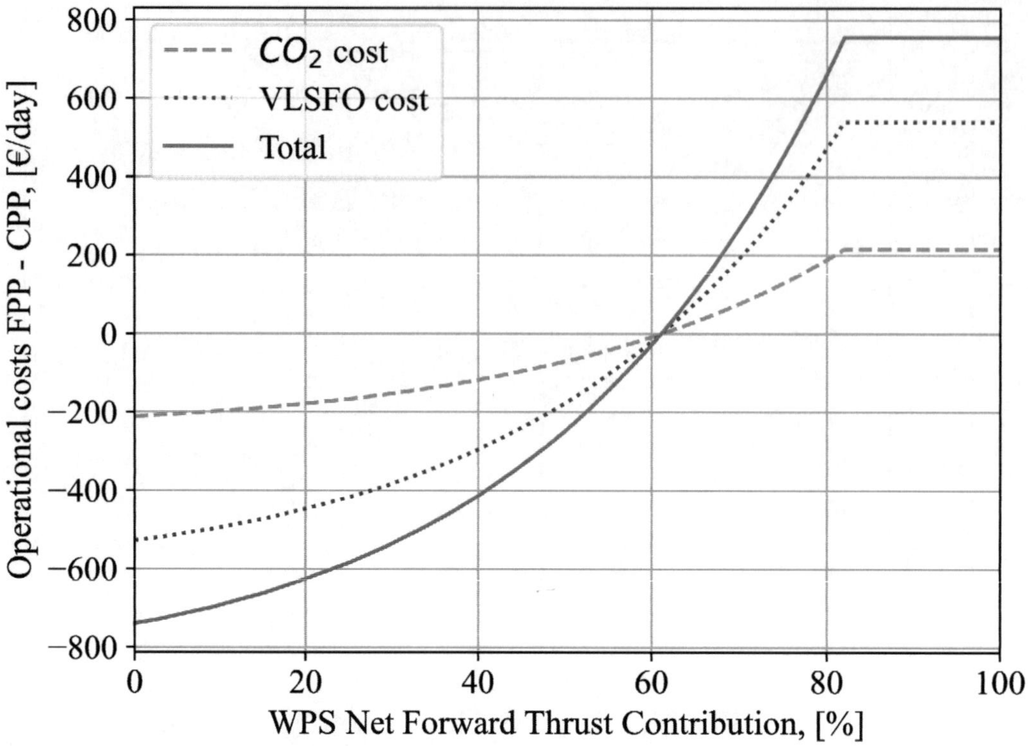

Figure 6: Difference in operational costs as function of increasing WPS net forward thrust contribution in percentage for a FPP and a CPP.

To evaluate the cost-benefit of the CPP for wind-powered vessels, payback period was computed for a WPS installation that gives a net thrust contribution of 80%. Since it is a big WPS installation, we assume that a new ship capable to withstand all side forces and still operate efficiently is required. Thus, the price difference between the FPP and the CPP is assumed to be the difference between both installations not the full price of the CPP, which would be the case for a retrofitting installation where the existing FPP should be changed to a CPP. We assume that the ship is sailing 260 days per year. The payback period for such an installation is 2.5 years approximately. This shows that installing a CPP for wind-powered ships is a cost-beneficial solution. It is worth noting that the payback period for wind-powered ships using alternative fuels, whose prices are expected to be higher than the combined fuel cost and MBM for fossil fuels, will be shortened. In addition, despite of the money saved, if we require a hybrid configuration at constant speed with high WPS thrust contributions (higher than 50-60%), it may be advised to install a CPP to avoid having an operating propeller too close to the stall point. Also, the presented results are for calm water resistance. In real life, the ship would rarely operate at calm water sea state but will face waves and currents which would lead to added resistance. Thus, in real life the minimum allowed engine load, % MCR, and the peak of maximum propeller efficiency would be reached at higher percentages of WPS net contribution, which makes this study conservative.

7. CONCLUSIONS

Most Wind Propulsion Systems (WPS) operate in a hybrid mode alongside actual main propulsion units. As seen in the present study, this interaction significantly affects the propeller and engine operating conditions and thus, their performance. However, in actual performance prediction programs (PPPs) their performance is commonly assumed to be constant and equal to design conditions, which has been demonstrated in this paper to be inaccurate. The impacts are found to be considerable. For the studied vessel, the open-source oil tanker KVLCC2, accounting for the propeller and engine performance variation leads to up to 12% higher fuel savings and the minimum engine load reached for big WPS installations. Not accounting for these performance variations can greatly influence the actual savings of WPS installations and, consequently, their business case.

The generic tool developed and outlined in this paper can predict the propeller and engine performance variation of any commercial ship with WPS for constant speed operation and two different propeller plants: FPP and CPP. It can predict the performance variation for any WPS installation size in terms of its net thrust contribution. The results show that the additional wind propulsion unloads the propeller, which results in a beneficial increase in propulsive efficiency and a detrimental rise in the main engine specific fuel oil consumption (SFOC). However, both effects result in overall favourable propeller and engine performance variation. The CPP seems to be more suitable for big WPS installations (higher than 60% wind power), while keeping the existing FPP is satisfactory for smaller installations (lower than 60%). A main concern for wind-powered ships under hybrid conditions is reaching the minimum allowed operating engine load, which is 10% of MCR for most conventional marine engines. To solve this, these ships may be advised to install the minimum power allowed together with a CPP to be able to operate in heavy weather. Alternatively, they could increase speed when the limit is reached, or exchange the main engine with a diesel-electric one.

In conclusion, the results presented in this study show that WPS not only save fuel and pollutant emissions by reducing the engine power required but using the entire propulsive system more efficiently.

As future work, we aim at including this generic tool in a more detailed PPP where real-life factors such as added resistance, side forces, leeway effects, and curl flow seen at the propeller will be accounted for. With that, we will be able to generate more accurate results for a specific ship with WPS.

8. ACKNOWLEDGEMENTS

The technical information provided by the engine manufacturers MAN Energy Solutions (Copenhagen) and all the support and assistance received from Professor Poul Andersen are greatly appreciated. This work received funds from Innovation Fund Denmark and the Danish Maritime Fund.

9. REFERENCES

1. IMO, 'Fourth IMO Greenhouse Gas Study 2020. Full Report', *International Maritime Organization*, 2021.

2. UNITED NATIONS, 'Paris Agreement', *United Nations Climate Change Conference COP 21*, 2015.

3. UNITED NATIONS, 'Transforming our World: the 2030 Agenda for Sustainable Development', *A/Res/70/1. United Nations General Assembly*, 2015.

4. IMO, 'Adoption of the Initial IMO Strategy on Reduction of GHG Emissions from Ships and Existing IMO Activity Related to Reducing GHG Emissions in the Shipping Sector', *MEPC 72*, 2018.

5. MOLLAND, A. F., AND HAWKSLEY, G. J., 'An Investigation of Propeller Performance and Machinery Applications in Wind Assisted Ships', *Journal of Wind Engineering and Industrial Aerodynamics, 20. Elsevier Science Publishers B. V.*, 1985.

6. NAAIJEN, P., KOSTER, V., 'Performance of Auxiliary Wind Propulsion for Merchant Ships Using a Kite', *Second International Conference on Marine Research and Transportation,* 2007.

7. RECHE-VILANOVA, M., BINGHAM, H. B., FLUCK, M., MORRIS, D., PSARAFTIS, H. N., 'Propeller and Engine Performance for Windships: Benefits and Trade-Offs', 2023.

8. HOLLENBACH, K. U., 'Estimating resistance and propulsion for single-screw and twin-screw ships', *Ship technology research 45. Schiffstechnik, vol. 45, no 2, p. 72,* 1998.

9. BERNITSAS, M. M., RAY, D., AND KINLEY, P., 'Kt, Kq, and Efficiency Curves for the Wageningen B-Series Propellers', *Department of Naval Architecture and Marine Engineering. College of Engineering. The University of Michigan,* 1981.

10. OOSTERVELD, M. W. C., VAN OOSSANEN, P., 'Further computer-analyzed data of the Wageningen B-screw series', *International shipbuilding progress, vol. 22, no 251, p. 251-262,* 1975.

11. BRESLIN, J. P., ANDERSEN, P., 'Hydrodynamics of Ship Propellers', *Cambridge University Press*, 1992.

12. MAN. 'Basic Principles of Ship Propulsion', *MAN Energy Solutions*, 2018.

13. KRISTENSEN, H. O., PSARAFTIS, H. N., 'Energy demand and exhaust gas emissions of marine engines', *Project no. 2014-122: Mitigating and reversing the side-effects of environmental legislation on Ro-Ro shipping in Northern Europe Work Package 2.3, Report no*, 2015.

14. PSARAFTIS, H. N., 'Market-based measures for greenhouse gas emissions from ships: a review', *WMU Journal of Maritime Affairs*, 2012.

15. IMO, 'Guidelines for Determining Minimum Propulsion Power to Maintain the Manoeuvrability of Ships in Adverse Conditions', *MEPC.1/Circ.850/Rev.3*, 2021.

10. AUTHORS BIOGRAPHY

Martina Reche-Vilanova holds the current position of PhD Candidate at the Technical University of Denmark and at North Windship Technologies. Her previous experience includes design of autonomous sailing drones and development of performance prediction programs for wind-assisted ships.

Harry B. Bingham is the Professor of Marine Hydrodynamics at the Dept. of Civil & Mechanical Engineering, Technical University of Denmark. He has been teaching and doing research on theoretical and numerical methods for ocean wave interaction with marine structures for more than 30 years.

Harilaos N. Psaraftis is a Professor at the Department of Technology, Management and Economics at DTU. His areas of interest are maritime logistics, intermodal logistics and port logistics. He has a PhD from MIT, where he was a faculty member from 1979 to 1989. He was a professor at NTUA from 1989 and until he joined DTU in 2013.

Manuel Fluck holds the current position of R&D Engineer at North Sails. He works on consulting projects for racing teams and the development of new technologies including North's contribution to the Windship industry. He holds a PhD in wind turbine aerodynamics and was involved in the performance analysis of DynaRig motor sailing vessels.

Dale Morris is the Technical Manager for North Windship Technologies, as well as Group Head of Global Design for Southern and Hall Spars. He has 12 years of experience as a rig designer for large superyachts and race yachts. He holds a MSc in Naval Architecture from Chalmers University of Technology.

WIND PROPULSION TOWARDS ZERO-EMISSION CRUISE VESSELS

N Abiven, **V Vautier**, **F Cany**, **J B Fabre**, and **L Rouxel-Duval**, Chantiers de l'Atlantique, France

SUMMARY

Wind propulsion for ships has gained new momentum due to the stringent GHG reduction requirements set up by the International Maritime Organization and further more within the European Green Deal. Indeed wind propulsion is a key enabler in this decarbonizing journey as it is one of the very few single solutions enabling energy saving beyond 10%, but still limited with current wind propulsion solutions and wind-assisted vessel applications. In this challenging context, Chantiers de l'Atlantique, through its pioneering Ecorizon® R&D Program, has developed novel Solid Sail wind propulsion solution suited for medium/large ships and hybrid wind-powered vessel architectures such as Silenseas® ship concept. Simulation/experimental demonstration results show great potential for hybrid wind-powered vessels with energy saving beyond 40% and more, especially for merchant vessels.

Keywords: Renewable energy ; Wind propulsion ; Zero-Emission Waterborne Transport

1. INTRODUCTION

Representing 90% of the global trade, the maritime transport is very dependent on fossil fuel energy and known as a large contributor to global emissions: about 2,9% of the world Greenhouse Gas Emissions ("GHG") emissions in 2018 [1], and 13,4 % of the European Union (EU) GHG emissions from transport in 2017 [2]. To tackle this environmental issue, the EU aims at reducing GHG emissions in the maritime transport by at least 40% by 2030 in accordance with its Green Deal strategy.

In this context, Chantiers de l'Atlantique (CdA), the largest shipyard in Europe and a leader in energy and environment efficiencies through its dedicated and pioneering Ecorizon® R&D Program (started 2008), has considered from the very beginning wind propulsion as a key technology to be developed. Indeed it is one of the very few single solutions enabling energy saving beyond 10% and more. Wind-propulsion has not been used for the last 100 years in modern shipping, due to strong drawbacks such as safety, crew-intensive solution, limited fabric sail lifespan, inherent wind versatility compared to low-cost steam/internal combustion engine powered vessels. Thus, to overcome these drawbacks and allow large deployment of such solution a well as significant energy saving even on large ships, CdA has considered several "must-have" requirements (i.e. safety, automation, cost/power-efficiency, durability and scalability for large vessels) in its wind propulsion specifications.

Based on these key stringent success requirements, CdA has made a thorough analysis of potential wind propulsion solutions. State of the art conducted with renowned Dijkstra naval architect team (designer of numerous Dynarig sailing mega-yachts) based on earlier version of the Figure 1 (see § 2) has shown that no existing/potential solution would fit these key success requirements and has pushed CdA in 2008 to develop in-house an innovative wind propulsion solution, i.e. Solid Sail solution. This rather unique and unconventional solution is based on large flat sail composite panels hinged together which deform under wind pressure thus generating aerodynamic trust, and mounted on a free-standing 74m canting aerorig.

This solution is safe, automated, cost/power-efficient, durable and overcomes both current wind powering limitations (i.e., ship size and energy saving). Indeed, together with relevant ship architectures such as CdA large sailing cruise ship concept design (named Silenseas®), this wind propulsion system would allow cruise vessels to reduce their GHG emissions up to 45% compared to similar conventional ships and is transferable to other kind of large ships (e.g. bunker vessels, tankers) with even larger potential energy savings while ensuring their operational capabilities. Achieving such GHG emission reduction level with one single solution is rather unique for shipping, thus confirming CdA R&D long term investment to develop such technology.

The today paper describes the methodology deployed to develop and demonstrate such solution, the results achieved and further developments foreseen before large market uptake.

2. STATE OF THE ART

Numerous wind propulsion technologies for application on low emission vessels are currently under development. Solution performances, cost efficiencies and Technology Readiness Levels (TRL) are quite different and may lead to different applications in terms of energy saving (from wind-assisted vessels with low energy saving to wind-powered vessels with very large energy saving), ship types, sizes and routes.

For wide market uptake, especially covering the medium/large vessel market, stringent requirements are to be fulfilled, as presented in section 1 above. Among them, solution scalability to power large vessels (i.e. equivalent sail area exceeding 1000 sqm) motivates all these developments, as standard fabric/soft sail is limited to max 600 sqm due to limited fabric strength at clew load, thus inducing use of numerous sails and further complexity. Another important criterion to consider is the ability to perform well upwind, as high-sea wind-powered vessels will mainly encounter headwind conditions (i.e., apparent wind angle vs the ship heading << 90°), even when choosing the most wind-wise favorable routes and reducing ship speed.

Up-to-date state of the art, with respect to this medium/large vessel market, in figure 1 below summarizes the pros and cons of different technologies on a 1 to 4 notation/merit index. Evaluation of the standard soft sail with Marconi rigging solution demonstrates all limitations of such solution for the medium/large ship market. On the other hand, CdA Solid Sail shows a very high overall potential, which confirms CdA initial motivation to develop such solution. Specific developments that have led to such evaluation will be further developed in this paper.

	Soft Sail Marconi	Dynarig	Flettner rotor	Flapped wing	Kite	Inflated wing sail	Telescopic hard sail	Solid Sail / AeolDrive
Setting, lowering sails	1	2	-	2	1	2	3	3
Trimming	2	3	3	2	2	2	3	3
Tacking	2	2	4	4	1	3	4	4
Gybing	1	4	4	3	4	3	4	4
Clew loads	1	4	-	3	4	4	4	3
Scalability	2	4	3	2	2	3	3	4
Air draft limitation	1	2	3	2	4	3	3	3
Maintenance	3	2	2	1	2	1	2	4
Durability	2	3	3	2	1	2	3	4
Comfort	3	4	1	4	4	3	4	4
Head wind AWA 25°	2	2	1	4	1	3	2	3
Crew intensiveness/expertise	1	2	4	2	2	3	2	3
Techno. readiness (eq. "TRL" in 1-4 scale)	**4**	**4**	**4**	**2**	**1**	**2**	**3**	**3**

Relative merit index

1	Complex, fragile, prone to failure, low performance, high crew-intensiveness & expertise, low TRL
2	⬇
3	⬇
4	Simple, elegant, robust, easy to use, inexpensive, high performance, low crew intensiveness & expertise, high TRL
-	Not applicable, not necessary

Figure 1: Pros and cons of wind propulsion technologies for medium/large vessels.

In terms of wind powering commercial ship applications, state of the art is limited to a few wind-assisted vessels not specifically designed for wind powering and corresponding low energy savings. Beyond providing performance/cost effective wind powering solutions, a change of paradigm in ship architecture is needed to develop novel wind-powered vessel architectures with large energy savings, and corresponding design and operation assessment methods and tools. Some of them have been already developed and are presented below.

3. METHODOLOGY DEVELOPED AND MAIN RESULTS

Since the very first concept design of Solid Sail solution and giant sail application on cruise ship (i.e. Eoseas concept release in 2009), CdA has conducted intensive R&D work on Solid Sail development and validation as well as wind-powered ship architecture design. These have accelerated starting 2016 with 1st small scale demonstrators and new Silenseas® cruise vessel concept design. The figure 2 below summarizes the different development steps carried out so far that will be further developed in this article.

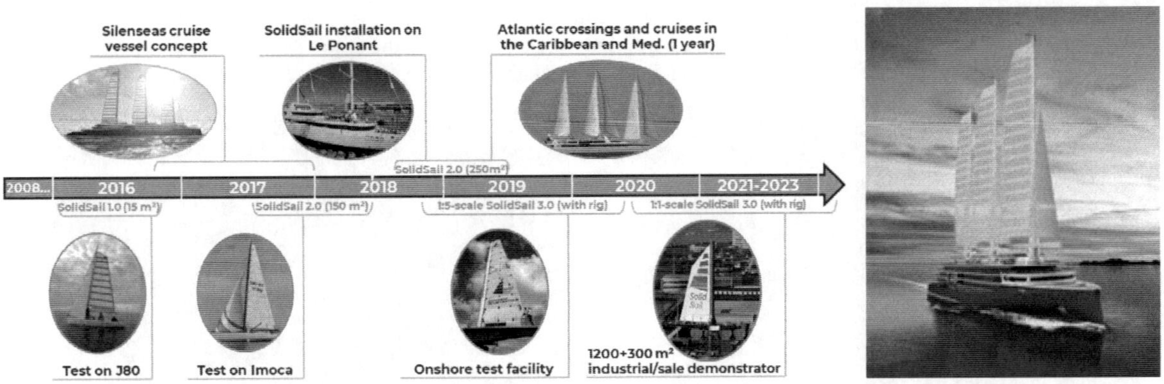

a *b*

Figure 2: (a) overview of Solid Sail validation scheme; (b) Silenseas® cruise vessel concept.

3.1 SOLID SAIL SOLUTION DEVELOPMENT AND DEMONSTRATION

Along with a thorough design to weight and cost approach, the product development and demonstration have concerned four main areas:

- Aerodynamic/structure simulation and/or experimental testing of the sail under wind loading with aerodynamic thrust assessment (aiming at classical soft sail performance equivalence),
- Cinematic/structure simulation and/or experimental testing of Solid Sail during transient operation conditions,
- Aerodynamic/cinematic/structure simulation and/or experimental testing of the full Solid Sail rig in all service conditions (hoisting and lowering the sails, sailing, tackling, gybing, rig tilting…) with actuator and automation operation and validation,
- Manufacturing process development/optimization of the outstanding composite mast (30-180mm thick, 74m high, 2m chord).

The first three development steps have made use of simulation as much as possible for design optimization correlated with real demonstration tests, at lower scales first until final validation at full scale started in 2021 (as shown in figure 2 above), in parallel to overall vessel aerodynamic simulation performances (see section 3.2). Basic solution principles (sail panels folding/unfolding and scantling, flat panel deformation under wind loading to the required flying shape up to overall behavior and performance) have then been validated in 2016 on a 1st-generation Solid Sail 1:10-scale main-sail prototype (through "Silenseas Project" supported by the French Agency for Ecological Transition (ADEME).

In 2017, with the support of Bretagne Region ("Solid Sail 2.0 Project"), a 2nd-generation 1:2-scale main-sail prototype was tested on Jean LE CAM's IMOCA (18m Vendée-Globe monohull racing yacht). After large sea trial validation, an extended version of the 2nd-generation sail prototype was installed in 2018 on Le Ponant (an 88m sailing cruise ship) for one-year operation and 2 Atlantic crossings. This sail proved its robustness as well as its ease of use and allowed better anticipation of all scale effects that could affect the full-size product (1200 sqm objective).

These extended tests gave ways to further optimization, especially for the hinge feature (from metallic hinges to dyneema hinges), sail panel deformation control in correlation to sail panel detail design and load bearing/transfer from panel to panel. Aerodynamic calculation based on Fluid-Structure Interaction (FSI) were successfully correlated globally and used to carry such optimization: see Morvan et all (2021) [3], Sacher et all (2017) [4]. Difficulties remain for local strain/stress analysis as the Solid Sail solution, despite its name, is not a fully solid body, as it is a combination of large semi-rigid carbon/glass epoxy panels linked by dyneema hinges whose stiffness is non-linear by essence and thus difficult to describe in the model.

Cinematic/structure simulation of the sail during transient operation conditions (hoisting and lowering, tackling, gybing…) have been tested and encountered even more difficulties, due especially to its Solid Sail "semi-solid" behavior. Experimental real testing on the different demonstrators has then been used extensively here to overcome these difficulties. Solid Sail cinematic behavior under these transient operation modes is now sufficiently monitored so that we can engineer it in a reliable way.

a

b

Figure 3: (a) 1:5 scale demonstrator; (b) 1:1 scale demonstrator.

All these developments were assessed and demonstrated starting 2019 using an onshore test facility at 1:5-scale of the whole wind propulsion system (i.e. 3rd-generation Solid Sail with a 360° rotating self-supported aerorig), through "LeanShips Project" supported by Horizon 2020 R&D Program (see figure 3a above). Positive results obtained allowed CdA in 2021 to launch fabrication and outfitting of a full scale demonstrator, with all expected features (tilting feature, sensors and automation system…), in 2 steps: first a 500 sqm main sail with a 40m mast successfully tested in 2022, see figure 3b above, and this year 2023 the full 1500 sqm sail area and 74m high mast (under progress). All handling procedures have been tested in the first version (automation of sails and rig, vibration loads, emergency procedures, fatigue tests...). Feedbacks of these last tests showed an expected Solid Sail service life span of 25 to 30 years versus 1 to 2 years only for a premium classic sail. Finally, Solid Sail has received in March 2022 an Approval in Principle (AiP) from Bureau Veritas (BV), based on BV Rule Note for WIND PROPULSION SYSTEMS (WPS) – NR 206.

In terms of weight and aerodynamic thrust, initial objectives for Solid Sail were to achieve close performances compared to classical fabric sail (in a 20% range difference). Solid Sail weight has been dramatically reduced along its different versions, from early 2016 design (1:4 ratio) to current 3.0 version achieving a 1:2 ratio. Further improvements, especially in the membrane definition and fitting equipment, are foreseen which could lead below a 1:1.5 ratio (meaning that the SOLID SAIL will be 1.5 heavier than a classic soft sail at most). It has to be noticed that this comparison is skewed as it is made for similar sail area in the two sail technologies (i.e. maximum 600 sqm for classical sail due to clew load limitation) but is not available for 1200 sqm Solid Sail area objective as classical soft sail reference does not exist.

In term of aerodynamic thrust, the results were even more promising. Speed tests made on J80 racing yachts, based on performance comparison while sailing, gave a 20% speed loss for a low Apparent Wind Speed (AWS) under 5kn (while the Solid Sail version tested were designed for 25 kn AWS), reducing to 10% at 10 kn AWS and becoming equal at 12 kn AWS. On an IMOCA platform, speed tests were made with respect to the polar tables of this offshore racing yacht, with a conversion rule to deal with the difference of sail area. The results in the [10-18 kn] AWS range obtained show Solid Sail achieving 92% of the polar table speed reference based on membrane racing sails. These results were very enthusiastic. Further tests on our 1:5 scale rig demonstrator were made with both Solid Sail and a classical soft sail of the same surface, measuring the torsion efforts in the mast and the efforts on the clew, cunningham and head. Both results were also compared to CFD calculation. Again the results gave similar values between both Solid Sail and classical soft sail.

All these positive results and accumulated 7-year experience in Solid Sail aerodynamic and folding mechanism have been used to develop a Solid Sail jib version ("Jib Sea" project with the support of Bretagne Region). Such a design faced new challenges, as the folding cinematic is made on a non-vertical and flexible forestay instead on a rigid vertical mast for the main sail version and flying shape control is even more performance-critical on a fore-sail. But the test results conducted on a 1:5 scale Solid Sail jib have turned very successful both in terms of aerodynamic performance as well as cinematic and ease of use. Thus decision has been made to launch fabrication of a full scale 450 sqm Solid Sail jib to equip our full scale rig demonstrator for final validation in 2024.

After 7 years of intensive test and demonstration, Solid Sail solution now achieves a 5-6 Technology Readiness Level (TRL). In parallel design to cost and manufacturing developments have been intensively conducted, on all systems involved, such as Solid Sail panels and fittings, carbon/epoxy composite mast, balestron, rig fittings and sensor/automation system. Tremendous results have been obtained so far, such as Solid Sail €/sqm cost divided by 2-fold, composite mast cost by 1.5-fold, and the overall Solid Sail solution rig cost by 1.5 fold. Additional technical optimization and pre-manufacturing activities are foreseen to reduce further this solution cost, improve its Return of Investment ratio much below 15 years (based on current MGO fuel price) and ease future market uptake.

3.2 WIND-POWERED CRUISE SHIP AERODYNAMIC SIMULATIONS AND CORRELATIONS

Beyond Solid Sail product simulation, development and demonstration, performance assessments at ship level have been carried out to assess and validate novel wind-powered vessel architectures using our wind powering solution to aim at large energy savings. They are based on Computational Fluid Dynamics (CFD) simulation coupled with ship model experimental testing to insure proper correlation. The CFD simulation tool developed here enables fast assessment of ship performance compatible with versioning optimization and project phase development requirements, avoiding lengthy and costly experimental tests campaign in wind tunnel, while providing additional local information for further optimization (e.g. aerodynamic flow around each rig, air pressure distribution on each side of rig, forces on each rig…): see figure 4 below. ©starccm+ CFD tool is used here with a specific search of right balance between calculation time and accuracy.

The mesh and geometry are coarse about four millions of cells. The calculation model is steady Reynolds-Averaged Navier Stokes (RANS) Volume of Fluid (VOF), less time consuming than unsteady models. The number of calculation iteration is kept small (i.e. 200), while still insuring proper convergence. The turbulence model has only one parameter (Spallart Almaras) giving a gain of twenty percent reduction on time calculation compared to models of two parameters, such as ambient model k-ε or k-ω. In parallel, an optimization loop on the adjustment of the sail angle has been developed, with respect to 2 optimization variables (driving force and Lift/Drag ratio). As an example, an optimization loop of a 3-rig vessel at one apparent wind angle takes one hour on a 32-core computer.

As shown in table 1 below, correlation achieved between ship model wind tunnel test and CFD simulation is excellent, as differences between wind tunnel test and CFD simulation results are in the 2% range in general, except for the driving force (Fx) which is at 4%.

Table 1: Wind tunnel and CFD simulation results for a 1:60-scale ship model.

35° Apparent Wind Angle (AWA)	Driving force F_x (N)	Heeling force F_y (N)	Roll moment M_x (N.m)	Yaw moment M_z (N.m)	Effort center position with respect to rudder reference (m)	
					$X=M_z/F_y$	$Z=-M_x/F_y$
Wind tunnel test	14.25	-39.72	21.51	8.41	-0.212	0.54
CFD simulation	14.85	-39.10	21.10	8.50	-0.217	0.54
Correlation (%)	104,2	98,4	98,1	101,1	102,4	100,0

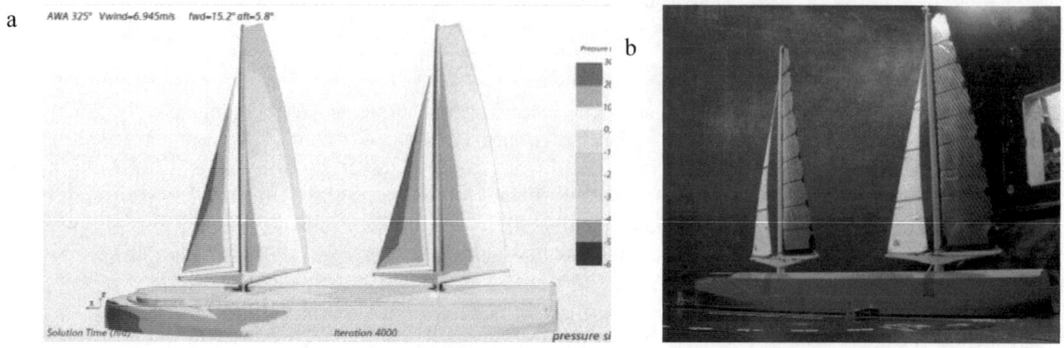

Figure 4: 35° Apparent Wind Angle (AWA): (a) CFD simulation; (b) wind tunnel test.

Based on such good correlation, a complete simulation set of various wind speeds & directions can be conducted to obtain a Velocity Performance Program (VPP), giving the wind speed contribution to the ship for all these wind conditions and ship headings.

3.3 WIND-POWERED CRUISE SHIP HYDRODYNAMIC SIMULATIONS AND CORRELATIONS

But to reach significant wind powering contribution to the overall ship powering, thus enabling large energy savings, novel wind-powered ship architectures need to be developed and especially assessed through hydrodynamic simulations and correlations, taking into account the peculiar impact of wind thrust to the ship equilibriums and behavior.

Wind thrust may generate significant ship heeling and drift. The former can be addressed through standard stability tools and solutions (i.e. adapted ship beam and water ballasts). On the other hand, impact of the latter on ship hydrodynamic performances induces a new era of ship architecture, as such behavior is usually not an issue on conventional propeller-powered vessels. Therefore CdA has developed and conducted series of drift hydrodynamic numerical/experimental simulations and correlations to assess induced lateral drag and impact on ship hydrodynamic behavior, in particular course stability and maneuverability. Figure 5 below shows related ship tank tests and CFD simulations performed.

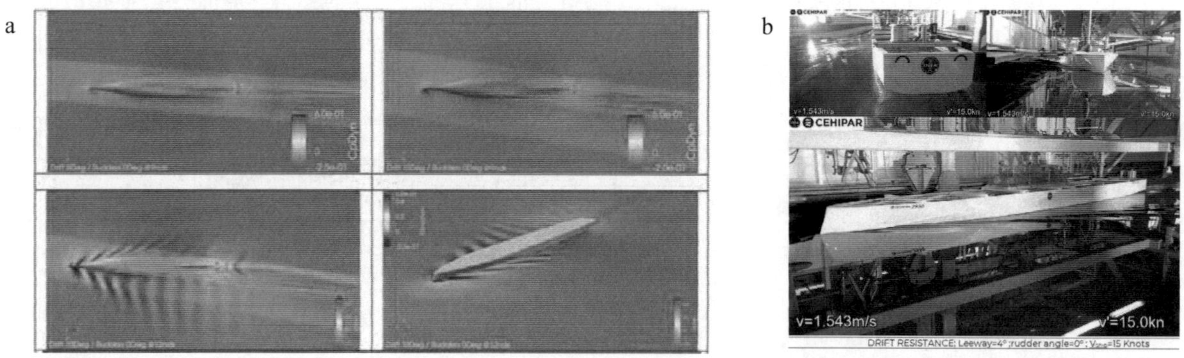

Figure 5: Hydrodynamic drift assessment (4° leeway, 0° rudder angle): (a) CFD simulation; (b) tank test.

Good correlation has been achieved that enabled CdA to conduct further CFD simulations to evaluate different hydrodynamic ship architecture and features (i.e. appendages) and develop related novel and optimized wind-powered ship architectures.

3.4 HYBRID WIND-POWERED SHIP ARCHITECTURE AND ENERGY PERFORMANCE ASSESSMENT

Based on above performance assessments for both Solid Sail component/global rig system and hydrodynamic ship behavior, a preliminary assessment tool of potential energy saving induced by the hybrid wind propulsion mode has been developed. It is based on the following methodology, data and simulation tools:

- Velocity Performance Program (VPP), obtained through CFD simulation, according to section § 3.2 above,
- Ship architecture design definitions in hybrid propulsion mode (standard speed/power curve, drift angle impact, appendage contribution…),
- Development of a Power Prediction Program (PPP) for different ship speeds & wind speeds & directions, giving in a hybrid propulsion mode the additional propeller propulsion power potentially needed to achieve the required ship speed, based on above data,
- Development of a preliminary energy digital twin simulation tool in hybrid propulsion mode, extrapolated from CdA current energy digital twin model used for standard vessels,
- Development of a preliminary routing additional feature in this energy digital twin simulation tool, based on wind and current statistics.

CdA current energy digital twin simulation tool has been developed in the course of the EU funded JOULES project and has been extensively used and further refined by CdA for all its following ship designs. High levels of accuracy and energy savings have been tested through this tool and further confirmed, as proven on the energy-award winning Harmony of the Seas (2017 Fuel Efficiency Seatrade Award).

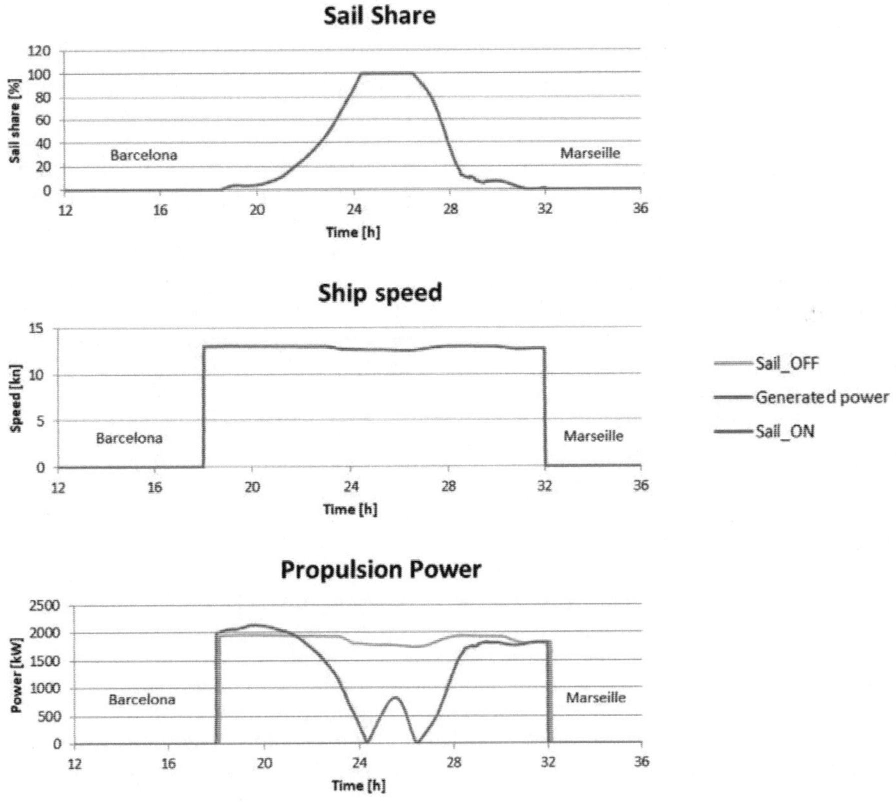

Figure 6: Hybrid wind powering simulation on Barcelona-Marseille cruise profile

This preliminary assessment tool of energy saving for hybrid powered vessels has then been tested on several representative cruise sailing profiles for Silenseas® sailing cruise vessel concept. The following results obtained demonstrate the relevance of hybrid propulsion mode for such application:

- Ability to be in full wind powering mode for ship speed about 10 kn, for wind speed about 15 kn, with ability to generate additional electric power through hydrogeneration for wind speed beyond,

- 34% propulsion energy saving obtained on Barcelona-Marseille cruise sailing profile without weather routing (with an average ship speed of 13 kn and an average true wind speed of 17 kn), including 2 hours in full wind powering mode plus power hydrogenation, out of a 14-h sailing cruise from port to port (see figure 6 above),
- 20% about overall energy saving on this cruise sailing profile, taking into account relative hotel load and propulsion shares on such cruise vessel,
- Additional contribution on above energy savings using weather routing (5-10% minimum propulsion energy saving even at wind speeds about 10 kn).

Such energy saving levels and equivalent GHG emission reductions obtained above are very promising, even on Mediterranean cruises which are usually not favorable for wind powering.

4. CONCLUSIONS AND WAY FORWARD

CdA development, simulation and assessments of its Solid Sail wind powering solution and related hybrid wind powered ship architecture presented above demonstrate the great potential contribution of hybrid wind-powered vessels with respect to the International Maritime Organization (IMO) and European Green Deal decarbonizing objectives. Solutions developed here make wind-powered vessels possible with energy saving much beyond current wind-assisted vessels (i.e. 40% energy saving and more, especially for merchant vessels where the additional energy needs beyond propulsion are limited).

Indeed CdA Solid Sail technology and Silenseas® wind-powered concept ship have recently convinced both ACCOR and NEOLINE ship owners to use these solutions for their future Orient Express Silenseas cruise vessel [5] and Neoliner ro-ro [6] to be delivered in 2026 and 2025 respectively. Both ships should achieve significant energy savings and GHG reductions: -40/-50% and up to -80% respectively, depending on wind conditions and routes.

To achieve such challenging objectives and thus demonstrate the large potential of wind powering solution for maritime decarbonization, additional R&D developments are under way by CdA with its engineering, composite structure and maritime equipment partners, such as:

- Detail developments of novel hybrid wind-propulsion ship architectures for different ship applications with specific new features to address heel and drift during navigation and maneuvering (e.g. development of anti-drift equipment and potential additional appendages), and corresponding ship architecture design assessments and validations: stability, power management strategy, coupled aerodynamic / hydrodynamic CFD simulation and experimental tests at large scales on standard ship operation conditions (sailing, maneuvering) as well as extreme cases (safety issues and rules), energy digital twin simulation. These developments will also benefit from results of the on-going Horizon Europe funded R&D "Optiwise" Project.
- Solid Sail jib and global rig full scale validation,
- Solid Sail process and pre-manufacturing optimization analyses (especially for the composite sail panels and the composite mast).

These R&D developments together with Orient Express Silenseas and Neoliner wind-powered vessel applications will provide a thorough assessment of wind-powered vessels and should confirm 40-80% range energy-savings expected for medium/large vessels, thus providing key leverage towards wider EU Zero-Emission Waterborne Transport co-Programmed Partnership objective to demonstrate and potentially launch very first zero-emission vessel before 2030 and overall IMO/European Green Deal decarbonizing objectives. Indeed, Bell et all (2019) [7] forecast wind propulsion market to count for roughly 40-45% of the global fleet by 2050, thus achieving about ¼ of the total GHG emission abatement worldwide, based on the above energy saving objectives. Such potential market share would represent several hundreds to thousands of ships per year to be equipped with wind systems by 2050, thus inducing a tremendous market potential for wind powering solutions.

5. ACKNOWLEDGMENTS

This R&D work has been made possible through several major contributions, from which the authors gratefully acknowledge their technical or financial support:

- Nicolas ABIVEN, Stéphane CORDIER, Vianney VAUTIER and Chantiers de l'Atlantique's R&D teams for the development and validation of the Solid Sail solution,
- AVEL Robotics, CDK, LORIMA, MULTIPLAST and SMM as main industrial partners to the industrial development and prototype production of the composite sails and mast,
- ENSTA Bretagne, GSEA Design and MECA, as main academic and engineering partners,
- Bretagne and Pays de Loire Regions, The French Agency for Ecological Transition (ADEME), and EU Horizon 2020 R&D Program (i.e. Leanships project).

6. REFERENCES

1. 'Fourth IMO GHG Study 2020', *International Maritime Organization*, 2021

2. 'Greenhouse gas emissions from transport in Europe', *European Environment Agency*, 2019

3. MORVAN A., SACHER M., NÊME A., LEROUX J.-B., JOCHUM C., ABIVEN N., 'Efficient jib-mainsail fluid-structure interaction modelling based on an extended lifting-line theory validations with semi-rigid sails experiments', *7th High Performance Yacht Design Conference, HYPD, Auckland, New Zealand*, 2021.

4. SACHER M., HAUVILLE F., DUVIGNEAU R., LE MAÎTRE O., AUBIN N., DURAND M., 'Efficient optimization procedure in non-linear fluid-structure interaction problem: Application to mainsail trimming in upwind conditions', *Journal of Fluids and Structures, Elsevier, 69, pp. 209 – 231*, 2017.

5. ORIENT EXPRESS, 'Orient Express Silenseas: the legendary train heads out to sea as the world's largest sailing ship is unveiled', *Orient Express website press release:* https://www.orient-express.com/high-life/orient-express-silenseas-the-legendary-train-heads-out-to-sea-as-the-worlds-largest-sailing-ship-is-unveiled/, 2023

6. NEOLINE, 'Neoline turns its Neoliner pilot project into a reality and launches the construction of its first 136m sailing cargo ship', *Neoline website press release:* https://www.neoline.eu/en/neoliner-under-construction/, 2023.

7. BELL M., DEYES K., FITZPATRICK N., WILSON C., CHASE A., LEWIS C., BATES G., 'Reducing the maritime sector's contribution to climate change and air pollution', *Frontier Economics Report for the UK Department for Transport*, 2019.

7. AUTHORS BIOGRAPHY

Nicolas ABIVEN holds the position of Senior R&D Project Manager at Chantiers de l'Atlantique. He is responsible for the development of SOLID SAIL technology since 2016. His experience is based on professional yacht racing and ship building.

Vianney VAUTIER holds the position of Senior Naval Architect at Chantiers de l'Atlantique. Specialized in passenger vessels, he developed the Silenseas Concept back in 2017. He graduated from Ecole Centrale de Nantes in 2010.

François CANY holds the current position of Hydrodynamics and Aerodynamics Engineer at Chantiers de l'Atlantique. He is responsible of hydrodynamics and aerodynamics performances: comfort, plume behaviour and efforts of ship. He graduated from Ecole Nationale Supérieure de Techniques Avancées in Paris majoring in Naval Architecture.

Jean-Baptiste FABRE holds the current position of Energy Efficiency Project Manager at Chantiers de l'Atlantique. He conducts R&D developments of a broad spectrum of energy and environment efficiency solutions and assesses their ship energy and emission impacts using digital twin modelling. He graduated from Ecole Nationale Supérieure Mines-Télécom Atlantique.

Laurent ROUXEL-DUVAL is R&D Vice President at Chantiers de l'Atlantique and is a Board member of the EU Zero-Emission Waterborne Transport co-Programmed Partnership. His previous experience at Chantiers de l'Atlantique includes the launch and management of ship concept approach and proposals which led to the Silenseas Concept among others. He graduated from University of Illinois at Urbana-Champaign, USA and Université de Technologie de Compiègne, France.

SYNERGIES OF WIND-PROPULSION AND GATE-RUDDER™ TECHNOLOGY

N Bulten, Wärtsilä, The Netherlands

SUMMARY

To reduce vessel GHG-emissions, new developments are observed based on both hydrodynamics and aerodynamics. Implementation of wind-propulsion is getting full attention, where the energy of the free available wind is harvested to propel the vessel. On hydrodynamic side a novel Gate-Rudder concept is introduced, which works as an energy saving device. In addition, the Gate-Rudder improves the vessel course-keeping capability. The known side-forces from wind-propulsion can therefore be better controlled with the Gate-Rudder. With the successful application of wind-propulsion, the actual propeller loading will reduce. The effectiveness of the rudder at the lower propeller MCR conditions has been investigated. The Gate-Rudder generates a constant side-force, which brings a clear synergy with wind-propulsion concepts. The following step in the search to reduce GHG-emissions as much as possible is to analyse the vessel drive line. The gains in power reduction of the propeller should not be lost due to increased specific fuel oil consumption due to a shift of the engine operating point. Given the variation in weather and thus in the contribution of wind-propulsion, a more flexible propulsion concept needs to be considered.

NOMENCLATURE

A_R	Rudder Area (m^2)
β	Vessel leeway (drift) angle (rad)
δ	Rudder angle (rad)
ρ	Density of water (kg m^{-3})
d	Vessel draft (m)
Fx	Vessel surge force (N)
Fy	Vessel sway force (N)
LPP	Length between perpendicular (m)
Mz	Vessel yaw moment (Nm)
r	Vessel turning rate (-)
v_{ship}	Vessel speed (m s^{-1})

1. INTRODUCTION

To reach the challenging EEDI targets significant steps in application of Energy Saving Technologies (EST) need to be made. Given the variety of the different concepts, it is important to identify which combinations give additional synergies. Both wind-propulsion and Gate-Rudder technologies have a proven track record of significant energy saving potential in operation. The effective wind-assisted propulsion is found for sails, rotor-sails and suction-sails. Gate-Rudder technology has shown clear fuel saving on numerous vessels as well [1].

The uncertainty of the wind (-assisted) propulsion contribution impacts the conventional propeller hydrodynamics. In case of favourable wind (in speed and direction), the propeller loading will reduce quite a lot. On the other hand, the vessel needs to be able to sail in case of unfavourable winds as well, which determines the maximum load on the propeller. The expected interaction of conventional hydrodynamic propulsion and wind-propulsion is shown in Figure 1. In this approach the contributions from aerodynamics and hydrodynamics are decoupled to get a first estimate of the actual vessel operation and the possible achievable reduction in fuel consumption and GHG emissions.

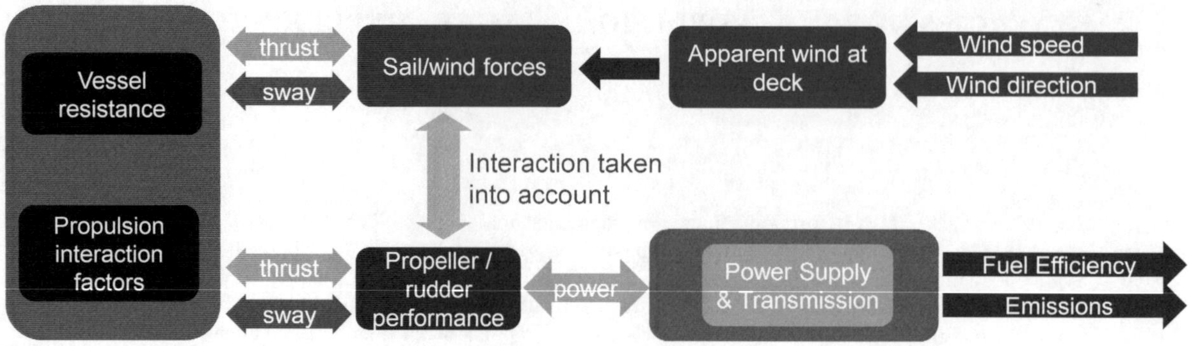

Figure 1: Sketch of integrated conventional hydrodynamic propulsion and wind-propulsion

Wind-propulsion is known to come with additional sway forces, due to the varying direction of the wind-speed. This increases the demands on the vessel course-keeping capability. With effective wind (-assisted) propulsion the loads on the conventional propeller are reduced, which come with reduced effectiveness of the rudder downstream of the propeller. Since the Gate-Rudder is not located in the wash of the propeller, this wash is not expected to be the main driving factor. Given the positions of the two rudder blades, the vessel speed seems to be a more appropriate candidate, which should be maintained with successful wind-propulsion. To quantify this rudder effectiveness, the course-keeping capability of the Gate-Rudder is analysed for different propeller loading conditions.

Fixed Pitch Propellers (FPP) are the common type as used for tankers and bulkers. However, when considering modern wind-propulsion devices on deck, which are designed to provide a significant part of the required thrust, then the change to a Controllable Pitch Propeller (CPP) should be considered. Even though the hydrodynamic efficiency gain might be moderate, when the CPP is compared to the FPP, the specific fuel consumption of the Diesel engine needs to be considered in the overall evaluation as well.

To quantify the different aspects of the Gate-Rudder performance a full-scale CFD simulation model is created, including the detailed hull, propeller and rudder geometries. The applied methodology will be discussed in section 3. This CFD model is used to determine the straight sailing operation, but also the situations where rudder steering angles and vessel leeway angles are applicable. Outcome of the CFD simulations are overall hull-forces and moments, similar to the MMG-manoeuvring methodology.

External forces on the vessel can originate from the environment (windage) or they can be a result of the wind-propulsion devices placed on deck. Forces from the aerodynamics and hydrodynamics are to be combined to get the overall vessel performance. A methodology is applied where both aspects are initially decoupled. As a result, the hydrodynamic approach provides the required knowledge and information for any wind-assisted propulsion device on deck, being rigid sails, rotor-sails or suction sails.

Part of this research work is executed within the framework of the EU-project CHEK, where the concept of wind-assisted propulsion on a bulker-vessel is investigated. Some more background about this project is given in section 2. In addition, CFD simulations are made for a container vessel, where the contribution of the wind-propulsion will be stretched even further.

2. CHEK PROJECT

The EU-funded project CHEK (deCarbonising sHipping by Enabling Key technology symbiosis on real vessel concept design) is to develop and demonstrate a first-of-a-kind vessel concept design at full-scale for a Kamsarmax bulker vessel based on real operational profiles and equipped with a combination of innovative energy saving technologies to reduce the greenhouse gas (GHG) emissions. For the bulker vessel an air-lubrication system (ALS), wind-assisted-propulsion, ultrasonic anti-fouling methods and a Gate-Rudder are considered. In addition to this, the vessel drive-line concept is adapted to meet the larger operational range of the propeller.

The combination of several drag-reduction technologies (ALS and anti-fouling) and the contribution of wind-propulsion result in a clear shift in propeller operating point. To keep optimum hydrodynamic performance of the propeller a Controllable Pitch Propeller (CPP) is selected. The power-supply is based on two 4-stroke Dual Fuel engines, which drive a twin-in-single-out gearbox. This gearbox is designed in such way, that the optimum RPM (and diameter) of the CPP can be selected. As a result, the vessel will sail at the optimum fuel efficiency point of the engines, meaning low RPM, close to the engine torque limit. Both geometries of the propellers and rudders are shown in Figure 2.

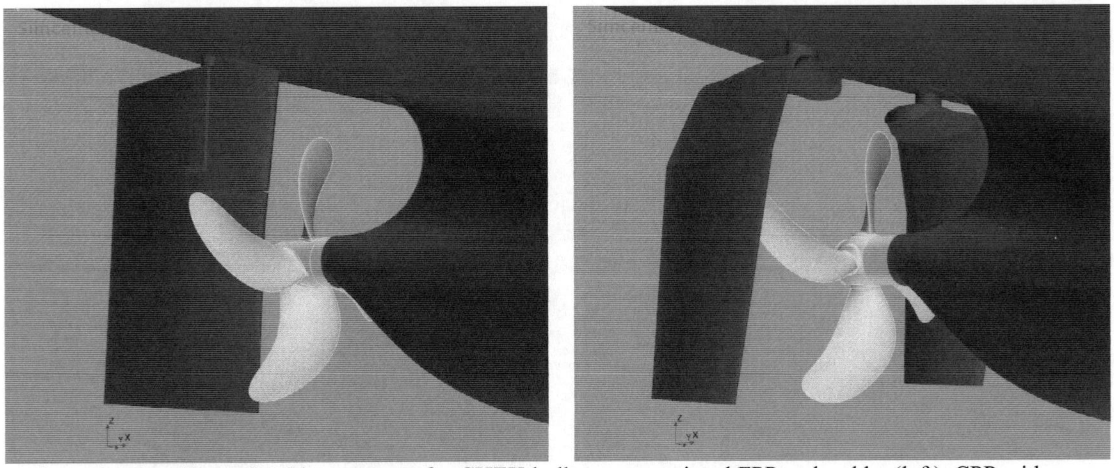

Figure 2: Propeller and rudder concepts for CHEK bulker, conventional FPP and rudder (left), CPP with Gate-Rudder (right)

3. CFD METHODOLOGY FOR HYDRODYNAMIC PERFORMANCE COEFFICIENTS

3.1 PURPOSE OF CFD SIMULATIONS

In the journey towards maturity of full scale CFD simulations the (obvious) focus has been on the calm water straight sailing conditions. This is in line with the towing-tank experiments, and it has been the approach to validate the numerical results with experimental data. Next step in the vessel operation simulations is to consider the effects of vessel manoeuvring due to rudder actions and yaw-motions. The MMG-method [2] is suitable for such evaluation, though this method requires the manoeuvring coefficients of the vessel. For a conventional rudder, equations or modelling approaches are developed to implement the performance of the rudders. However, for the Gate-Rudder these coefficients are not available and therefore a series of full-scale CFD simulations is executed.

To create sufficient data points for the MMG-coefficients a matrix of rudder angles, leeway (drift) angles and turning circle radii needs to be analysed. The results for the vessel surge and sway forces and yaw-moment can be captured in the response surfaces coefficients. With these coefficients all typical vessel manoeuvring actions, like zig-zag, turning circle and pull-out manoeuvre can be calculated.

3.2 CFD METHODOLOGY MAIN FEATURES

When defining the set-up of the CFD model the operating aspects of rudder angle, leeway angle and turning radius need to be covered. The rotating arm set-up, as proposed in already 1964 by Abkowitz [2], is applied. This method suffered from limitations in experimental facilities, though it is well suited for a numerical approach, where the turning radius of the rotating arm can be set to any desired value. This planar rigid motion is implemented in the commercial CFD software of Siemens Simcenter Star-CCM+. In the current approach the straight sailing operation is actually based on a turning circle diameter of 100km, which should represent the straight sailing operation. In this set-up the vessel will execute a rigid-body motion which means that the surrounding water is actually at zero velocity. Figure 3 shows the top view of the rotating arm set-up and the rigid body motion of the vessel. The water velocity is indeed zero at the boundaries of the numerical domain when looking in the stationary frame. In the vessel-reference frame the velocity distributions show the same behaviour as in the conventional applied CFD set-up, where the water gets the inflow velocity according to the vessel speed and the vessel remains at fixed position.

To solve the various interaction effects between hull, propeller and rudders in the disturbed flow field properly, it has been decided to model the actual 3D propeller geometry and to use the sliding mesh transient method. The propeller rotation is a superposition on the rigid vessel motion. In this way the actual blade load fluctuations at blade-pass frequency and higher harmonics are captured as well. The final time-step represents a 2°-propeller rotation.

For the simulations with vessel leeway and/or turning rate, the inflow to the propeller is expected to have some nett pre-rotation. The impact of this effect will be captured with the actual 3D propeller geometry. In case the set-up would be simplified by a virtual disk methodology, the impact of the pre-rotation will probably be missed in the simulation. The geometry of the aft-ship with propeller and Gate-Rudder can be seen in Figure 3)

The rather large matrix of vessel operating conditions (vessel speed, propeller loading, rudder angle, leeway angle, turning radius) require many calculations. To limit the overall calculation effort to acceptable proportions, it is decided to use the double-body approach, which means that the impact of the free-surface effects is neglected in this way. Effects of vessel heel angle are excluded from the simulations as well in this approach.

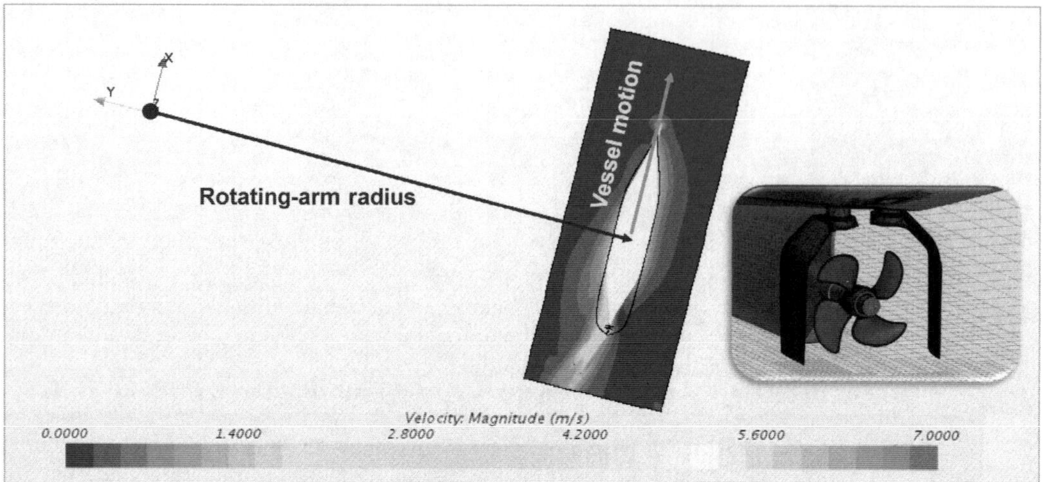

Figure 3: applied CFD set-up for the rotating arm set-up

3.3 SPECIFIC GATE-RUDDER CFD DETAILS

It can be seen in Figure 2 that the Gate-Rudder blades are placed to a large extend outside of the hull-boundary layer in the undisturbed flow along the vessel. One aspect of the working principle of the Gate-Rudder is to produce thrust (or minimize the drag) in normal transit operation. The flow near the leading edge of the Gate-Rudder blades needs to be solved accurately and the actual shape of the rudder blade geometry need to be captured well within the CFD simulations. The combination of Advancing Layer Meshing (ALM) and polyhedral cells provide means to control the mesh in the boundary layer. To capture the strong curvature of the leading edge of the Gate-Rudder blades, the Anisotropic Advancing Layer Meshing (AALM) is used. Figure 4 shows the final rudder surface mesh near the leading edge as well as the mesh of the extrusion layer from rudder surface to the core polyhedral mesh.

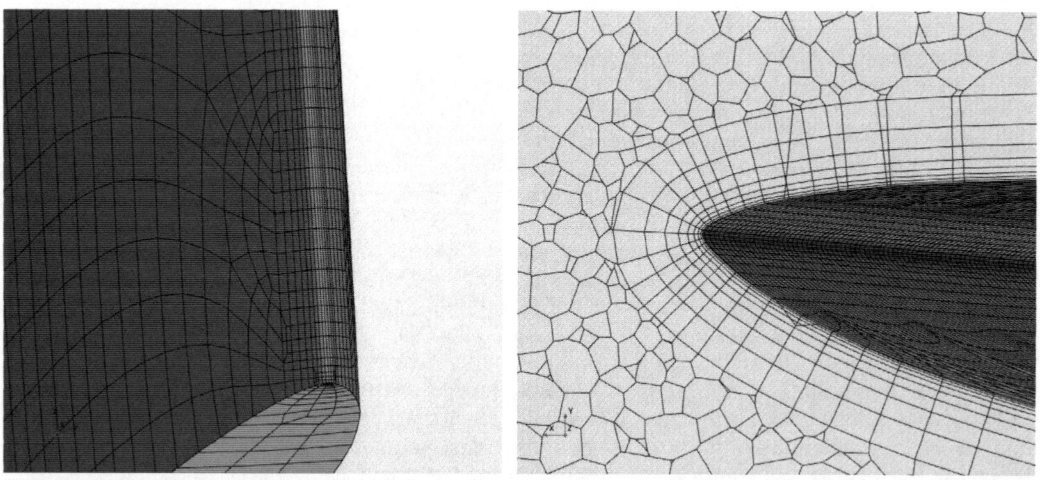

Figure 4: mesh of Gate-Rudder blade based on anisotropic advancing layer meshing and polyhedral meshing

3.4 SCOPE OF CFD SIMULATIONS

The scope of CFD simulations covers the different aspect as described before. First runs are based on straight sailing at constant vessel speed with propeller load variations. The CFD simulations provide the forces on the

different components and the shaft power. The effects of the actual free-surface can be incorporated once the actual power consumption of the vessel is known. In this approach to match the actual propeller power, other aspects like roughness and fouling and added resistance due to waves can be included as well.
In line with the theory of vessel manoeuvring, the impact of rudder angles is determined based on straight sailing direction, where the rudder steering angles are varied in discrete steps. In the current CFD set-up this requires remeshing of the numerical domain, though it could be handled by over-set meshing as well. Vessel leeway runs are obtained by a shift of the rotating arm mid-point, and therefore no remeshing of the domain is required. Change of the radius of the rotating arm is based on the relocation of the rotating arm mid-point, which is another action without remeshing.

3.5 INITIAL RESULTS OF CFD SIMULATIONS

The work in the CHEK project is currently executed and for the Gate-Rudder several design cycles will be made to investigate the impact of design changes on the vessel performance and manoeuvrability. The CFD simulations for the reference case of the semi-spade rudder, as shown in Figure 2, has been completed already. The results of the yaw-moment and the sway-force as function of rudder angle variation and leeway (or drift) angle variation are shown below. The forces and moments are presented in non-dimensionless representation based on:

$$CY = \frac{Fy}{1/_2 \cdot \rho \cdot Lpp \cdot d \cdot v_{ship}^2} \tag{1}$$

$$CN = \frac{Mz}{1/_2 \cdot \rho \cdot Lpp^2 \cdot d \cdot v_{ship}^2} \tag{2}$$

These dimensionless results show the calculated hydrodynamic yaw-moment and sway-force of the complete vessel, thus including hull, propeller and rudder geometry. In this approach all possible interaction effects are taken into account. The trends of the results are clear from the calculated data points.

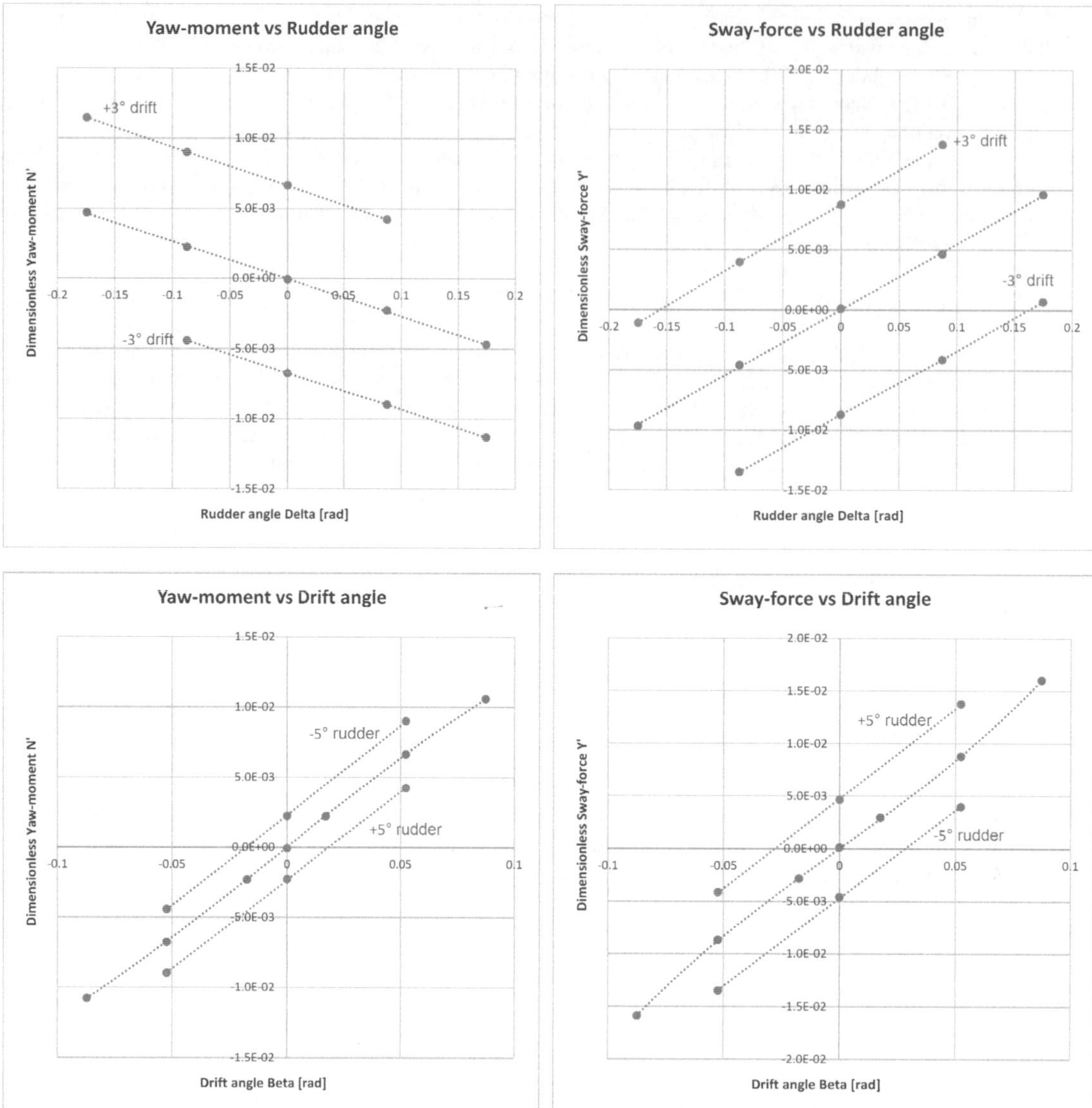

Figure 5: Vessel yaw-moment and sway-force as function of rudder angle and leeway (drift) angle variation

3.5 TRANSFORMATION OF CFD RESULTS INTO RESPONSE SURFACES

Review of the CFD results, as shown in Figure 5, reveals that the sway-force and yaw-moment give the expected anti-symmetrical behaviour. This can be fitted properly with 1st and 3rd order coefficients. The CFD results from rudder angle δ, vessel drift angle β and vessel turning rate r variations are captured in a single formula representing a multi-variable response surface:

$$CY = Y_\beta \beta + Y_{\beta\beta\beta}\beta^3 + Y_\delta \delta + Y_{\delta\delta\delta}\delta^3 + Y_r r + Y_{rrr} r^3$$
(3)

$$CN = N_\beta \beta + N_{\beta\beta\beta}\beta^3 + N_\delta \delta + N_{\delta\delta\delta}\delta^3 + N_r r + N_{rrr} r^3$$
(4)

For the surge force it has been found that the vessel leeway angle introduces a pre-swirl into the propeller. Consequently, the propeller loading decreases or increases depending on the leeway direction. Other aspects of the surge-force, like the rudder angle and turning-radius, show a symmetrical behaviour. To capture the hull resistance and the propeller thrust forces properly the overall curve-fit formula for surge-forces is defined as follows, where n is the propeller rotation rate:

$$CX = c + X_{\beta}\beta + X_{\beta\beta}\beta^2 + X_{\beta\beta\beta}\beta^3 + X_{\delta}\delta + X_{\delta\delta}\delta^2 + X_r r + X_{rr}r^2 + X_n n + X_{nn}n^2 \quad (5)$$

Once the coefficients have been determined, based on simple linear algebra, the actual vessel operating condition can be determined for any given sway-force and yaw-moment, either coming from the environment or from wind-propulsion devices on deck.

4. VESSEL COURSE KEEPING CAPABILITY

4.1 GATE-RUDDER SEA-TRIAL RESULTS

End of 2020 a dedicated sea trial was executed of two sister vessels, one equipped with conventional flap-rudder and one with a Gate-Rudder. Both vessels sailed the same route at the same day with only limited time in between. In this way both vessels experienced the same weather, which was actually not favourable. Rather strong side winds impacted the course-keeping of the vessels. More details about the seat trial campaign have been reported by Sasaki *et al.* [1].
One aspect from the joint sea trials which will be analysed in more detail are the vessel motions due to the wind and waves, as shown in table 1. Both the vessel roll and yaw amplitude differ significantly when flap-rudder and Gate-Rudder values are compared. Roughly speaking the amplitudes for roll and yaw are halved for the vessel with Gate-Rudder. This is attributed to the larger lateral area of the two combined Gate-Rudders.

Table 1: vessel motions during joint sea-trial

	Roll	**Yaw**	**Freq.**
Flap-rudder	**3-5 deg**	**0.6-1.0 deg**	**Abt. 30sec**
Gate-rudder	**1-3 deg**	**0.3-0.5 deg**	**Abt. 30sec**

Based on these findings the beneficial course-keeping capability of the Gate-Rudder has triggered follow-up research actions. Question is whether it is a simple matter of more lateral area, or that additional hydrodynamic effects play a role.

4.2 CFD ANALYSIS OF GATE-RUDDER WITH LEEWAY ANGLE

To get better understanding of the occurring flow phenomena when the Gate-Rudder sailing with leeway angle, results of the CFD simulations need to be analysed in more detail. Impact of the leeway angle can be expected based on a simple theoretical approach, as shown in Figure 6. This picture shows the expected contributions to the inflow to the rudder profile. The vessel is oriented in such way, to represent the leeway angle, as indicated by the vessel-centerline. Due to the leeway, the angle of attack of the flow towards one of the two Gate-Rudder blades will increase. Second aspect which influences the angle of attack to the Gate-Rudder is the magnitude of the radial velocity due to the suction of the propeller. With an increased angle of attack the rudder profile section pressure distribution will change which can result in a nett positive thrust force.
Figure 7 shows the pressure distribution on the Gate-Rudder blades from two different viewing directions. In this condition the rudders are in straight-ahead condition (0° steering) and only vessel leeway is applied in the simulations. When looking at the innerside of the starboard rudder blade, the low pressure region increases with larger leeway (drift) angles. This is in line with the theoretical concept. However, a second effect can be observed at the port-side rudder blade, where a larger low pressure region at the shoulder-part section can be seen. This low pressure indicates a change in angle of attack to the shoulder-section. The main flow is obstructed by the hull geometry, or in other words the Gate-Rudder is located in the shadow of the vessel. Reduction of the axial flow component in combination with the remaining radial component results in a change of the angle of attack and therefore a different pressure distribution. The consequence of the two above described phenomena result in a positive thrust-force in the vessel axial direction.

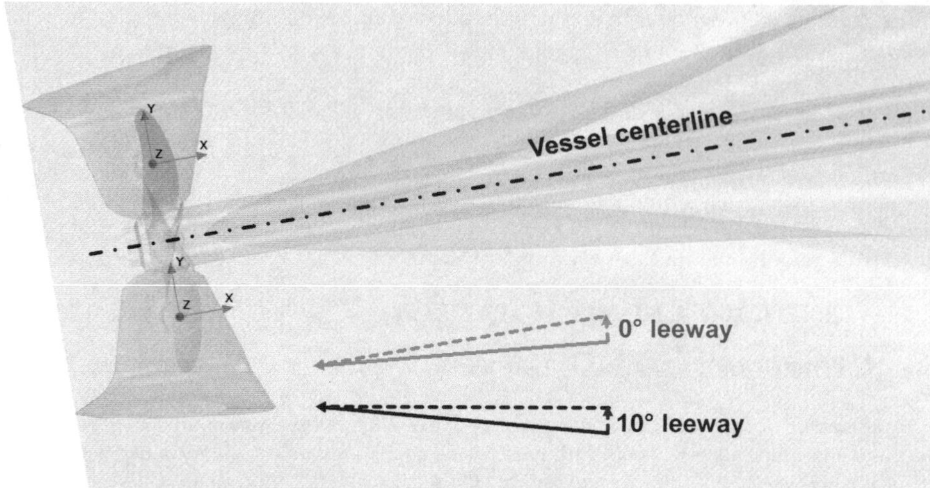

Figure 6: Theoretical review of Gate-Rudder inflow angle when sailing with leeway angles

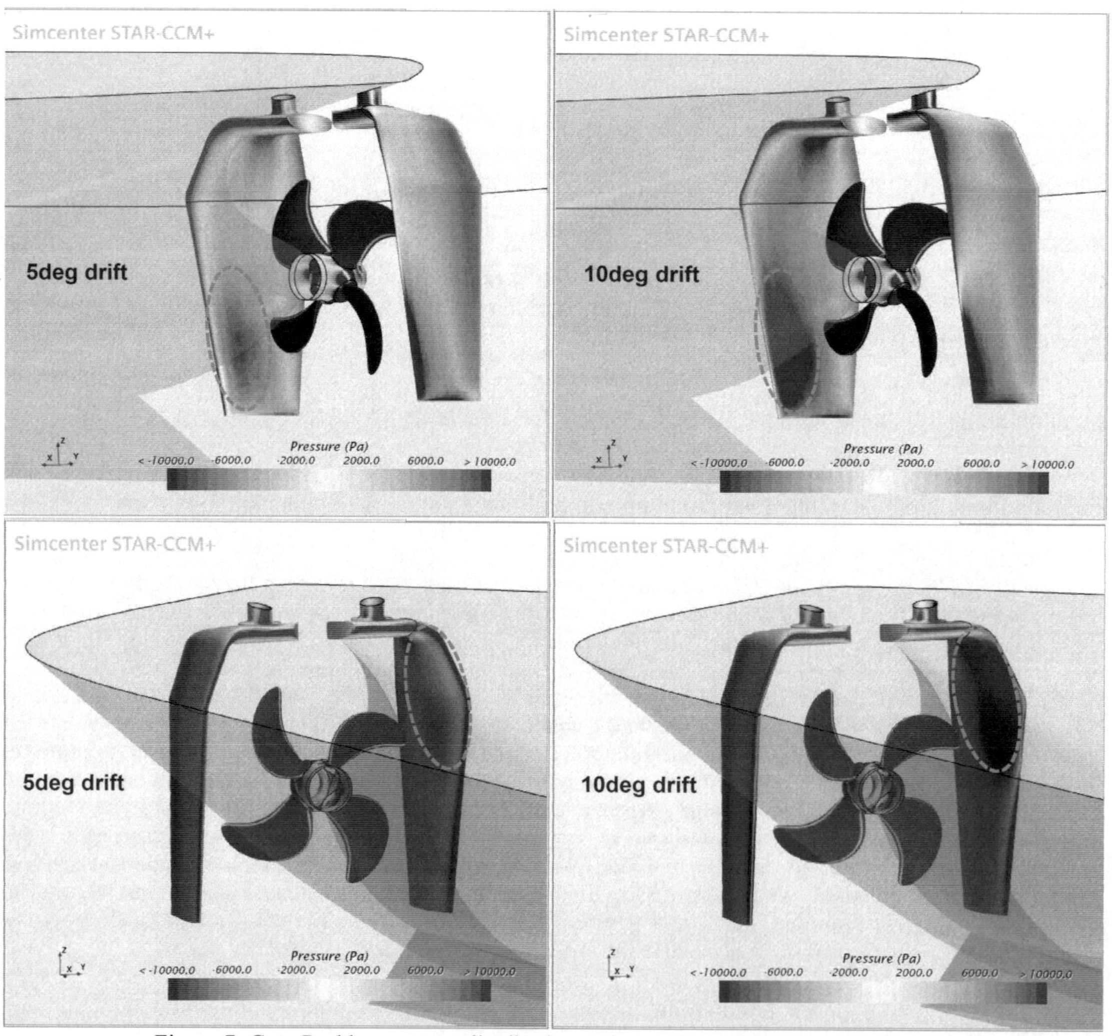

Figure 7: Gate-Rudder pressure distributions when sailing with leeway angles

5. GATE-RUDDER SYNERGY WITH WIND-PROPULSION

5.1 CONVENTIONAL RUDDERS AT REDUCED PROPELLER LOAD

The theory of vessel manoeuvring indicates that the effectiveness of a conventional rudder depends on the propeller wash [3]. Figure 8 shows the velocity field from the propeller along the rudder, when the rudder is at straight sailing condition. The red contour indicates high velocity and the green colour indicates the far-field velocity relative to the vessel. The propeller accelerates the water and this flow with higher momentum is passing the rudder to create the side-forces. Increased propeller loading gives higher velocity and thus a more effective rudder. On the other hand, in case the propeller loading can be reduced, which is the main purpose of applying wind-propulsion, the effectiveness of a conventional rudder is reduced.
It should be kept in mind that sailing at reduced power (like slow steaming) has a subtle difference compared to wind-propulsion concepts. When slow-steaming is applied, both the propeller loading and the vessel speed will drop and therefore the vessel course-keeping and manoeuvring capabilities remain acceptable for the operating vessel speed. However, when a significant part of the propulsion comes from wind, then the vessel speed remains high with a significantly lower propeller load.

Figure 8: Velocity field around conventional rudder with active propeller

5.2 GATE-RUDDER EFFECTIVENESS AT REDUCED PROPELLER LOADING

Where the theory is rather clear for a conventional rudder in case the propeller loading is reduced, a quite different story applies to the Gate-Rudder. As mentioned in section 4, a large part of the flow along the Gate-Rudder originates outside the hull boundary layer. The speed of the vessel through the water could be the main driving factor for the rudder performance. Moreover, the impact of the propeller loading (reduced MCR) could be significantly lower compared to the conventional rudders, as discussed before.
To quantify the effectiveness of the Gate-Rudder a series of CFD simulations is made for different combinations of rudder-angle and vessel leeway angle. For each combination of rudder and leeway angle the propeller loading has been varied for the same (constant) vessel speed. Change of propeller loading can be achieved with change of RPM or, in case of Controllable Pitch propeller, with a change of blade pitch. Since a change of blade pitch requires remeshing of the numerical domain, the option to vary the RPM has been applied in this research. The CFD results provide data for the vessel surge and sway force as well as the yaw-moment. These response surfaces can be coupled to any aerodynamic calculation method, which can predict the forces from wind-propulsion devices. Coupling the two data sets, will give the equilibrium condition of vessel leeway, rudder angle and propeller RPM to achieve a target vessel speed for given environmental condition of the wind (magnitude and direction).
Figure 9 shows two example cases of the velocity field and the pressure distribution of the propeller and rudder blades. At very low MCR, there is hardly any acceleration of the flow observed downstream of the propeller. Nevertheless, the angle of attack to the rudder blades still looks reasonable. At moderate MCR a clear acceleration of the flow downstream of the propeller is observed. In addition to this, the pressure distribution on the propeller blades has been changed (lager part of the blade surface below the threshold colour of the legenda). When comparing the pressure distribution of the rudder blades and the flow field near the rudder leading edges, rather small differences are found.

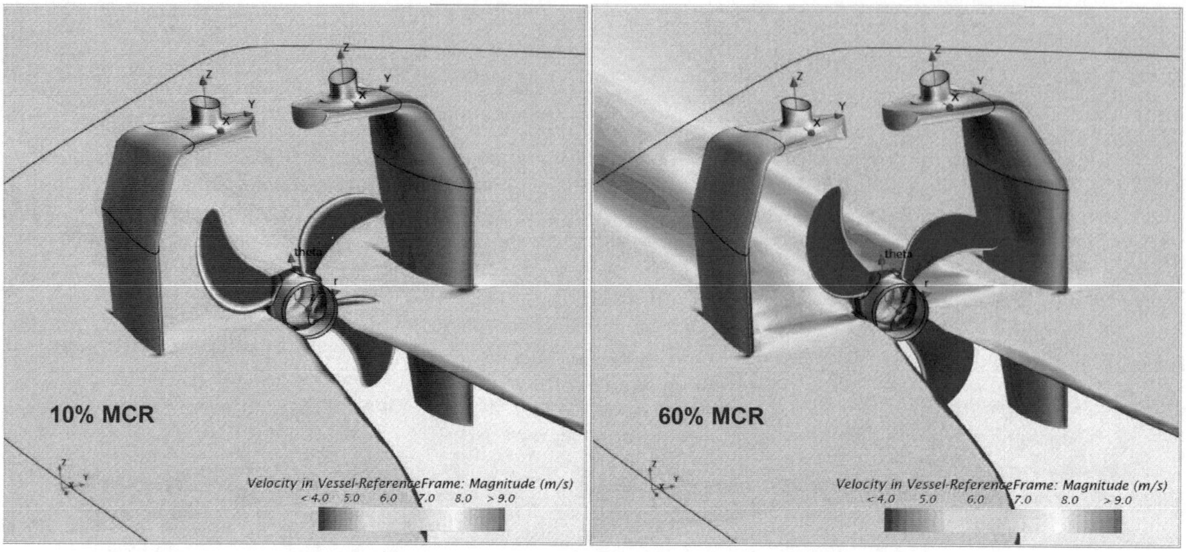

Figure 9: Comparison of velocity field and pressure distribution on Gate-Rudders for 10% and 60% MCR condition

The combined rudder sway-forces of the Gate-Rudder as function of the propeller MCR are shown in Figure 10. The rudder sway-forces of both blades have been added together and made dimensionless based on:

$$CFY = \frac{Fy_1 + Fy_2}{{}^1/_2 \cdot \rho \cdot v_{ship}^2 \cdot A_R} \tag{6}$$

This diagram clearly shows the stable performance of the Gate-Rudder over the range of analysed MCR conditions. For a moderate rudder angle (5°) and leeway angle (5°) the dimensionless sway-force remains almost constant. At larger rudder and leeway angles there seems to be even a positive trend towards lower MCR conditions. More detailed analysis of the flow field is required to explain this bonus.

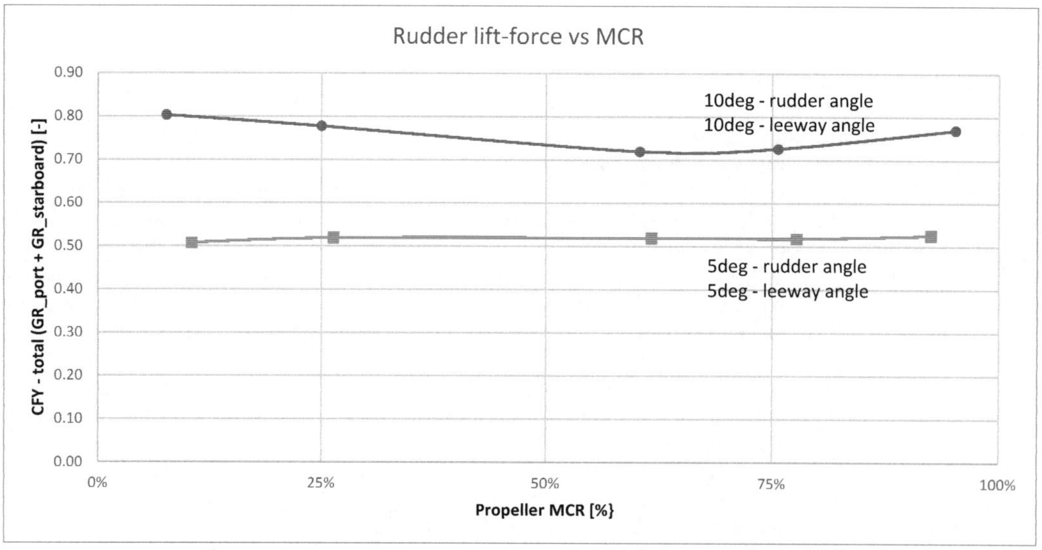

Figure 10: Gate-Rudder sway forces as function of propeller MCR

6. PROPELLER – DRIVE-LINE CONFIGURATIONS

As mentioned before, the purpose of applying wind-propulsion is to reduce the propeller loading and to reduce therefore the engine load to reach a certain operating speed. Reduction of engine power has been used for a longer period, either driven by high fuel prices or by regulations like EEXI. When a vessel is equipped with a Fixed-Pitch Propeller (FPP) the dimensionless working point of the propeller (Kq) remains more or less constant. Therefore, the vessel will remain operating along the so-called propeller curve when power is being reduced.

When wind-propulsion is considered the story changes, since the vessel speed remains constant, while the required shaft power reduces. In such situation the dimensionless propeller working point shifts to a lower loading. In the well-known propeller open-water diagram the working point moves to the right, resulting in reduced torque for same propeller RPM. This effect of reduced propeller loading at same RPM can be transferred into the engine diagram. Figure 11 shows a sketch of an engine diagram of a typical 2-stroke engine. The design point of the Fixed Pitch Propeller is based on the maximum available power and a light running margin (LRM) of 5%. The corresponding power-RPM curve is shown in the diagram as FPP curve.

The impact of the wind-propulsion is twofold: the propeller RPM-power relation changes which brings the lines downwards with increasing wind-propulsion. Moreover the required propeller RPM reduces to reach a certain vessel speed and therefore the actual working point moves to the left.

To make a proper evaluation of the benefits of wind-propulsion, the focus should not be on the required shaft power levels, but on the actual specific fuel oil consumption (SFOC) at the found operating point. To get a first impression of the SFOC behaviour at the different engine operating points, some estimated lines are included in Figure 11. For normal FPP operation the FPP fuel consumption improves when power is reduced from maximum power to the lower, operational levels. When wind-propulsion starts playing a role, the benefits of improved fuel consumption will become different.

Based on Figure 11, for 55% MCR the propeller RPM is about 74.6 in this example case, when following the FPP curve. However, with sufficient wind-propulsion this could change to 81.7 RPM at 55% MCR. The change in propeller operating point will have an impact on the fuel consumption which should therefore not be overlooked in the overall performance evaluation.

An alternative approach would be to consider a Controllable Pitch Propeller (CPP) to keep the propeller working in the regions with better fuel consumption as indicated in the right figure and discussed by others [4]. With a proper balance of hydrodynamic efficiency and the engine performance, improved overall vessel performance can be found over the large range of operating conditions when implementing wind-propulsion.

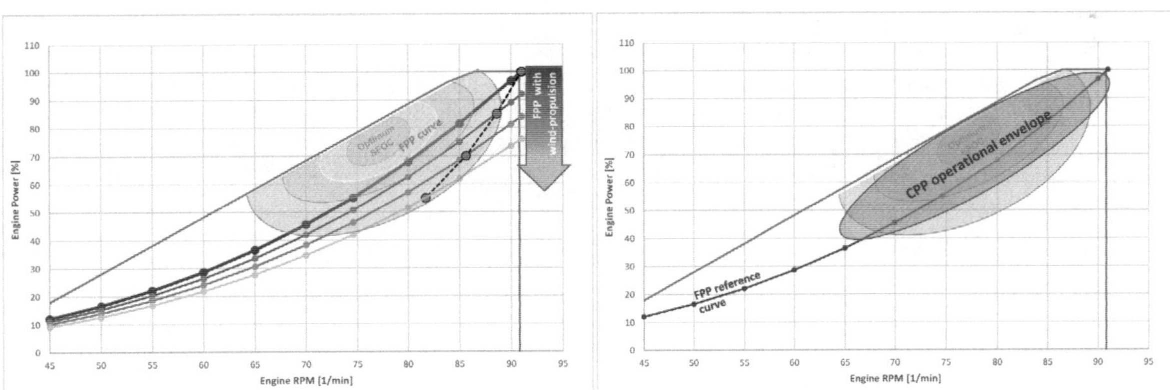

Figure 11: Engine diagram with operating points for FPP (left) and CPP (right) for reduced propeller loading due to wind-propulsion

7. CONCLUSIONS

The challenging targets for GHG-emission reductions require significant changes in the vessel design. The conventional bulker vessel concept, based on 2-stroke engine and Fixed-Pitch Propeller, will not be able to reach the future targets. In CHEK-project a number of new concepts are combined, which should provide

synergies in the overall vessel performance. Reduction of the required propeller thrust is obtained by using air-lubrication systems (ALS), ultrasonic anti-fouling and wind-propulsion. The effectiveness of the propulsion system is enhanced with a Gate-Rudder.
The contribution of the wind-propulsion gives a sway-force component, which needs to be compensated. The performance of the Gate-Rudder on this aspect has been analysed, not only for normal propeller operation, but also for significantly reduced propeller loading. The concept of Gate-Rudder, where the vessel speed is the driving factor for the side-force generation, gives a constant side-force over the whole range of propeller loading.
Given the variation in weather and therefore in the contribution of the wind-propulsion, a much larger propeller operating envelope is found. In order to maintain good fuel consumption for all operating conditions, the vessel drive-line needs to be considered in the future design as well. The working point of the conventional FPP with 2-stroke might shift so far from its design point, that a major part of the power-saving will not be reflected in fuel saving. Therefore, when the complete lay-out of propulsion system and drive-line is considered, a solution based on CPP is worthwhile to consider.

8. ACKNOWLEDGEMENTS

Part of the work presented in this paper has been executed in EU-project CHEK, which has received funding from the European Union's Horizon 2020 research and innovation programme under grant agreement No 955286.

9. REFERENCES

1. SASAKI, N., KURIBAYSHI, S., BULTEN, N., YAZAWA, M., 'Joint sea trial of ships with Gate-Rudder and conventional rudder', RINA Full Scale Ship Performance online conference, February 2021

2. ABKOWITZ, M.A., 'Ship hydrodynamics - steering and manoeuvrability', 1964

3. YASUKAWA, H., YOSHIMURA, Y., 'Introduction of MMG standard method for ship maneuvering predictions', J Mar Sci Technol, 2015

4. GYPA, I., JANSSON, M., GUSTAFSSON, R., WERNER, S., BENSOW, R., 'Controllable-pitch propeller design process for a wind-powered car-carrier optimising for total energy consumption', Ocean Engineering 269, 2023

10. AUTHORS BIOGRAPHY

N. Bulten holds the current position of Senior Product Performance Manager at Wärtsilä – Business line Propulsion. He is responsible for hydrodynamics within Propulsion business. He has more than 25 years of experience in the field of CFD and marine propulsion systems. In his current assignment he focusses on development and implementation of system simulations and advanced CFD-methodologies to include various Energy Saving Devices into the performance predictions.

KEY FEATURES OF *'WIND READY'* RETROFIT OF A KAMSARMAX BULK CARRIER

N Contopoulos, BEng, CEng, Anemoi Marine Technologies, UK
B Jiang, BEng, MSc, MRINA, CEng, Anemoi Marine Technologies, UK
H Jiang, MEng, Lloyd's Register Classification Society (China) Co. Ltd, China
R Tustin, BSc, MRINA, CEng, Lloyd's Register Classification Society (China) Co. Ltd, China

SUMMARY

A retrofit to install three Anemoi Marine Technologies (AMT) Rotor Sails with rail deployment to a Lloyd's Register (*LR*) Classed Kamsarmax bulk carrier '*TR Lady*' is being executed in two phases:

- **1st Phase retrofit to a *'Wind Ready'* state**: completed in Chengxi Shipyard in November 2022
- **2nd Phase for Rotor Sail installation**: expected to be completed for *'TR Lady'* in 2Q2023

The paper describes:

- Configuration studies and contract for Rotor Sail retrofit of Kamsarmax bulk carrier '*TR Lady*'
- Execution of modifications for structural and electrical integration as well as updates and modifications for ship regulatory compliance, as part of a 1st phase retrofit of '*TR Lady*'

The paper is intended as a possible reference document for ship owners when considering readiness of a ship for installation of Rotor Sails.

NOMENCLATURE

ACP	Panama Canal Authority (*Autoridad del Canal de Panama (ACP)*)
AMT	Anemoi Marine Technologies
BCU	Bridge Control Unit (BCU) a bridge console for Rotor Sail control
BV	Bureau Veritas Classification Society
CII	Carbon Intensity Indicator
COLREGS	Collision Regulations of SOLAS Convention of IMO
Composite Rotor	Composite construction rotating sleeve of Rotor Sail
DnV	Det Norske Veritas Classification Society
EEDI/EEXI	Energy Efficiency Design Index (*EEDI for new ships*) (*EEXI for existing ships*)
Flettner Rotor	Common phrase for Rotor Sails, in etymology, named after Anton Flettner, who pioneered, and patented, original application of Rotor Sails for ship propulsion in 1920s
FSAM	Anemoi Marine Technologies Fuel Saving Assessment Model
GL	Germanischer Lloyd Classification Society (*merged with DnV in 2013*)
IMO	International Maritime Organization
Kamsarmax	Bulk carrier type of about 80,000 tonnes deadweight designed to size limits of Panama Canal (*32.3 metres beam limit*) and Port Kamsar (*229.0 metres length overall*) in Guinea, Africa
KR	Korean Register Classification Society
LR	Lloyd's Register Classification Society
Magnus Effect	Lift and thrust generation by a rotating cylinder named after experimental scientist, Heinrich Gustav Magnus, who observed and reported the phenomenon in 1852
MARPOL	The International Convention for the Prevention of Pollution from Ships of IMO
MEPC	Marine Environment Protection Committee of IMO
MSB	Main switchboard
MSC	Marine Safety Committee of IMO
RO	Organisation, typically a Classification Society, recognised and authorized by the ship's flag state to verify compliance of the ship with flag state shipping laws and regulations
Rotor Sail	Assembly comprising composite rotor, fixed steel tower, drive system and bearings

RPM	Revolutions Per Minute
SAT	Shipboard Acceptance Tests
SEEMP	Shipboard Energy Efficiency Management Plan
SELMA	Ship Electric Marine Automation; AMT sub-contractor for electrical integration design
SOLAS	Safety of Life at Sea Convention of IMO
Tufton	Tufton Investment Management
'Wind Ready'	State of completion of ship modifications prior to installation of Rotor Sails

1. INTRODUCTION

Environmental regulations and policies that incentivize fuel saving and higher ship energy efficiency is driving the adoption of wind assistance for commercial ships. There are three main types of rigid wind assistance rigs being installed on commercial ships currently including:

- Rotor Sails
- Suction Wing Sails
- Rigid Wing Sails

At the time of writing this paper, and based on public domain record, fourteen ships have been equipped with a total of twenty-six Rotor Sails. See Appendix 1. On average, over the last five years, two Rotor Sail installations are completed per year.

A Rotating or Rotor Sail [1] sometimes also called a Flettner Rotor [2] from original ship installations of Rotors pioneered by Anton Flettner generates lift, and thrust, for fuel saving in operation by rotation of a vertical cylinder by electric motor.

1.1 MAGNUS EFFECT AND 'KUTTA-JOUKOWSKI' LIFT EQUATION

The observed phenomenon of lift and thrust generation by an AMT Rotor Sail is called the '*Magnus effect*' [3] named after experimental scientist, Heinrich Gustav Magnus, who observed the phenomenon in 1852 [4].

For prediction of Rotor Sail thrust a simplified model, '*Kutta-Joukowski'* lift equation [5], offers a good first order prediction of thrust compared with experimentation and is incorporated in AMT publications [6] and LR Guidance [7].

1.2 APPLICATION OF ROTOR SAILS ON BULK CARRIERS BY ANEMOI MARINE TECHNOLOGIES

A first case of wind readiness *(characterised as 'Wind Ready')* for a future expected installation of AMT Rotor Sails was completed at newbuilding stage in 2017 for an LR Classed Kamsarmax bulk carrier, *'Axios'*, including structural integration of four sets of transverse rails as well as electrical integration.

A first case of a newbuilding bulk carrier utilizing Rotor Sails for wind assistance was completed in 2018 for an LR Classed Ultramax *'Afros'* utilizing four first-generation AMT Rotor Sails with longitudinal rail deployment on starboard main deck. See Figure 1 image.

Figure 1: Rail deployed AMT Rotor Sails installed on Bulk Carrier *'Afros'* in 2018 *(Image courtesy of AMT)*

1.3 ANEMOI MARINE TECHNOLOGIES SECOND GENERATION ROTOR SAIL

Figure 2 cutaway view shows components [8] of AMT second-generation Rotor Sail including:

- **Rotor:** rotating component which generates lift and thrust. Rotor is of composite construction and comprises a sleeve or skin, a top disc and an internal diaphragm connecting Rotor to tower upper bearing.
- **Tower:** steel static component which supports Rotor and incorporates bearings and drive system.
- **Upper Bearing:** standard roller bearing which connects Rotor to Tower that carries all axial and most radial loads.
- **Drive System:** electric motor and drive shaft. Depending on design the drive system may be fitted at either upper or lower bearing.
- **Lower Bearing:** roller bearing. Carries radial loads (*no axial loads*). Wheels run on racetrack/base ring.
- **Pedestal:** at tower base with corner leg interfaces with upper deck foundations on ship.

Ship's electrical system supplies each Rotor with power. A safety control system regulates speed of the Rotor in operation.

Interface of corner legs with upper deck foundation on a ship may also be seen as a design and engineering interface for Rotor Sail structural integration.

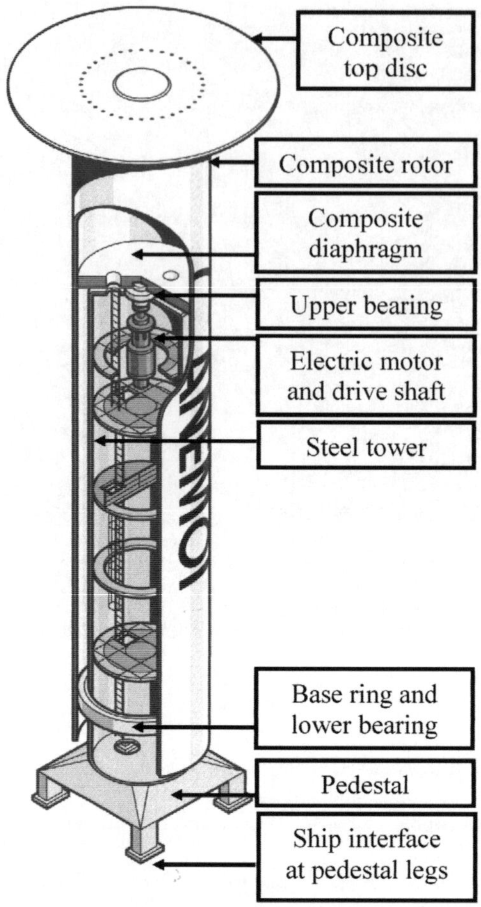

Figure 2: Cutaway showing main components of AMT second-generation Rotor Sail *(image courtesy of AMT)*

2. STUDIES OF ROTOR SAIL CONFIGURATION FOR KAMSARMAX BULK CARRIER *'TR LADY'*

A Rotor Sail installation for a 2017 built Kamsarmax bulk carrier, *'TR Lady',* was studied as part of co-operation agreement reached between AMT and LR in October 2020 [9].

2.1 FUEL SAVING INVESTIGATION AND SELECTED ROTOR SAIL CONFIGURATION

Different Rotor Sail configurations for Kamsarmax bulk carrier, *'TR Lady',* were investigated [10] using performance predictions of (*AMT proprietary*) Fuel Saving Assessment Model (*FSAM*) with following data inputs:

- **Rotor Data:** utilizing performance data accumulated from a land-based AMT Rotor test installation in Blyth
- **Wind Data:** from the latest version, at the time of investigation, of IMO *'Global Wind Probability Matrix'* [11] to define wind speed and direction probabilities
- **Ship Route Data:** from IMO Global Wind Matrix trading routes database
- **Vessel Data:** from ship-owner/operator

Estimated fuel savings and payback periods *(based on a range of fuel prices)* for two ship speeds and two loading conditions for each Rotor Sail configuration were calculated. Based on maximized fuel saving, and associated shortest payback period, a final retrofit configuration was selected with three Rotor Sails on centreline between hatches.

Investigation results [10][12] are summarized in Figure 3 showing a comparison of FSAM predicted average % fuel saving for each Rotor Sail configuration examined for an optimal low ship speed operating profile.

	Rotor Sail position				Rotor Sail installation						Fuel saving comparison
	Cross Deck Strip (*Between Hatches 5 & 6*)	Cross Deck Strip (*Between Hatches 4 & 5*)	Cross Deck Strip (*Between Hatches 2 & 3*)	Forward of Hatch 1	Rail Deployment (*Y/N*)	Number of Rotor Sails	Diameter, metres	Height, metres	Rotor Sail projected area (*m^2 per Rotor Sail*)	Total Rotor Sail projected area (*m^2*)	Average % fuel saving (*Low speed operating profile*)
Option 1	✓	✓	✓		Y	3	4	28	112	336	12.9
Option 2	✓		✓		Y	2	4	28	112	224	8.7
Option 3				✓	N	1	5	35	175	175	7.0
Final Design	✓	✓	✓		Y	3	5	24	120	360	14.4

Figure 3: Summary of FSAM predicted average % fuel saving for each Rotor Sail configuration option examined

2.2 PORT OPERATIONS INVESTIGATION FOR ROTOR SAIL INSTALLATION

Loading and unloading operations, with Rotor Sails installed, were investigated [10] for air draft and port equipment operational limitations and clearances. Information and data for this investigation was sourced from:

- **Published data on ports:** in public domain
- **Ship-owner supplied information:** based on port visits during 2020 by bulk carrier *'TR Lady'*
- **Surveys of ports:** carried out with those ports supportive of application of wind assistance solutions on shipping

Loading port investigations, with rail deployed Rotor Sails stowed outboard, indicated sufficient clearance for operations of shore-side cargo loading equipment in all ports. Unloading port investigations indicated that luffing of shore-side unloading gantry cranes may be required for crane movement.

3. CONTRACT FOR ROTOR SAIL RETROFIT OF KAMSARMAX BULK CARRIER '*TR LADY*'

Collaborative design studies led to a decision in May 2021 by Tufton Investment Management (*Tufton*), on behalf of the ship-owner, to contract [14] Anemoi Marine Technologies for supply and retrofit of Kamsarmax bulk carrier '*TR Lady'* with three rail deployed AMT Rotor Sails as illustrated in Figure 4 image.

Figure 4: Illustration of AMT Rotor Sails to be installed on Bulk Carrier *'TR Lady'* in 2023 *(Image courtesy of AMT)*

3.1 AMT SCOPE OF SUPPLY AND RETROFIT DESIGN FOR BULK CARRIER *'TR LADY'*

Retrofit contract for Kamsarmax bulk carrier '*TR Lady'* included following scope of AMT Rotor Sail system supply [13]:

- Three assembled, and tested, 5.0-metre-diameter by 24.0-metre-tall second-generation AMT Rotor Sails
- Rail deployment system elements including capstan winch, rope, deck sheaves, deployment system control cabinet
- AMT Rotor Sail control system including bridge control unit and anemometers

In addition, AMT were contracted to provide scope of design and engineering services [13] shown in Figure 5:

Integration design for Rotor Sail retrofit		
Integration design task	Scope of work for integration design	
Structural integration	a.	Rotor Sail foundation
	b.	Rails and rail beam structure, sheaves, and capstan winch
	c.	Structural analysis calculations
Electrical integration	a.	System diagrams for ship power and Rotor Sails
	b.	Electrical load balance and short circuit calculation update
	c.	Cabling and cable tray layouts
	d.	Main switchboard modifications
Ship regulatory compliance design updating on Rotor Sail retrofit		
Regulatory compliance task	Scope of design updating for ship regulatory compliance	
EEXI assignment	a.	Rotor Sail performance calculation
	b.	Net effective power calculation
	c.	EEXI technical file update
Safety of navigation	a.	Bridge visibility
	b.	Radar blind sectors
	c.	Arrangement of lights *(COLREGS & ACP compliance)*
Stability and loading	a.	Mass and CoG (*VCG/LCG*) report *(SOLAS 2020 criteria)*
	b.	Report of inclining experiment
	c.	Updated stability and loading manuals

Figure 5: AMT scope of design for Rotor Sail integration and ship regulatory compliance updating

For Figure 5 scope of engineering and design two parties were sub-contracted by AMT:

- **CS Marine Ltd (*CSM*):** the original ship designer of '*TR Lady*' for drawings, and documentation, for structural integration design for Rotor Sail and rail deployment system as well as updating ship's drawings, and documentation, and design of modifications for ship regulatory compliance
- **Ship Electric Marine Automation (*SELMA*):** for drawings, and documentation, for electrical integration design.

In co-operation with AMT and CS Marine Ltd the project execution document [13] was further developed by LR to define expected scope of submission and Classification design review [15] [16].

3.2 TIMELINE TO 'WIND READY' RETROFIT OF BULK CARRIER *'TR LADY'*

For '*TR Lady*' Kamsarmax bulk carrier retrofit, in descending date order, notable project milestones are shown below:

- **October 2020**: '*Joint design co-operation agreement, Rotor Sail installed energy efficient Kamsarmax Bulk Carrier design development*' [9] for a study to create a specification and design of a Kamsarmax Bulk Carrier with *AMT* second generation Rotor Sails.
- **December 2020**: kick off meeting including Tufton, AMT, CS Marine, LR.
- **January 2021**: presentation of three options for wind assistance configurations [10] utilizing AMT's Rotor Sails for retrofit of '*TR Lady*' including fuel saving analysis and an examination of port operation limitations.
- **April 2021:** initial project execution scope [13] defining AMT and Shipyard equipment supply and AMT scope of design for Rotor Sail integration and ship regulatory compliance
- **May 2021:** contract [14] between AMT and Tufton for supply and retrofit of '*TR Lady*' with three AMT second generation Rotor Sails and rail deployment system
- **June 2021:** Tufton decision on final retrofit configuration adopting 24 metre height Rotor Sails based on air draft limitation of Huey P. Long Bridge in Jefferson Parish, Louisiana and to access grain terminals on Mississippi River.
- **November 2022**: 1st Phase retrofit to *'Wind Ready'* state at Chengxi Shipyard including completion of structural and electrical integration as well as modifications for ship regulatory compliance.

Rotor Sail retrofit of bulk carrier '*TR Lady'* is organised in two phases. 2nd Phase installation, commissioning, and shipboard acceptance tests (*SAT*) of three rail deployed 5.0-metre-diameter by 24.0-metre-tall AMT Rotor Sails is expected to be executed in 2Q2023.

4. 1st PHASE 'WIND READY' RETROFIT OF KAMSARMAX BULK CARRIER *'TR LADY'*

Figure 6 montage shows photographs of some externally visible, and distinct, modifications completed in Chengxi Shipyard for 1st Phase '*Wind Ready*' retrofit of bulk carrier '*TR Lady*'. The modifications comprise:

- **Rail deployment system integration** on cross-deck strips between hatches.
- **Ship regulatory compliance modifications** including navigation light modifications and new forward radar antenna.

In the following short descriptions, for each photograph, modifications are described to assist reader visualize the scope of completed for '*Wind Ready*' retrofit of bulk carrier '*TR Lady*':

- **Top and centre photographs** are two aerial drone camera views looking along ship's centreline with one view looking forward from aft and one view looking downward from directly overhead. These aerial views show newly integrated rails and rail beam structures and foundations (*red coating colour*) and capstan winches and sheaves (*with green colour covers*) arranged in cross-deck strips between hatches 2 and 3, hatches 4 and 5, and hatches 5 and 6.
- **Bottom right photograph** shows forward mast modifications including incorporation of Port and Starboard blue steering lights for compliance with Panama Canal bridge visibility requirements [17] as well as a new forward radar antenna to address aft radar antenna blind sectors with Rotor Sail installed.
- **Bottom left photograph** shows aft mast modifications including incorporation of Port and Starboard aft mast head lights to comply, as far as is practicable, with requirements for main mast visibility in accordance with COLREGS and for issue of a certificate of exemption by LR on behalf of, and with prior agreement of, Marshall Islands flag.

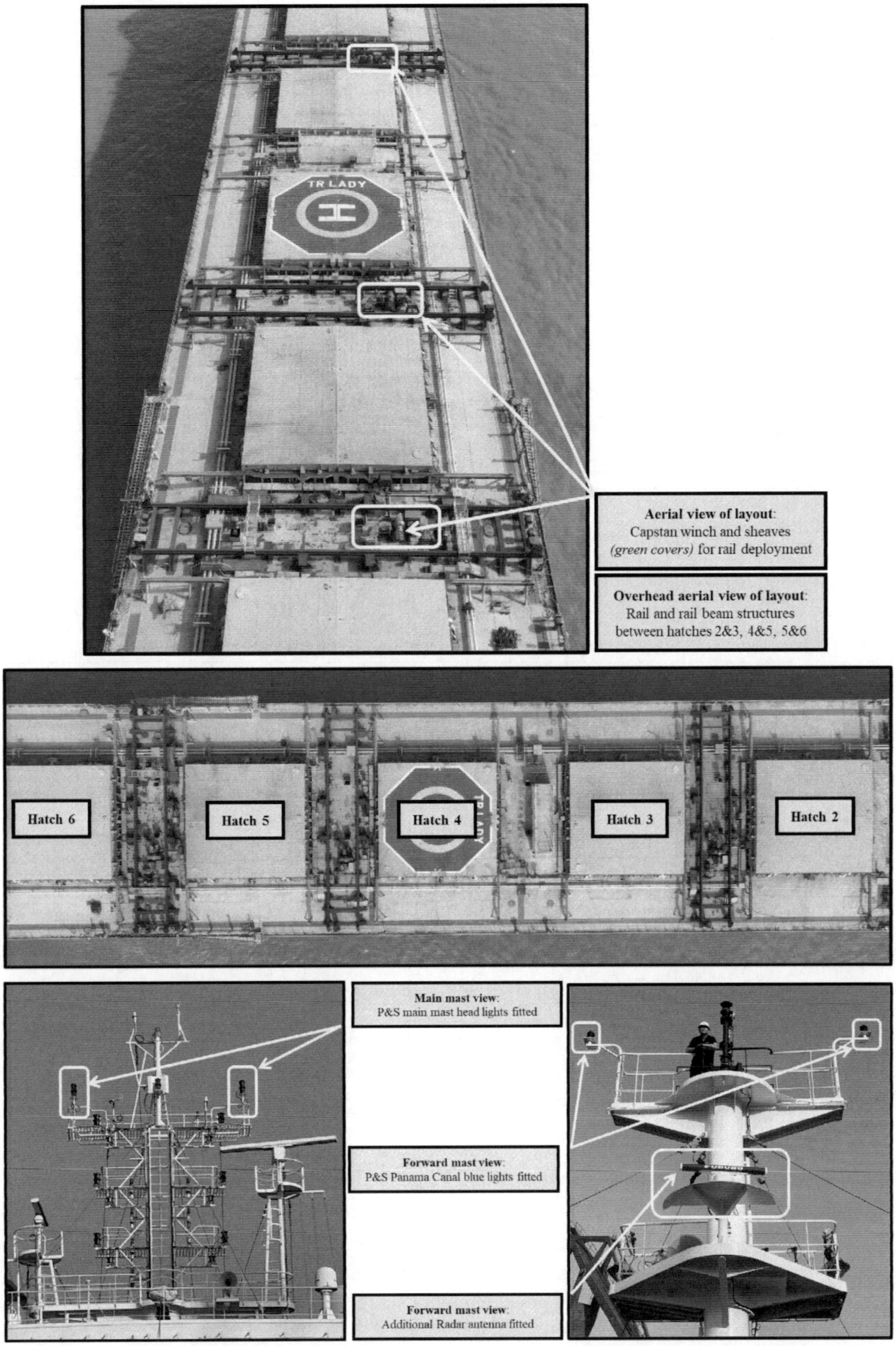

Figure 6: photo montage of modifications to *'TR Lady'* from 11-2022 *'Wind Ready'* retrofit *(Images courtesy of AMT)*

Scope of work completed in 1st Phase '*Wind Ready*' retrofit of '*TR Lady*' and planned for 2nd Phase retrofit is summarized in Figure 7 for structural and electrical integration and key aspects of ship regulatory compliance.

STRUCTURAL AND ELECTRICAL INTEGRATION		
Retrofit work packages	**Shipyard fabrication and/or supply**	
	1st Phase scope of work: '*Wind Ready*'	**2nd Phase scope of work**: Rotor Sails installed
Scope of Structural Integration		
Cross-deck Rotor Sail foundations	• Structural seats arranged adjacent to centreline *(For Rotor Sail in operational condition)*	• Rotor Sail installation on foundation structure • Mechanical fixing of Rotor Sail to foundation
Cross-deck Rotor Sail rail systems	• Rails and rail beams • Capstan winch including foundation • Accumulator including foundation • Sheave ancillary foundations	• Rail brake unit installation and set up onto rails including link beams • Winch rope installation and set up • Hydraulic piping connection for accumulator • Hydraulic buffer installation at rail beam ends • Commissioning SAT for rail system
Minor underdeck reinforcement and cross-deck modifications	• Under cross deck reinforcement for Rotor Sail foundation and rail system • Modifications for foundation and rail system including relocation of manholes, vent/sounding pipes, access walkways etc.	
Scope of Electrical Integration		
Main deck cabling for Rotor Sails	• Cabling for Rotor Sail power and control • Main deck conduit for new cabling • Main deck cabling connection boxes	• Flexible power cables to install and connect between deck boxes and Rotor Sails
Control system for Rotor Sails Electrical system modifications	• Anemometer installations *(fwd & aft masts)* • Bridge Control Unit *(BCU)* installation • Installation of capstan winch control panel • Installation of remote I/O station to collect ship equipment/sensor signals • MSB modifications with additional circuit breakers *(Rotor Sails and capstan winches)*	• Commissioning SAT for Rotor Sail

SHIP REGULATORY COMPLIANCE			
Regulatory Compliance Issue	**Shipyard scope of modifications, installation, and test**		
	1st Phase scope of work: '*Wind Ready*'	**2nd Phase scope of work**: Rotor Sails installed	**Certificates & Manuals**
Safety of Navigation			
ACP [17] **Panama Canal steering light bridge visibility**	• Foremast modification • Two blue Panama Canal steering lights installed (P&S) (*NOT functioning)*	• Two blue Panama Canal steering lights (P&S) installed at 1st Phase '*wired in*' and functioning	-
IMO SOLAS Radar antenna blind sectors	• Foremast modification • Additional radar antenna on foremast	-	-
IMO COLREGS Aft mast head light visibility	• Main mast modification • Two aft mast head lights (P&S) installed *(NOT functioning)*	• Two aft mast head lights (P&S) installed at 1st Phase '*wired in*' and functioning. • Existing centreline aft mast head NOT '*wired in*' and functioning *(or removed)*	**Safety Equipment Certificate** • 2nd Phase: flag state exemption
Stability and Loading			
IMO SOLAS [18] **Lightship weight & VCG change**	• Inclining experiment	• Mass and VCG of Rotor Sails to re-confirm	**Loading Manuals** • 1st Phase inclining experiment • 2nd Phase update of manuals

Figure 7: completed scope of work at 1st Phase retrofit and planned scope of work for 2nd Phase retrofit of '*TR Lady*'

4.1 *'TR LADY'* 1st PHASE RETROFIT LESSONS FOR FUTURE ROTOR SAIL INSTALLATION CASES

From practical experience with 1st Phase retrofit of '*TR Lady*' the following aspects deserve special attention for efficient execution of future Rotor Sail installations:

- **Detailed consideration of scope of integration and ship compliance design** including, if appropriate, a definition of expected completed work at each phase in case of a '*Wind Ready*' retrofit.
- **Detailed preparatory inspection of the ship in advance of retrofit** including dimension survey of existing structures as well as a 3D scan to identify conflicts of installed Rotor system with existing outfitting.
- **Assessment of compatibility of rotor sail control and vessel performance monitoring systems** to avoid duplication of data acquisition.

4.2 BENEFITS OF ADOPTING A TWO-PHASE RETROFIT WITH 1st PHASE '*WIND READY*' RETROFIT

By adopting two-phase retrofit beneficial possibilities arise including, for example,

- Timing of retrofit phasing to align with scheduled dry-docking as well as Rotor Sail availability for installation.
- Phasing may allow for contracts with similarly phased investment decisions, for example, where a Rotor Sail installation may be phased to align with energy efficiency improvements for regulatory compliance.
- Finally with a 25-year design life the re-deployment of a Rotor Sail between ships is also a possibility in a scenario where an aged Rotor equipped ship is being sold or scrapped and equipped with residual operational Rotor Sail life.

In steps a running order of execution of a two-phase, or even possibly a three-phase retrofit could be contemplated, with each step corresponding to execution of a retrofit phase:

- **1st Step:** preparing ship for Rotor Sail installation e.g., 1st Phase 'Wind Ready' retrofit adopted for '*TR Lady*' case. Timing for a '*Wind Ready*' retrofit to be executed at time of scheduled dry-docking and ship repair in the normal ship maintenance and repair cycles.
- **2nd Step:** installation of Rotor Sails timed to cater for availability of Rotor Sails and possibly also timed to meet a regulatory threshold for compliance action such as improvement of energy efficiency.
- **Optional 3rd Step:** to increase fuel saving potential of installation by adding Rotor Sails as an action to meet future regulatory thresholds for energy efficiency.
- **Another optional 3rd Step:** to re-deploy Rotor Sails to another *'Wind Ready'* ship before planned out of service of an aged Rotor Sail equipped ship.

5. EEXI CALCULATION FOR KAMSARMAX BULK CARRIER 'TR LADY'

Energy Efficiency Design Index, EEXI, is being implemented from 1st January 2023 to ensure existing ships catch-up with energy efficiency level of new ships complying with EEDI.

For calculation of EEXI for *'TR Lady'* AMT's method of calculation of Net Effective Power [19] contribution of Rotor Sail utilizes IMO Guidance for Innovative Energy Efficiency Technologies [11][20] in following calculation steps:

- **Importing of wind data**: based on latest version of IMO Global Wind Probability Matrix [11][20]
- **Correction, and conversion, of imported wind data**: for Rotor Sail height and apparent wind direction
- **Importing of Rotor Sail performance data**: calculated from Anemoi land-based test site in Blyth [21]
- **Calculation of Net Effective Power ($f_{eff} P_{eff}$) contribution of Rotor Sails**: for EEDI/EEXI Technical File

In case of Kamsarmax Bulk Carrier *'TR Lady'* a Net Effective Power calculation for Rotor Sail contribution was completed by AMT [22] in support of ship-owner submission of Technical File for assessment of EEXI on 2nd Phase retrofit installation of Rotor Sails.

6. IMO REGULATION OF ATTAINED OPERATIONAL CARBON INTENSITY OF A SHIP

Initial (*April 2018*) IMO GHG strategy [23] includes a target to reduce Carbon Intensity of shipping by 40% (*compared to 2008*) by 2030 with a target of 70% reduction by 2050.

From 1st January 2023, data will be collected to verify attained carbon intensity for all ships over 5000 gross tonnes. A ship rating will be assigned from 2024, and following a yearly reduction factor trajectory thereafter, by comparing (*attained*) operational carbon intensity of a ship with a required annual operational carbon intensity indicator (*CII*). All ships are required to achieve an 11% reduction in operational CII by 2026. When a ship attains a low rating (*E rating in a single year, D rating for 3 consecutive years*) corrective measures are required to adjust attained Carbon intensity to achieve a better attained rating against required annual operational Carbon intensity.

A key element of maintaining a ship rating are measures to improve energy efficiency as set out in the Shipboard Energy Efficiency Management Plan (*SEEMP III*). The SEEMP III, implemented by 31st December 2022, requires ship operators to articulate a 3-year plan for maintaining or improving the CII of a ship through energy efficiency measures to maintain pace with the downward trajectory slope of the CII rating bands to 2026.

The significance of implementation of these regulations from 1st January 2023 is a likely encouraging of adoption of wind assistance solutions as one practical measure to achieve energy efficiency improvements necessary to improve a ship's rating to maintain compliance.

IMO will review progress in 2026, at end of this period of short-term GHG reduction measures, and it is expected that more stringent CII reduction factors will apply from 2026 to reach IMO's stated goal of 40% reduction by 2030.

6.1 ILLUSTRATION OF ROTOR SAIL RETROFIT TO MAINTAIN SHIP RATING

In Figure 8 [24] the required Carbon Intensity Indicator *(CII)* for an example Bulk Carrier is shown. It is assumed that CII reduction factors will be applied from 2026 with a required 21.5% (*also assumed*) reduction in operation CII by 2030 in order to meet IMO 40% carbon intensity reduction target. Colour shading, and key, in Figure 8 shows rating bands for attained carbon intensity.

In Figure 8 an illustration of the impact of a Rotor Sail retrofit is shown with grey font arrow indicating a reduction in attained carbon intensity on retrofit. This is an idealised illustration of Rotor Sail impact as, in practice, a progressive measurement of attainment of carbon intensity is made during an annual measuring period rather than an abrupt step change in measured carbon intensity.

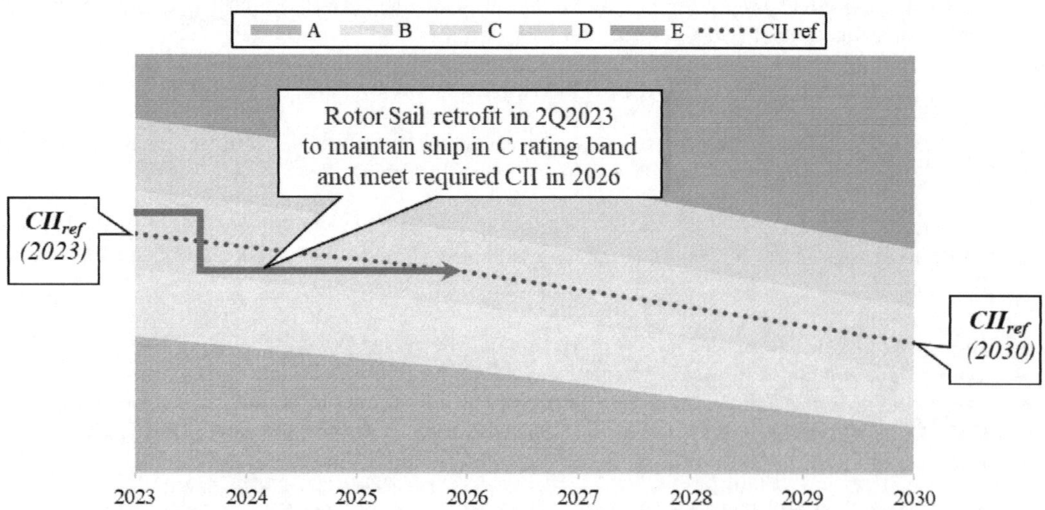

Figure 8: Idealized illustration of Rotor Sail retrofit to maintain ship rating and to meet required CII in 2026

In Figure 8 the installation of Rotor Sails in a 2Q2023 retrofit of a bulk carrier is illustrated as an energy efficiency improvement measure with a target, in this idealized example, of utilizing Rotor Sail retrofit to maintain the ship within C-rating band and also meeting required 11% reduction in carbon intensity.

7. CONCLUSIONS

The intention of the authors, in offering a paper for publication, is to share experience and offer insights on the preparation of an existing ship – in this case a Kamsarmax bulk carrier - for a retrofit installation of Rotor Sails. In doing so the aspiration is to create a useful reference for ship owners when considering wind readiness of a ship.

To conclude our paper, and based on experiences and learning of the co-authors from 1st Phase *'Wind Ready'* retrofit of *'TR Lady',* the following aspects including definition of design and engineering scope at early project stages are deserving of emphasis, and attention, by ship-owners when considering readiness of a ship for a Rotor Sail installation:

7.1 DEFINITION OF SCOPE OF DESIGN AND ENGINEERING SCOPE OF WORK

Early detailed definition of scope of design and engineering was made on the two key retrofit aspects:

- Rotor Sail integration
- Ship regulatory compliance

Significant project scoping effort [13] [15] [16] was made by both AMT and LR immediately before and, within a short time, after the May 2021 retrofit contract for the retrofit of '*TR Lady*'. This time, and effort, at early project stages ensured smooth execution of design and engineering scope of work by AMT sub-contractors and for subsequent Classification review and approval activities.

7.2 ANTICIPATING, AND PREPARING FOR, SHIP REGULATORY COMPLIANCE IMPLICATIONS

In co-operation with AMT the original project execution document [13] was further developed by LR to define an expected scope of re-examination of ship regulatory compliance aspects. '*TR Lady*' is Marshall Islands registered ship and LR is a recognized organisation (RO) acting on behalf of the Republic of Marshall Islands for ship regulatory compliance aspects.

LR developed [16] ship regulatory compliance scope of work incorporating references to regulatory interpretations, as well as text extracts, for example definitions of limits of lightweight and centre of gravity changes of SOLAS 2020 [18] defining requirement for an inclining experiment on Rotor Sail installation.

In addition, ship regulatory compliance scope of work was validated with Deputy Commissioner of Maritime Affairs for Republic of Marshall Islands in a July 2021 meeting and, at same time, expected process for issue of a certificate of exemption by LR on behalf of Marshall Islands flag was discussed in expectation that full regulatory compliance for some aspects of Safety of Navigation may not be achievable after installation of Rotor Sails.

7.3 REGULATION OF CARBON INTENSITY AND IMPROVEMENT OF ENERGY EFFICIENCY

One insight, to infer from the idealised illustration, in Figure 8 in Section 6,1, is that it may be possible to anticipate an optimal timing of wind assistance retrofits in an energy efficiency improvement plan (*as part of SEEMP III*).

Wind readiness may also be appropriate to consider in order to prepare the ship for later installation of Rotor Sails. In such a case a '*Wind Ready*' retrofit can also be planned based on ship dry-docking schedules and timed so as to ensure that future retrofit of Rotor Sails meets required timing of energy efficiency improvement for ship rating on measurement of attained carbon intensity.

7.4 IN CLOSING

It will be appreciated that not all experience, and insight, from the *'TR Lady'* Rotor Sail two-phase retrofit is transferrable as each Rotor Sail installation will have its own unique, or unusual, characteristics and these differences include for example ship type, ship design, ship configuration, operating profile, and trade. This notwithstanding a possibly obvious, but nevertheless worth stating, remark may be appropriate for closing that **prior preparation when considering the wind readiness of a ship is a pre-requisite for a smooth execution of a Rotor Sail installation**.

8. ACKNOWLEDGEMENTS

The authors wish to thank colleagues in AMT and LR in North Asia and UK for their support.

One person deserves special acknowledgment and mention:

- Anton Flettner who created the first experimental prototype Rotor Sail ship '*Buchau*' in 1924.

And also deserving mention are two motivated COOs and a '*meeting of minds*' that ended with signing of a co-operation agreement [10] in October 2020 that has culminated in AMT and LR involvement in a project to retrofit '*TR Lady*':

- Nick Contopoulos from Anemoi Marine Technologies
- Mark Darley from Lloyd's Register

Any views and opinions expressed are the authors alone and do not necessarily represent those of Anemoi Marine Technologies or Lloyd's Register

9. REFERENCES

1. FREEHILL-MAYE, L., 'Rotating Sails help to revive Wind-Powered Shipping' [*Rotating Sails Help to Revive Wind-Powered Shipping - Scientific American*], *Scientific American*, December 2020

2. HERZOG, H., 'More facts about the Flettner Rotor Ship' [*More Facts About the Flettner Rotor Ship - Scientific American*], *Scientific American*, February 1925

3. NUTTALL, P. & KAITU, J., 'The Magnus Effect and the invention of the Flettner Rotor'[*The Magnus Effect and the Flettner Rotor: Potential Application for Future Oceanic Shipping (researchgate.net)*], *The Journal of Pacific Studies (Page 165, Volume 36, Issue 2)*, January 2016

4. LLOYD'S REGISTER (LR), 'Guidance Notes for Flettner Rotor Approval, Background, Section 1.2.3' [*Guidance Notes for Flettner Rotor Approval (lr.org)*], *Lloyd's Register website*, May 2015

5. NATIONAL AERONAUTICS AND SPACE ADMINSTRATION (NASA), 'Lift of a Rotating Cylinder' [*https://www.grc.nasa.gov/WWW/K-12/airplane/cyl.html*], *NASA Glenn Research Center website*, 13th May 2021

6. ANEMOI MARINE TECHNOLOGIES (AMT), 'Using the Magnus Effect to generate thrust' [*Anemoi Marine - Rotor Sails for a sustainable future*], *Anemoi Marine Technologies website*, Undated

7. LLOYD'S REGISTER (LR), 'Guidance Notes for Flettner Rotor Approval, Design, Section 3.3 Theroretical Maximum Thrust' [*Guidance Notes for Flettner Rotor Approval (lr.org)*], *Lloyd's Register website*, May 2015

8. ANEMOI MARINE TECHNOLOGIES (AMT), 'Rotor Sail Systems: introductory presentation to Lloyd's Register', *Anemoi Marine Technologies*, 21st December 2020

9. ANEMOI MARINE TECHNOLOGIES (AMT) & LLOYD'S REGISTER (LR), 'Joint design co-operation agreement, Rotor Sail installed energy efficient Kamsarmax Bulk Carrier design development', *Anemoi Marine Technologies/Lloyd's Register (restricted circulation: not for*

publication), October 2020

10. ANEMOI MARINE TECHNOLOGIES (AMT), 'Concept Configuration Proposal: Kamsarmax TR Lady', *Anemoi Marine Technologies (restricted circulation: not for publication)*, 14th January 2021

11. IMO MARINE ENVIRONMENT PROTECTION COMMITTEE (MEPC), '2013 Guidelines on treatment of innovative energy efficiency technologies for calculation and verification of the attained EEDI' MEPC.1/Circ.815 [Session (imo.org)], *International Maritime Organisation*, 2013

12. JIANG, B., 'Savings – Global Trade' summary slide from AMT presentation of Fuel Savings Assessment for '*TR Lady*', *Anemoi Marine Technologies (restricted circulation: not for publication)*, 1st February 2023 email

13 ANEMOI MARINE TECHNOLOGIES (AMT), 'Project Execution Scope, Version A', *Anemoi Marine Technologies (restricted circulation: not for publication)*, 20th April 2021

14. ANEMOI MARINE TECHNOLOGIES (AMT), 'Tufton inks deal with Anemoi for installation of Rotor Sails on 82k DWT bulk carrier' [Tufton inks deal with Anemoi for installation of Rotor Sails on 82k DWT bulk carrier - Anemoi (anemoimarine.com)], *Anemoi Marine Technologies website*, 26th May 2021

15. LLOYD'S REGISTER (LR), 'TR Lady Structural and Electrical Integration: scope of design reviews/deliverable documentation and testing', *Lloyd's Register (restricted circulation: not for publication)*, 27th August 2021

16. LLOYD'S REGISTER (LR), 'TR Lady Compliance Checks: scope of design reviews/deliverable documentation and testing', *Lloyd's Register (restricted circulation: not for publication)*, 27th August 2021

17. PANAMA CANAL AUTHORITY (ACP), 'OP Notice to Shipping No. N-1-2021Vessel Requirements' *[Notices to Shipping - Autoridad del Canal de Panamá (pancanal.com)]* 4. Navigation Bridge Features Required of Transiting Vessels i. Steering Light (1), *Panama Canal Authority website*, January 2021

18. IMO MARINE SAFETY COMMITTEE (MSC), 'Revised explanatory notes to the SOLAS 2020 Chapter II-1 subdivision and damage stability regulations' Resolution 429 (98) [MSC 429 98 (imo.org)] Intact Stability Regulation 5.4 Page 9, *International Maritime Organisation*, 2017

19. ANEMOI MARINE TECHNOLOGIES (AMT), 'RPT-AMT-9004-0001A - Net Effective Power Calculation Method', *Anemoi Marine Technologies (restricted circulation: not for publication)*, 8th October 2021

20. IMO MARINE ENVIRONMENT PROTECTION COMMITTEE (MEPC), '2021 Guidelines on treatment of innovative energy efficiency technologies for calculation and verification of the attained EEDI and EEXI' MEPC.1/Circ.896 [Microsoft Word - MEPC.1-Circ.896 (imo.org)], *International Maritime Organisation*, 2021

21. ANEMOI MARINE TECHNOLOGIES (AMT), 'RPT-AMT-9004-0002A - Rotor Sail Performance Report', *Anemoi Marine Technologies (restricted circulation: not for publication)*, 8th October 2021

22. ANEMOI MARINE TECHNOLOGIES (AMT), 'RPT-TFT-2001-0020A – Rotor Sail Net Effective Power', *Anemoi Marine Technologies (restricted circulation: not for publication)*, 16th March 2022

23. IMO MARINE ENVIRONMENT PROTECTION COMMITTEE (MEPC),'Initial IMO strategy on reduction of GHG emissions from ships'[250_IMO submission_Talanoa Dialogue_April 2018.pdf (unfccc.int)], *International Maritime Organisation*, 13th April 2018

24. LLOYD'S REGISTER (LR), 'MEPC 78: CII & SEEMP Part III' slide 12, *Lloyd's Register,* 2022

10. AUTHORS BIOGRAPHY

Nick Contopoulos holds the current position of Chief Operating Officer at Anemoi Marine Technologies (AMT). He is one of the founding members of Anemoi Marine Technologies which he took a lead role in setting up. He is responsible for execution, and delivery, of projects to manufacture and install Anemoi Rotor Sails on ships at new construction and in service. He is a Chartered Engineer with a degree in Engineering from the University of Exeter. He has 15 years' experience across multi-disciplinary engineering and shipping sectors, with a proven track record of delivering complex engineering projects in Europe and the Far East.

Baiqian Jiang holds the current position of project engineer at Anemoi Marine Technologies, UK. He is responsible for technical management of projects for integration of Anemoi Rotor Sail equipment onto ships, from pre-contract planning to final project delivery. He is a Chartered Engineer and practising Naval Architect with a bachelor's degree in offshore engineering and a master's degree in naval architecture from Newcastle University. He has 7 years of engineering consulting experience in the shipping, oil & gas, and renewables industries.

Hao *(Henry)* Jiang holds the current position of Hull Section Specialist in Shanghai Technical Support Office at Lloyd's Register Classification Society (China) Co. Ltd, China. He is responsible for Classification plan approval work of hull structures of Bulk Carriers, Oil Tankers, and Chemical Carriers to the requirements of Harmonized Common Structural Rules (HCSR) of IACS. He is also responsible for steel structure approval of wind assistance propulsion systems on commercial ships for projects in China. Henry is a Naval Architecture and Ocean Engineering master from Shanghai Jiao Tong University. He has 10 years' experience of plan approval for Classification of new construction and existing ships.

Rob Tustin holds current position of North Asia New Construction Manager at Lloyd's Register Classification Society (China) Co. Ltd, China. He is responsible for delivery of pre-tender and pre-contract consultancy advice for non-standard and special ship type projects in three major shipbuilding countries in North Asia. He is a Chartered Engineer and practising Naval Architect with a degree in Ship Science from the University of Southampton. He has 38 years' experience of Classification of ships with Lloyd's Register including survey and plan approval of new construction and existing ships.

APPENDIX 1: TRACK RECORD OF ROTOR SAIL EQUIPPED COMMERCIAL SHIPS

At the time of writing this paper, based on public domain records, fourteen ships have been equipped with a total of twenty-six Rotor Sails as shown in Figure 9 below.

Rotor Sail installations on '*E-Ship 1*', '*Afros*' and '*Sea Zhoushan*' were executed at newbuilding stage with remaining eleven installations of Rotor Sails executed by retrofit in service.

Rotor Sails have been installed on two commercial ships per year, on average, in the last 5 years.

					Rotor Sail Installation							
		Key dates			Number	Dimensions		Deployment			Ship Compliance	
Ship name	Ship type	Year of build	Ship integration	Rotor Sails installed	Rotor Sails *(To be installed)*	Diameter, metres	Height, metres	Fixed	Folding	Rail	Flag state	Class Society
E-Ship 1	Ro-Ro	2010	2010	2010	4	4	27	✓			Germany	GL
Estraden	Ro-Ro	1999	2014	11-2014 11-2015	2	3	18	✓			Denmark	LR
Axios	Kamsarmax Bulk Carrier	2017	2017	-	*(8)*	*(3)*	*(25)*			✓	Marshall Islands	LR
Viking Grace	Ro-Pax	2013	2018	2018-21	1	4	24	✓			Finland	LR
Fehn Pollux	General Cargo	1996	2018	2018	1	3	18	✓			Antigua & Barbuda	DnV
Maersk Pelican	Aframax Tanker	2008	2018	2018	2	5	30	✓			Singapore	LR
Afros	Ultramax Bulk Carrier	2018	2018	2018	4	2	16			✓	Marshall Islands	LR
Copenhagen	Ro-Pax	2012	2020	2020	1	5	30	✓			Denmark	LR
Annika Braren	General Cargo	2020	2020	2020	1	3	18	✓			Germany	BV
SC Connector	Ro-Ro	1997	01-2021	01-2021	2	5	35		✓		Malta	DnV
Sea Zhoushan	Very Large Ore Carrier	2021	05-2021	05-2021	5	5	24		✓		Panama	KR
Berlin	Ro-Pax	2012	05-2021	05-2022	1	5	30	✓			Germany	LR
Delphine	Ro-Ro	2018	02-2023	02-2023	2	5	35		✓		Malta	DnV
TR Lady	Kamsarmax Bulk Carrier	2017	11-2022	*(2023)*	*(3)*	*(5)*	*(24)*			✓	Marshall Islands	LR
14 ships with Rotor Sails installed *(or to be installed)*					26 Rotor Sails installed *(11 Rotor Sails to be installed)*						Flag & Class Society on installation date	

Figure 9: Rotor Sail equipped commercial ships

APPENDIX 2: SUMMARY OF BULK CARRIER *'TR LADY'* AND ROTOR SAIL RETROFIT PARTICULARS

		Kamsarmax Bulk Carrier '*TR Lady*' Rotor Sail retrofit	
		Data	Comments
	Date of Build	2017	Shipyard: *Jiangsu New Yangzi*
Principal Particulars			
	LOA (metres)	229.00	
	LBP (metres)	225.30	
	B moulded (metres)	32.26	
	D moulded (metres)	20.00	
	T_d design (metres)	14.45	
	Lightweight (tonnes)	13,410	
	Deadweight at T_d (tonnes)	82,020	
Main Engine			
	Engine Type	6S60ME-C8.2	
	Engine Output (MCR)(KW)	9801	
	Speed (T_d @ NCR) (knots)	14.3	
Rotor Sail Installation			
	Date of Rotor Sail Installation	2Q-2023	
	Number of Rotor Sails	3 units	
	Rotor Sail Diameter (metres)	5	
	Rotor Sail Height (metres)	24	Air draft restriction for Rotor Sail height
Construction	Rotor	Composite	
	Tower and Pedestal/Bogie	Steel	
	Maximum RPM	185	Anemoi product brochure data
	Estimated Thrust (KN)	265	
	Transverse rail deployment	3 rail sets	Cross deck strip in-between hatches 2/3, 4/5 and 5/6
	Air Draft (metres)	40.1	LBWL to top of lightning conductor on Rotor Sail
	Lightship weight increase (%) with Rotor Sails installed	2.8%	
***(Ship)* Statutory Compliance on Rotor Sail Installation**			
Safety of Navigation	Arrangement of Navigation Lights	Exemption	Flag exemption as aft mast head light obscured on retrofit
	Radar Blind Sectors	Compliant	Additional forward radar set installed on retrofit
	Bridge Visibility	Compliant	
Tonnage	International Tonnage	Re-measured	Gross tonnage increase on retrofit
	Panama Canal Tonnage	Re-measured	Gross tonnage increase on retrofit
	Suez Canal Tonnage	Re-measured	Gross tonnage increase on retrofit
Loading & Stability	Inclining Experiment	Re-measured	Inclining experiment due lightship & VCG change
	Loading & Stability Manuals	Re-examined	Re-examined, and approved, after inclining experiment
	Energy Efficiency	EEXI assigned	EEDI Phase 1 (*2017 DoB*) => EEXI (2Q *2023*) on retrofit

Figure 10. Summary of particulars of Kamsarmax Bulk Carrier '*TR Lady*' and Rotor Sail retrofit

WIND PROPULSION AND UNDERWATER RADIATED NOISES MITIGATION

P Cordier, XP SEA, France
C Noël, Semantic TS, France

SUMMARY

From various scientific studies, we know that vessels' Underwater Radiated Noises (URN) have significant negative impacts on marine life. As the worldwide maritime traffic increased over decades, oceans are getting noisier. Since a few years, part of the shipping industry is working to mitigate URN, and most maritime organizations are preparing future rules to protect marine life from noises. URN emitted by vessels mainly come from propeller cavitation and machinery equipment. Studies show a clear reduction of URN with ship speed decrease as propulsion power is reduced.
Wind propulsion is the most silent means to propel a vessel. Its contribution to ship propulsion is an opportunity to reduce URN to acceptable levels for marine life, but it implies ship speed reduction. The question is therefore "What would be the maximum achievable speed, combining wind and conventional propulsion, keeping the URN to an acceptable level for the marine life?"

UNITS
All values are given using the International System of Units.

ABBREVIATIONS

ANL	Ambient Noise Level
AWA	Apparent Wind Angle
AWS	Apparent Wind Speed
DT	Detection Threshold
OSS	Optimum Ship Speed
RPM	Round per Minute
SL	Source Level
TL	Transmission Loss
TWA	True Wind Angle
TWS	True Wind Speed
URN	Underwater Radiated Noises
WASP	Wind Assisted Ship Propulsion

1. INTRODUCTION

"Imagine you live in a dark place, and you must rely on your hearing the same way you would rely on your vision. And that space is loud and getting louder. It's certain to cause you stress." This statement from CLEAR SEAS Centre for Responsible Marine Shipping (Canada) clearly illustrates the issue of growing URN levels in the oceans and their impacts on marine life.

While several global entities like the EU are continuously working on measures to reduce URN emitted by ships, there is no state today currently imposing legally binding requirements. But they will certainly come someday.

As the shipping industry is working on URN mitigation measures for the existing fleet and newbuilding, wind propulsion is an additional opportunity to reduce noises along with other benefits.

To quantify the benefits of wind propulsion for URN mitigation and marine life protection, this article presents the following aspects:

- Sources and levels of URN emitted by vessels, and the effect of ship speed
- Sensitivity of marine mammals to URN depending on the level and frequency band
- Principles of Wind Assisted Ship Propulsion (WASP)
- An example of URN mitigation on a WASP cargo vessel.

2. URN MITIGATION INITIATIVES

The below main initiatives regarding URN mitigation show how active this topic is at a worldwide level. Many other initiatives exist in this field (national and regional).

2.1 IMO SDC 8: LATEST UPDATES

In a bid to address the issue, the IMO Sub-Committee on Ship Design and Construction (SDC 8, January 2022), began its work to review the 2014 "Guidelines for the reduction of URN from commercial shipping" (MEPC.1/Circ.833). Given the complexities associated with ship designs, the guidelines focus on primary sources of URN, namely on propellers, hull design, onboard machinery, and various operational and maintenance recommendations.

2.2 PIAQUO PROJECT

The European project AQUA laid the ground for the PIAQUO Project, an initiative led by EU gathering several international partners, aiming at reducing the acoustic impact of maritime traffic and adapting it in real time to the ecosystems.

2.3 THE ZERO POLLUTION ACTION PLAN

In the frame of COP 15 (Montreal, December 2022), the EU is taking action today to better protect marine life from URN. The new limits mean that no more than 20% of a given marine area should be exposed to continuous URN over a year. These URN pollution limits are part of the Zero Pollution Action Plan and are the first of this kind at global level.
These threshold values have been developed under the Commission's Marine Strategy Framework Directive.

2.4 CLASSIFICATION SOCIETIES' NOTATION

Seven classification societies have implemented voluntary class notations for ships related to URN limitations, motivated by environmental concerns. For other reasons, research and cruise vessels were the first type of vessels to obtain these notations. However, a tanker has been awarded the DNV's Silent-E class notation (DNV GL 2017) in 2021 being the first cargo vessel to do so for environmental motivations.

3. URN EMITTED BY VESSELS – EFFECT OF SPEED REDUCTION

3.1 SOURCES AND LEVELS OF URN EMITTED BY VESSELS

URN emitted by conventionally propelled vessels mainly come from propellers cavitation, hull appendages and machinery equipment.

The design of propulsion machinery and associated propellers usually leads to propeller cavitation at nominal speed. Cavitation is the formation and collapse of vapor-filled cavities in water due to low pressure in the fluid (vortices) and/or on the blade or appendage surface (bubble or sheet cavitation). This phenomenon creates significant underwater noise over a wide frequency range. Propeller cavitation appears above a certain propeller rotational speed (RPM) and ship speed (called cavitation inception speed) specific to each propeller and ship design. Avoiding or limiting cavitation is of interest for certain types of ship. For example, navy ships propulsion aims at acoustic discretion while research vessels need low URN for their acoustic equipment. Cruise vessels target passengers' comfort with low noise levels.
The cavitation of propellers mainly depends on the blade hydrodynamic loading. As the blade rotational speed increases, the internal side blade load increases while the external side pressure decreases until it reaches local pressures below which water vaporises. Cavitation may be mitigated by the combined use of low tip speed (low rpm/high pitch), larger blade area ratio, uniform inflow (hull design), optimized blade loading, and reduced tip loading. Cavitation may also be reduced if the blade loading is reduced at a given ship speed using an additional mean of propulsion (wind for example as we will see further).

Hull appendages, when properly designed, produce low levels of URN, and rarely cavitation.

Machinery equipment transmit noises and vibrations to the hull through the room air and their structural supports. The hull then radiates this noise in the sea water. The noisiest machinery equipment are:

- Propulsion diesel engines (low, medium, or high speed)
- Diesel Generators

Rotating machines and electrical motors may also contribute to URN at a lower level.

From report [2], the following noise levels, measured at 6 m depth, are given with respect to vessel sizes:
- Gross Tonnage from 500 to 50 000 UMS: 160 to 180 dB (typically general cargos)
- Gross Tonnage from 50 000 to 100 000 UMS: 170 to 200 dB (typically large tankers)

In terms of frequency band, the URN emitted by ships cover a wide range from 10 Hz to above 10 kHz depending on the type of vessel and speed. The major part of the world ocean shipping fleet is constituted by tankers, bulk carriers and containerships. They are therefore the major contributors to oceans URN levels. Report [5] shows that peak mean sound levels appear between 30 and 100 Hz for these vessels with levels ranging between 170 and 190 dB re 1 μPa m depending on the model (source at 6 m below sea level).

3.2 URN MITIGATION BY MACHINERY AND PROPULSION DESIGN

Modern ship with propeller propulsion can reduce significantly the URN levels by propulsion design. This can be done by optimizing the propeller for reduced cavitation by increasing the blade area ratio for example which involves a small loss in efficiency. It can also be achieved by reducing the propeller tip speed through RPM reduction which is made possible by low-speed electric propulsion motors. This implies typically a higher machinery weight and cost compared to mechanical transmissions. Il can also be achieved by improving the ship wake into the propeller, typically by going to twin shafts, or even better choosing thrusters or pods (more expensive).

Machinery equipment airborne and structure borne noises (and vibrations) can be reduced by resilient mountings and isolation of enclosures. Electrical propulsion achieves some of these objectives by allowing resilient mounting of the diesel generators.
It shall be noted that resilient mountings on main propulsion engines are adapted to 4-stroke medium to high-speed diesel engines, but not to heavy 2-stroke engines low speed engines (major part of the fleet) due to their weight.

These considerations have led to the propulsion architecture of cruise vessels for example, which have very strict requirements for noise and vibration onboard, thus implying low URN. These ships may obtain URN class notations at speeds exceeding 15 knots. This is similar for research vessels to avoid interferences with acoustic equipment.

Retrofit of existing vessels with the above design features may bring a noise reduction in the range of 6 to 8 dB in low and high frequency bands respectively [2] which is not negligeable. But retrofit costs are usually too high for cargo vessels.

3.3 URN MITIGATION BY VESSEL SPEED REDUCTION

Hydroacoustic studies show that URN levels increase exponentially with vessel speed. In view of the predominant source of URN from propellers' cavitation and the significant cost of retrofit, ship speed reduction (along with propeller rotational speed reduction) is still the most efficient measure.

Report [2] gives the result of a speed reduction experiment in large areas close to Vancouver harbour. The vessels were asked to slow down to 11 knots and the noise reduction was measured between 6 to 12 dB for tankers/bulkers and containers vessels respectively. These reductions look rather small in % but are significant for marine life since a 3 dB reduction is equivalent to divide the perceived noise level by a factor of 2. Hence 6 dB and 12 dB correspond to a noise power level reduction factor of 4 and 16 respectively.

Report [3] presents the results of URN measurements emitted by a total of 29 commercial ships transiting in the Santa Barbara Channel, southern California coast. A graph (see Figure 1) was derived showing the relationship between the ship speed and broadband source level for various types of vessels of which fastest container ships (20-22 kts) emit noise levels around 184-188 dB whereas the slowest chemical product tankers (11-12 kts) emit noise levels around 176-178 dB.

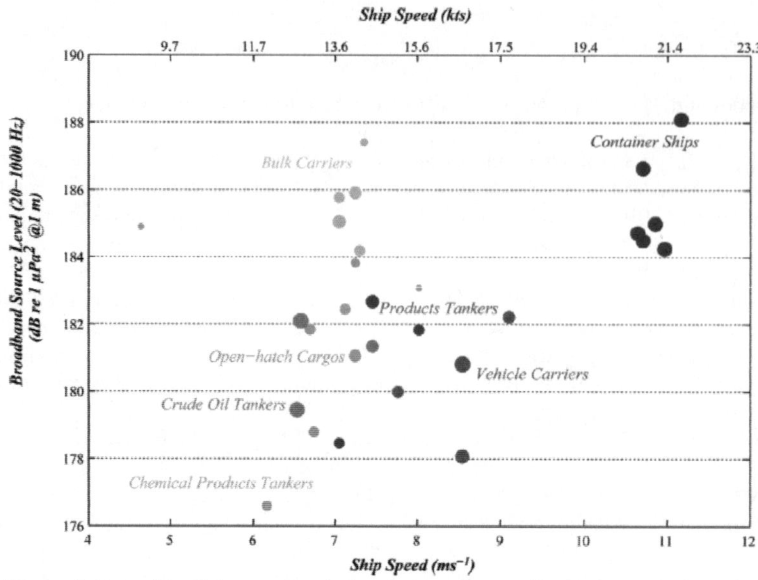

Figure 1: Broadband Source levels for various types of vessel vs ship speed[3]

Ship energy consumption analysis can identify the most efficient ship speed (minimum consumption for a given distance travelled) which is typically much lower than the design ship speed. Ship speed limitation (debated in IMO committees without decision yet) brings, on top of URN reduction, the following main advantages:

- Fuel consumption reduction
- CO2 emission reduction
- Pollutants and particles emission reduction
- Increased contribution of wind propulsion for WASP vessels

Note regarding electric propulsion on batteries: this configuration concerns few vessels in the world and used for small distances. Speed reduction is anyhow required to reduce URN.

3.4 PROPAGATION LOSS FACTORS IN THE SEA

It is important to estimate or take assumptions on URN propagation losses to assess the impact of ship noises on the marine life.
Several factors influence the URN propagation losses between the source and the receiver:

- URN levels and frequencies at the source
- Depth of the source (below sea surface)
- Water depth and type of seabed (as the noises are reflected on the seabed)
- Sea water characteristics (temperature and salinity) in various layers
- Distance of the receiver from the source

This large number of parameters makes it complex to assess accurately the noise level received at a certain point of the ocean. Unless live in-situ measurements are available, modelisation is required and usually involves certain simplifications to estimate the propagation losses.

The URN source is close to the propeller depth, bottom of hull depth, or draft (usually set at 6 m below the sea surface).

Shallow water depths increase the sound reflections which usually increase the sound levels close to sea surface. Reversely, deep waters will better absorb noises as they lead to less reflections.
On the other hand, in deep waters, low frequency energy (<100 Hz) may travel over tens of kilometres, which is why ship noise has the potential to mask the marine mammals' communication far away from the source [1].

Hard seabed like rock increases the sound reflections whereas sedimentary seabed partially absorbs noises. The shape of the seabed also influences the propagation pattern which is another source of complexity.

Sea water temperature and salinity have an impact on the sound celerity. When the water column is constituted by multiple layers with different characteristics, the boundaries are acting like reflection surfaces. For example, a hot layer of several meters near the sea surface (typically during summer) will trap the noises which will enhance URN in the top layer. Marine mammals living in this layer will obviously suffer from this phenomenon.

We can therefore conclude that the URN level at the source alone is not the only indicator for the impact on marine life.

Based on marine traffic statistics across the oceans, and using URN propagation models with various input parameters and assumptions, it is possible to derive sound maps. EU-funded projects generated numerous sound maps (BIAS, AQUO, SONIC, JOMOPANS and JONAS) based on predictions of sound pressure levels to show how noisy are the oceans, especially along the maritime routes.

4. MARINE LIFE AND URN FREQUENCIES AND LEVELS

4.1 MARINE MAMMALS' SENSITIVITY TO URN

This article does not aim at giving all aspects of marine life sensitivity to human activities. It only refers to study results focusing on the impact of URN emitted by vessels on certain specifies of marine mammals (which may cover the sensitivity of other species).
The marine mammals' frequency band of communication is commonly used as the band within which they are likely to be disturbed. However other frequency bands emitted by vessels may also be detrimental to them on other aspects.
Studies are usually comparing these communication frequency bands to those emitted by vessels to evaluate the potential impact on marine life.

Report [5] presents the vessels URN frequency broadbands depending on the source (propeller or machinery) versus marine mammals' communication frequency broadbands. Figure 2 shows an example of such comparison.

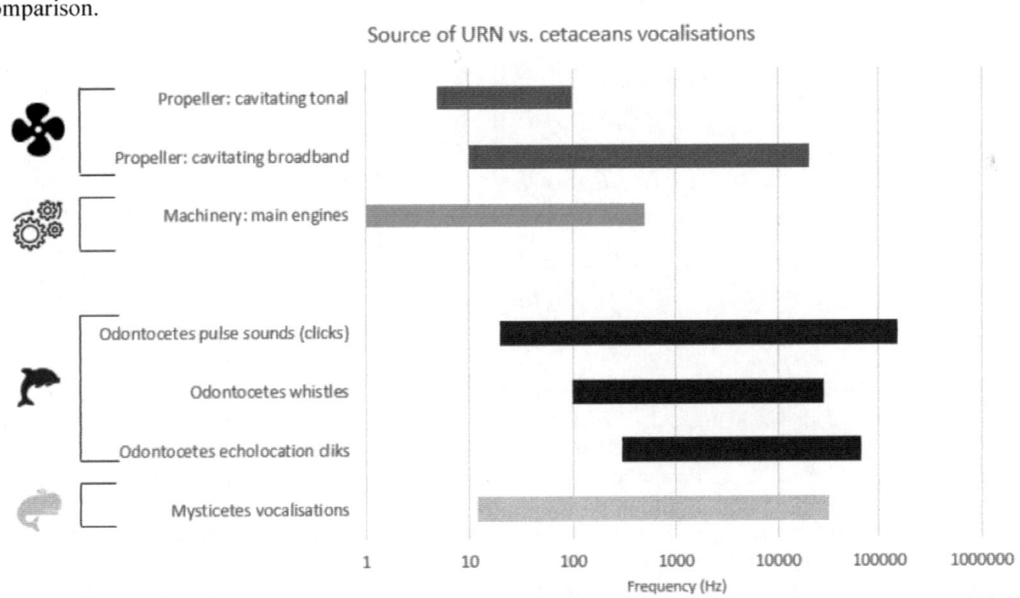

Figure 2: Frequency range of shipping noise and different types of vocalisations produced by cetaceans. The frequency ranges are based on the minimum and maximum value of frequency found in literature for the different types of vocalisations (red colour = high contribution / orange colour = medium contribution) [5]

It is clear that the propeller cavitation frequency range covers all species communication frequency ranges. We know from previous sections that peak sound levels from bulker, tanker and container ships occur between 30 and 100 Hz which band interferes at least with Ondontocetes clicks and Mysticetes vocalisations.

From [5], we understand that shipping noises potentially have the following impacts on these cetaceans (depending on the sound levels):

- Changes in vocal behaviour
- Changes in diving and swimming patterns
- Reduction in the communication range
- Foraging behaviour
- Physiological responses

Temporary or permanent hearing damages are unlikely from URN emitted by vessels.

Regarding fishes, we also learn from [5] that their communication frequency broadband ranges from 30 to 5000 Hz, also overlapping with shipping noises. There are also evidences of negative impacts from shipping noises on invertebrates (behavioural, morphological and physiological changes).

4.2 NOTIONS OF AMBIENT NOISE AND COMMUNICATION DISTANCE

An Ambient Noise Level (ANL) including shipping traffic can be defined by the Wenz statistical level curves [8].
In the case of marine mammals, in a certain ANL, two individuals (or groups) may communicate up to a certain distance.
If the ANL is high, the two individuals need to be close to communicate and if the ANL is low, the two individuals may be remote to communicate.
The first reaction of marine mammals when the ANL increases is to raise their source level (SL) gradually up to a certain physiological limit [1].

In the example of Figure 3, the maximum communication distance for frequencies around 100 Hz, in a Wenz type of ANL of 78 dB, is reached at R= 40 km. It represents the maximal range up to which the passive sonar equation is satisfied:
$SL – TL – ANL \geq DT$ with DT (Detection Threshold) = 10 dB.

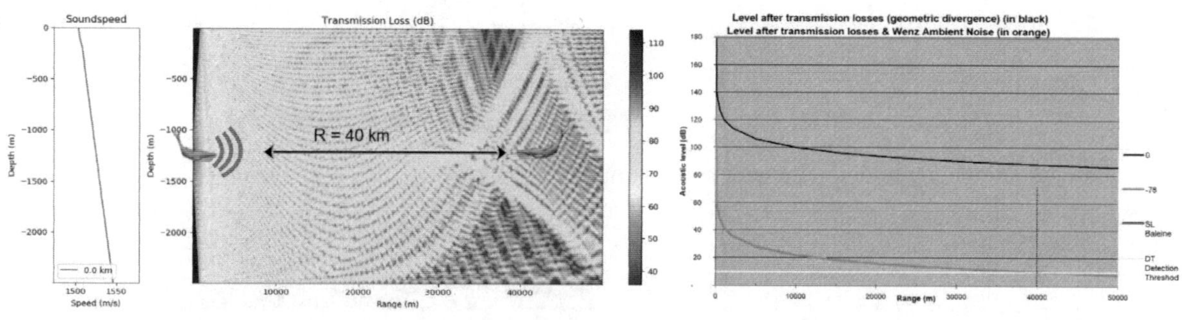

Figure 3: Illustration of communication distance limit between 2 marine mammals at 100 Hz with:
Water depth =2500 m, ANL= 78 dB (Wenz Traffic2), SLwhale =180 dB, Whale depth =1250 m
Model used to calculate TL between the source and the receiver separated by a range R:
Left: Parabolic model. Right: $TL = 20 \, Log_{10} (R)$ (Semantic-TS)

When a motor vessel comes close to any of the two individuals, the URN emitted by the vessel reaching the individual (after propagation losses) may come above or below the ANL. If the received vessel noise level is higher than the ANL, then it reduces the maximum communication distance, thus disturbing the animals' social life. This effect will be evaluated in the example presented at §6.

5. CONTRIBUTION OF WIND PROPULSION

5.1 BENEFITS OF WIND PROPULSION

URN emitted by 100% wind-propelled ships are expected to be below the natural underwater noise levels with careful machinery design. The major sources of URN on sailing ships are the following:

- Hull hydrodynamic noises which are very low for marine life

- Power generation and other machinery equipment noises which can be isolated
- Transient noises such as bow thrusters and anchor mooring noises

As shown in report [4], wind propulsion has been the subject of more attention recently, targeting to reduce shipping carbon footprint and air pollution, as well as fuel consumption as it is an important source of OPEX saving.
Modern wind propulsion concepts require fewer crew than in the past, using automation and electrical or hydraulic motorization in the rigging.

But even nowadays, targeting a high percentage of wind contribution involves lower ship speeds and lower reliability of Estimated Time of Arrival (ETA). Therefore, most wind propulsion concepts are considered as a WASP systems and try to find the best compromise between wind contribution and speed.

There is a large variety of modern wind propulsion concepts (main ones presented in [4]) which are usually mixed with conventional propeller propulsion. The optimum percentage of the mix between wind and conventional propeller propulsion mainly relies on the ship program, the selected wind propulsion technology and the size of the wind propulsion system.

The wind propulsive power depends on [4]:
- the true wind speed (TWS)
- the ship speed
- the angle between the true wind and the ship heading (TWA)
- the surface area of the propulsion system exposed to the wind
- the lift and drag coefficients provided by the selected technology

5.2 WIND THRUST CONTRIBUTION VERSUS SPEED

The typical relation between required propulsive thrust and ship speed is exponential and specific to each ship design and loading. The reason for having an exponential increase of the propulsive thrust is due to a hydrodynamic relationship between ship's drag (or resistance) and speed, specific to each vessel (several ship's parameters involved). At low speed, the relation of resistance to speed (V) is approximately to V^2 and it increases to V^3 at higher speeds. Typically, an increase of about 20% of the ship speed will increase its propulsive power demand by more than 50% [4].

On a sailing ship which does not use its propeller propulsion (100% wind propulsion), polar diagrams are commonly produced by Velocity Prediction Program (VPP) to show the sailing performances which combine the sail system and hull performances. The polar diagram is used to forecast the ship speed for given TWS and TWA. Typically, for a given TWS, the maximum ship speed is obtained for TWA varying from around 135° to 45° (angles with respect to ship heading).

As the ship speed increases, the AWS increases as well and the AWA decreases. The wind thrust and ship speed increase in a virtuous circle up to a certain equilibrium speed. For vessels with poor sailing abilities (high resistance and/or small sail system) using only wind propulsion (like cargo sail ships), this phenomenon is however minor.

As seen on most polar diagrams, for TWA varying from about 45° (TWA limit) to 0° (head wind), the ship speed drops rapidly down to 0. What determines the point where the wind thrust starts to decrease is the AWA limit. This limit is specific to each rigging and wind propulsion technology. An AWA limit around 35° is typical, smaller values being possible.

When combining conventional and wind propulsion, the ship speed may be selected by the captain, adjusting the propeller propulsion to the required level (RPM) depending on the complementary wind thrust. The driving parameters for wind thrust contribution are the AWS and AWA.

Above a certain ship speed, the AWA limit is reached and the wind thrust starts to decrease drastically as the AWA reduces towards ship axis (0 deg). This principle is applicable to most WASP systems. The ship speed limit associated to AWA limit (wind thrust start of decrease) is specific to each wind propulsion concept and vessel but it is clear that sail systems capable of producing thrust close hauled (high lift to drag ratio) will be more beneficial in terms of average wind power contribution.

Above a higher ship speed limit, the rigging starts to generate more drag than thrust: it is therefore bringing a negative contribution.

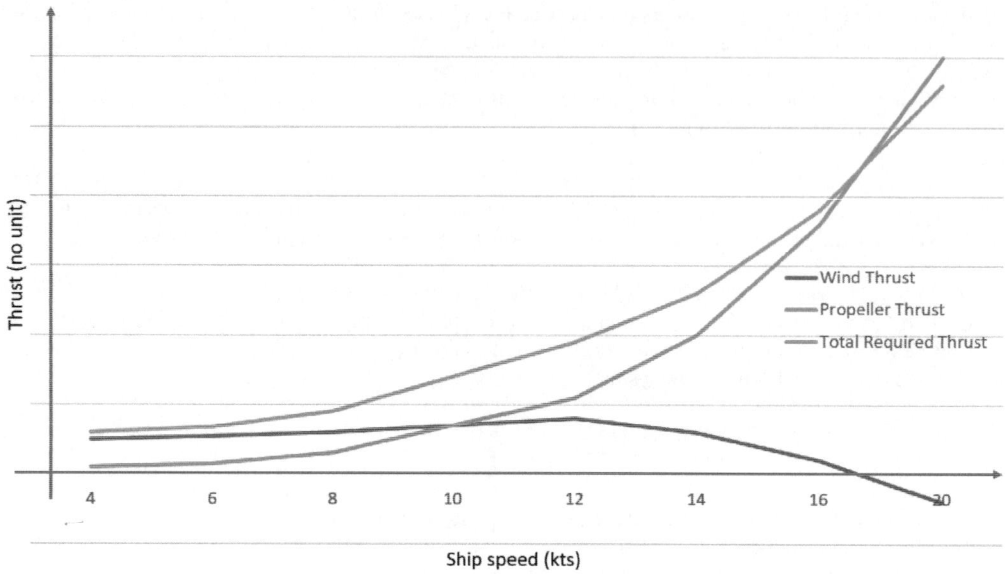

Figure 4: Typical wind thrust contribution for TWS of 15 knots and TWA of 60 deg (XP Sea)

The typical graph of Figure 4 represents the situation on a hypothetical WASP ship using specific wind conditions with TWS of 15 knots and TWA of 60 deg as an example.
An AWA limit of 35° is assumed in this example (angle below which wind thrust starts to decrease).
The ship speed associated to the AWA limit is around 12 knots. We may call this speed the "Optimum ship speed" (OSS) although it does not mean that it is the best speed for other criteria.

As the ship speed increases to the OSS, the AWS also increases up to about 23 knots. This drives the wind thrust up to its maximum at the OSS. Above the OSS, the wind thrust starts to decrease which requires increasing further the contribution of the propeller propulsion.

Above a certain ship speed (18 knots in this example), the AWA drops down below another limit (27 deg in this example) where the WASP system is expected to generate more drag than thrust, therefore a negative contribution.

5.3 OPTIMUM SHIP SPEED FOR ACCEPTABLE URN LEVELS

As it is clear that 100% wind propulsion at slow speed gives the lowest URN levels, and 100% conventional propulsion at maximum speed is the worst case for URN emission, the question is:
"What would be the maximum speed achievable combining wind and conventional propulsion at which the URN would be acceptable for marine life?"

The answer to that question is quite complex as it depends on the following main parameters:

- Vessel characteristics including hull dimensions, displacement, machinery and propulsion design
- Sail system characteristics
- Environmental conditions at any time (wind speed, wind direction, sea state, routing, etc.).
- URN propagation loss factors as seen previously
- URN levels and frequencies acceptable for the marine life (specific to each group of species).

Assuming hypothetical values or design options, we can focus on the relation between propeller RPM, ship speed, wind contribution and URN levels.

5.4 EXPECTED PROPELLER RPM AND URN REDUCTION

To quantify the contribution of wind propulsion to the URN reduction, we may compare the two following situations for the OSS ship speed of 12 knots as shown in Figure 4:

- Case A: 100% conventional propeller propulsion
- Case B: Mix between 60% conventional and 40% wind propulsion

In Case A, a propeller RPM_A is set with blades fully loaded for the given speed.
In Case B, a propeller RPM_B (< RPM_A) is set with blades less loaded.

At RPM_A corresponds URN_A where cavitation may occur
At RPM_B corresponds $URN_B < URN_A$. where cavitation is expected to be minimum or nil.

We are therefore looking at an estimate of the URN reduction = $URN_A - URN_B$.

When we look to the URN levels emitted by vessels at different RPM, they are associated to different vessel speeds, which means that in all cases, the propeller is fully loaded for the required speed in an equilibrium between thrust and drag.

Let us consider the case of a vessel under conventional propulsion only at a certain RPM0 and Speed0 associated to URN0. If we add wind propulsion to conventional propulsion keeping the same RPM0, the vessel will accelerate to Speed1, however the propeller loading will drop somehow which will reduce the cavitation and therefore the URN level to URN1<URN0. Keeping the wind propulsion contribution and same wind conditions, if we reduce the RPM from RPM1 to RPM2 and slow down the vessel back to Speed0, the cavitation will further be reduced and the URN level will further drop down to URN2<URN1<URN0. As a result, URN2 level potentially includes no cavitation, other sources of noises (machinery) being predominant.

6. EXAMPLE OF URN MITIGATION IMPACT BASED ON A CARGO VESSEL FITTED WITH WASP

6.1 METHODOLOGY

As stated previously, cargo vessels (bunkers, tankers and containerships) constitute the vast majority of ocean shipping traffic. These vessels are usually not designed with a machinery and propulsion for low URN levels (like navy, research and cruise vessels). URN reduction on these vessels through WASP retrofit and appropriate propulsion management would therefore bring a significant benefit to marine life worldwide.

The methodology considers the example of a cargo vessel defined in report [7]. Although this cargo vessel is not fitted with a WASP system, the purpose is to compare the two cases (A and B) as described in §5.3 assuming that the cargo vessel could be fitted with an appropriate WASP without changing other characteristics. The WASP system is not defined but should be compatible with the assumed performances.

Report [7] defines relations between RPM, ship speed and URN.
Then for each URN frequencies and levels A and B, site and environmental assumptions are taken to estimate the propagation losses. An assumption is taken on the ANL including shipping traffic (ANLshipping).

Ultimately, a comparison is made between the two cases to determine the impact of wind propulsion on the maximum communication distance between two individuals (or groups). The reasoning is that starting from a maximum communication range of 40 km (see Figure 3), two individuals separated from that distance should come closer by R_A in Case A and by R_B in Case B in order to communicate. R_A and R_B are therefore the minimum required gathering distance to be achieved by one individual (or a group) to communicate with another due to the presence of a vessel.

6.2 CARGO VESSEL CHARACTERISTICS, URN FREQUENCIES AND LEVELS

The cargo vessel characteristics in below tables and figures are extracted from [7].

Table 1: M/V Harriette characteristics [7]

Vessel name:	M/V OVERSEAS HARRIETTE
Date of construction	1977
Deadweight:	25 515 tons
Length overall	172.9 m
Beam:	22.8 m
Draft full load:	10.2 m
Draft ballasted:	7.9 m (condition of measurements)
Maximum speed:	15.9 kts (full load)
Associated rpm:	145 rpm
Propeller diameter:	4.9 m
Number of blades:	4
Engine type:	Direct-drive, two-stroke diesel (Hitachi/Babcock & Wilcox 6K67GF, 6 cylinders)
Engine power:	11 200 HP (low speed engine)
Diesel generator:	5 cylinders (Hitachi/B&W)

Table 2: Relation between Engine speed, ship speed and wideband source levels [7]

Propeller rotational speed (rpm)	Ship speed (knots)	Keel aspect wideband source level (*) (dB re: 1 mPa at 1 m)
68	8	178
86	10	180
105	12	184
122	14	190
140	16	192

(*) Peak values

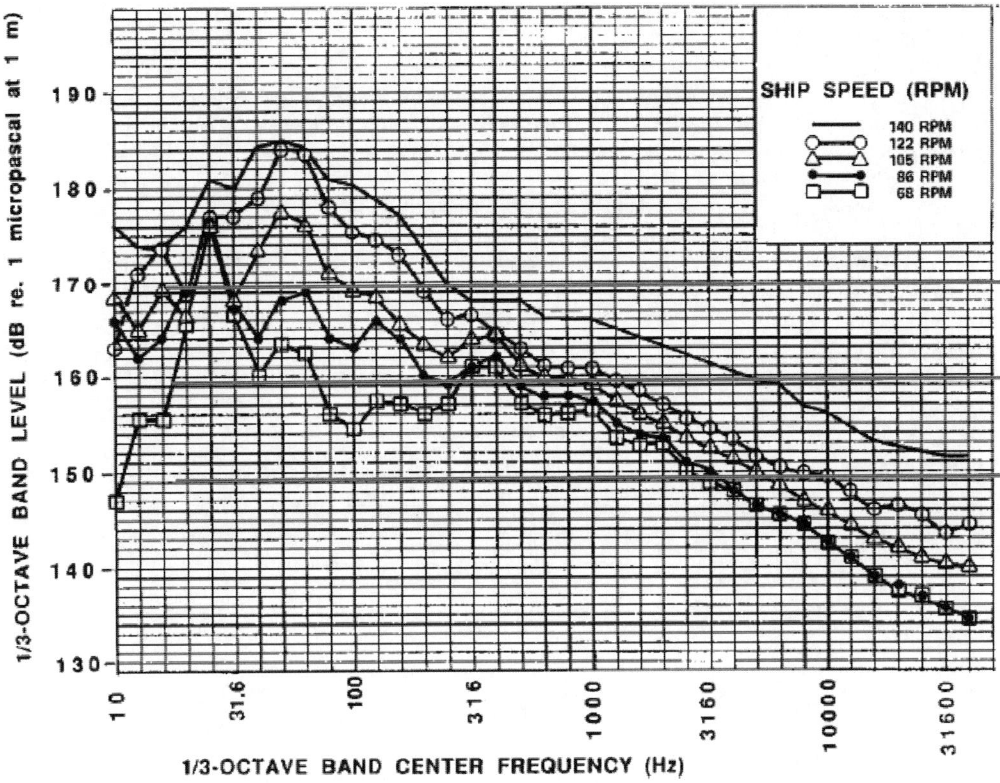

Figure 5: Keel-aspect third-octave bandwidth spectra of the vessel at various speeds [7]

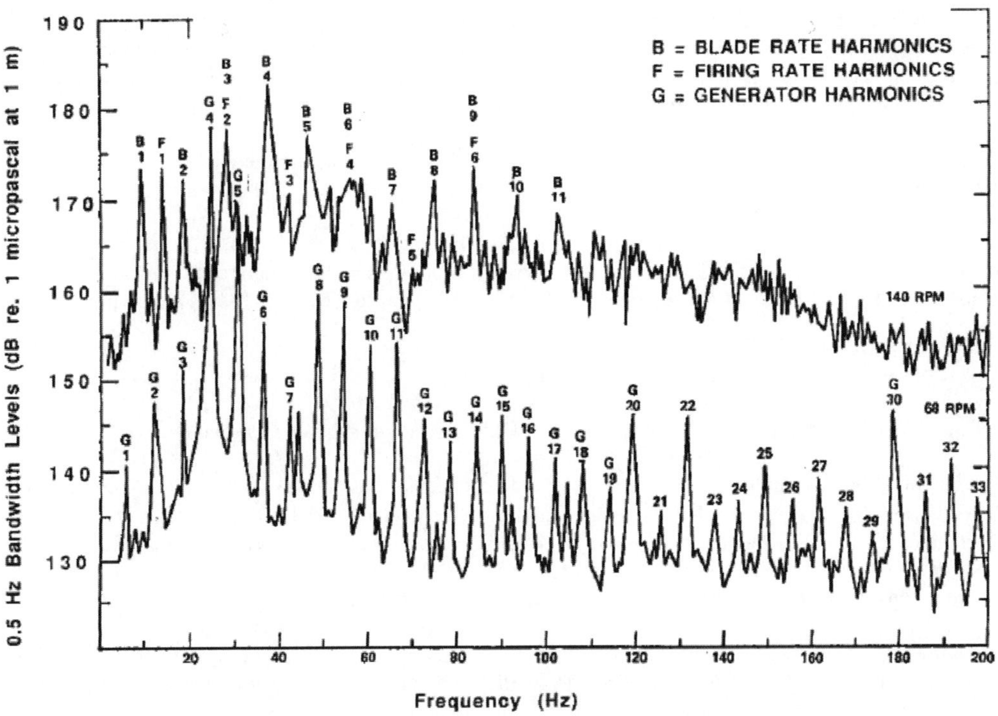

Figure 6: Keel-aspect narrow-band spectra of the vessel in 0.5 Hz bands at low speed (68 rpm) and maximum speed (140 rpm) [7]

B = Blade rate harmonics are related to cavitation
F = Firing rate harmonics are related to the propulsion engine noise
G = Generator rate harmonics are related to diesel generator noise

6.3 STUDIED CASES

Following the methodology described in §6.1, Table 3 presents cases A and B to be compared. The URN levels are maximum values of specified broadband ranges taken from Figure 5. It is to be noted that they are in line with the measured values presented in §3.2 and [2] [3].

Table 3: Studied cases (XP Sea)

Parameter	**CASE A 100% Conventional Propulsion**	**CASE B 60% Conventional 40% Wind**	**Difference**
Ship speed (knots)	12.0	12.0	
Thrust form conventional propulsion	100%	60%	
Thrust form wind propulsion	0%	40%	
Propeller rotational speed (rpm)	105	68	-37
URN maximum levels	Values in dB from Figure 5 except (*)		(dB)
10 – 31.6 Hz	178	165 (*)	-13
31.6 Hz – 100 Hz	179	165 (*)	-14
100 Hz – 316 Hz	170	161	-9
316 Hz – 1000 Hz	164	161	-3
> 1000 Hz	159	156	-3

CASE B uses the following wind conditions (same as in Figure 4):

- TWS = 15 knots / TWA = 60 deg
- AWS = 23 knts / AWA = 35 deg

(*) As stated in [7], it is understood that propeller cavitation is significant above a speed of 10 knots (86 rpm), which means that at 68 rpm, maximum URN levels are emitted by diesel generators (G) as shown in Figure 6. The URN peak levels for 68 rpm between 20 Hz and 40 Hz appear to go above 165 dB. This is not a realistic situation on modern cargo ships as resilient mountings and insulation in the machinery is common nowadays. This means that all URN emitted by diesel generators (G) should be below 165 dB at any frequency band. We therefore consider this value as a maximum in CASE B at 68 rpm.

For the evaluation of the impact on marine mammals, we will focus on frequencies around 100 Hz for which the reduction is from URN_A=169 dB to URN_B=155 dB, therefore -14 dB (see Figure 5).

In this example we may first give an equivalence in terms of noise contribution per vessel between Case A and Case B. Based on a logarithmic relation between noise levels and the number of emitting sources, in this example, with an offset of 14 dB, we may say that the URN_A emitted by one vessel is equivalent to 25 vessels emitting at URN_B ($10^{\frac{14}{10}} \cong 25$).
This result is coherent with the findings of report [9].

6.4 ILLUSTRATING THE IMPACT OF ADDITIONAL NOISE ON COMMUNICATION DISTANCE

In our illustrative example of Figure 3, below the ANLtraffic, the marine mammals may communicate to a maximum distance of 40 km. When a vessel sails near a receiving whale, it generates a source URN which propagates to the receiving whale. Close to the vessel, the local ANL is louder than the ANLtraffic. The local ANL decreases as it propagates and starts equalizing the ANLtraffic at a certain distance. Beyond this distance, we may consider that its contribution is "drowned" in the statistical noise.

In Figure 7 the first graph compares the propagation losses between Case A (orange curve) and Case B (green curve) crossing the ANLtraffic (yellow line at 78 dB). The orange and green curves cross the yellow line at 36 km and 8 km respectively.
The second graph shows the same information in terms of minimum gathering distance R_A and R_B. The two whales are considered distant from 40 km and need to gather by R_A = 36 km and R_B = 8 km to be able to communicate.

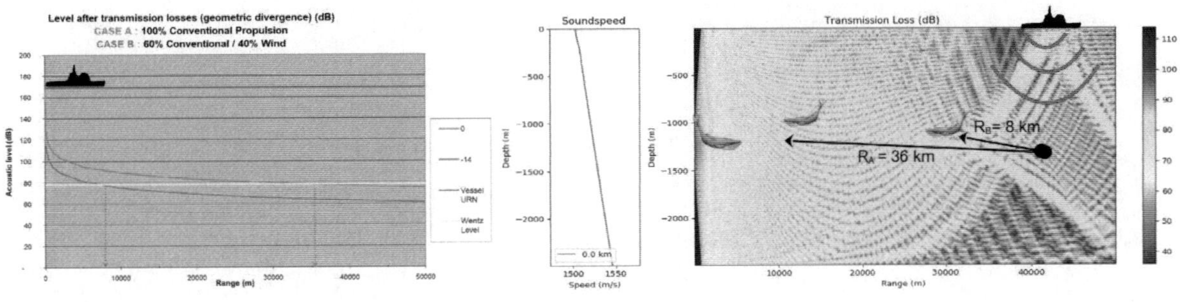

Figure 7: Gathering distance R_A (orange) and R_B (green) required to be achieved by marine mammals to communicate (Semantic-TS)

6.5 MACRO ESTIMATING THE IMPACT OF ADDITIONAL NOISE ON COMMUNICATION DISTANCE

To generalise the benefit of wind propulsion in terms of maximum communication range (R) between marine mammals, we start from a reference state without maritime traffic. Then we express the reduction of R as a function of the noise disturbance from a vessel.

We define:

- R_{ref} as the reference range, maximum communication range in an environment without maritime traffic (ANL_{ref}), satisfying the passive sonar equation:
  $SL - TL(R_{ref}) - ANL_{ref} + DT = 0$
- R as the maximum communication range in the disturbed ANL due to a vessel coming across the area:
  $SL - TL(R) - ANL + DT = 0$
- $\Delta_{ANL} = ANL \quad - ANL_{ref} > 0$ is therefore the disturbance of the noise, view as a strictly positive anomaly, when a ship is passing through.

Then $SL - TL(R_{ref}) - ANL_{ref} + DT = SL - TL(R) - ANL + DT$

implies $TL(R) - TL(R_{ref}) = -\Delta_{ANL}$

Assuming, as an order of magnitude, a classical model for the transmission losses: $TL(R) = A.\log_{10}(R)$

Then we find: $\log_{10}\left(\frac{R}{Rref}\right) = -\frac{\Delta_{ANL}}{A}$ or $\frac{R}{Rref} = 100\ x\ 10^{-\Delta_{ANL}/A}$ expressed in % of communication range reduction depending on the ANL degradation.

The following Figure 8 illustrates this relation and shows how fast the communication range is reduced with the increase ANL degradation.

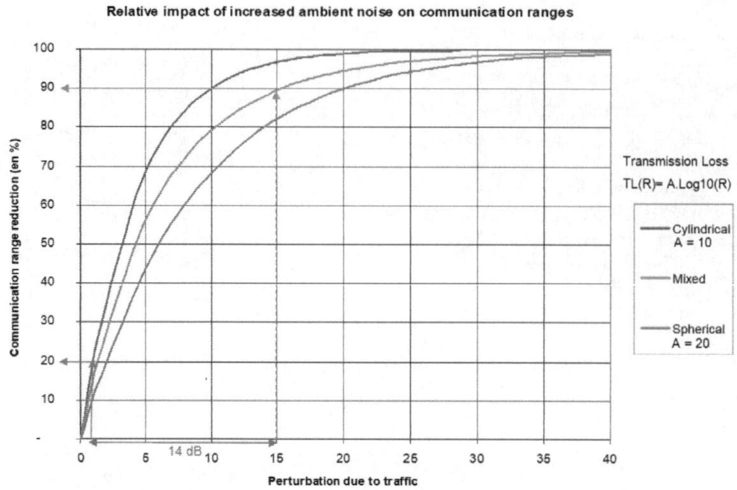

Figure 8: Loss of marine mammals' communication range vs ALN degradation due to shipping traffic (Semantic-TS)

In Case A, $\Delta_{ANL}= 15.3\ dB$, the communication range reduction is of 90% (36 km versus 40km)
In Case B, $\Delta_{ANL}= 1.3\ dB$, it drops down to 20% (8 km) which is a considerable saving

7. CONCLUSIONS

The above results show that the benefit of wind propulsion contribution to conventional propulsion can be very significant with regards to URN levels reduction and marine life preservation.

Many parameters influence the results. Parametric data of URN measurements, as a function of speed, RPM and ratio of conventional vs wind propulsion are currently critically lacking to be able to carry out a serious impact study. However, the presented example shows that a WASP cargo vessel bringing a wind contribution of 40% on the propulsion (Case B) will reduce significantly the disturbance of marine mammals' communication while keeping the same speed as with 100% conventional propulsion (Case A). In this example, the communication range between two individuals is decreased by 90% for Case A and by 20% for Case B compared to a reference ambient noise level.

The benefit is expected to be greater with stronger winds or broad reach point of sail, and reversely smaller with lighter winds or close reach point of sail.

Reducing URN levels to an acceptable level for marine life means setting WASP vessels' speed to an optimal balance between wind and conventional propulsion. This good practice should be applied especially when the vessel enters areas where URN sensitive marine species are present. This requires real time on-board monitoring of marine species position and, ideally, a software displaying live URN levels and maximum allowable URN levels depending on the distance to sensitive species.

It should also be highlighted that collisions between ships and marine mammals represent a significant source of decline in the species (difficult to distinguish from other sources). This risk and severity of such collisions are drastically lowered by reducing ships' speed.

All these benefits are obviously cumulated with carbon footprint reduction and fuel savings.

A recent order in France for two 220 m long sailing cruise vessels shows the way: project name "Silenseas".

Following this analysis, one can foresee that future regulations will push towards a reduction in vessel speeds and the use of wind propulsion in order to reduce fuel consumption, carbon footprint, and URN of shipping, all at the same time.

8. ACKNOWLEDGEMENTS

Many thanks to Yann Stephan (SHOM) for helpful advices in underwater acoustic analysis and to Stephane Cordier D. Sc., Naval Architect and Hydrodynamics expert, for his review.

9. REFERENCES

1. ERBE, C, DUNLOP, R.A, DOLMAN, S.J, "Effect of Noise on Marine Mammals", Chapter 10 of Springer Handbook of Auditory Research, 2018

2. DE JONG, C.A.F., HARMSEN, J, BEKDEMIR, C, HULSKOTTE, J.H.J, TNO report - 2020 R11855 "Reduction of emissions and underwater radiated noise for the Belgian shipping sector", 2020

3. MCKENNA, M.F, ROSS, D, WIGGINS, S.M, HILDEBRAND, J.A "Underwater radiated noise from modern commercial ships", 2012

4. Wind Ship Association white paper "Wind Propulsion for Ships – Technologies ready to decarbonise maritime transport. An industrial opportunity in France", 2022

5. EMSA report EMSA/NEG/21/2020 "Sounds: status of underwater noise from shipping – Study on inventory of existing policy, research and impacts of continuous underwater noise in Europe", 2021

6. BEST, P, FERRARI, M, GLOTIN, H, POUPARD, M, "Livrable_2.4 Impact of Anthropogenic Sounds on Marine Mammals-2019", 2019

7. ARVESON, P.T, VENDITTIS, D.J, "Radiated noise characteristics of modern cargo ship", 1999

8. WENZ, G. M. "Acoustic Ambient Noise in the Ocean: Spectra and Sources", 1962

9. PUTLAND, R.L, MERCHANT, N.D., FARCAS, A, RADFORD, C.A, "Vessel noise cuts down communication space for vocalizing fish and marine mammals", 2018

10. AUTHORS BIOGRAPHY

Pierre Cordier holds the current position of CEO at XP SEA (France) with 30 years of experience in the maritime domain.
XP SEA provides engineering and project management services to the maritime world. In particular in naval architecture, marine engineering and offshore wind. **Pierre Cordier** managed the basic design of Grain de Sail new cargo sailing vessel, GDS2, 50 m long, in cooperation with other naval architects. GDS2 is being built at Piriou shipyard.
Pierre Cordier previously worked at Chantiers de l'Atlantique and D2M Engineering.

Claire Noël holds the current position of Chief Scientist Officer and founder of SEMANTIC T.S. (France). Engineer in Mechanics and Microtechnics with a Phd in underwater acoustics, she has been working since 1993 on several research projects in the domain of acoustical oceanography & acoustic propagation modeling in random ocea

PERFORMANCE PREDICTION OF WIND PROPULSION TECHNOLOGIES VIA CFD

Dr R Azcueta and **H Ward,** Cape Horn Engineering, UK

SUMMARY

This paper presents a wide range of CFD analyses, along with verification and validation processes which aim to showcase the potential of CFD to support the shipping industry in the adoption of novel wind propulsion technologies. The presented case studies are based on the SINTEF Ocean bulk carrier benchmark for which good quality model testing results are available. Open water propeller simulations, resistance tests at model and full scale, propulsion tests with virtual disk and with rotating propeller, and finally 6 DOF simulations for motor-sailing conditions are presented and discussed. The importance of validation and verification practices for EEXI/EEDI calculations, in particular when introducing ESDs and WASP devices to improve vessel performance are highlighted. The requirement of demonstrating the qualification of the CFD solution by performing numerical uncertainty analysis following the ITTC and IACS recommendations is described. Finally, an investigation into the uses of Machine Learning with CFD for wing trim optimisation is presented.

1. INTRODUCTION

Global efforts to reduce the environmental impact of the shipping industry have accelerated rapidly in recent years, with new measures by the International Maritime Organisation (IMO), such as the Energy Efficiency eXisting Ship Index (EEXI), and Energy Efficiency Design Index (EEDI) for new designs, coming into force in January 2023. From 2022 the IMO accepts Computational Fluid Dynamics (CFD) as a mean to determine the vessel reference speed needed in the EEXI/EEDI calculation. Prior to this, only towing tank tests were accepted with numerical simulations seen as a secondary option "under special circumstances".

Cape Horn Engineering have been lobbying for many years for CFD simulations to achieve the accredited status necessary to become the preferred method to demonstrate the ship performance. There are many reasons why running a CFD programme is superior to utilising towing tanks. Not only are there huge time and cost benefits, fundamental to any project, but CFD also benefits from running simulations in a one to one scale, removing any errors or uncertainties when applying scaling effects from model scale to full scale. Notably, CFD results can be validated directly with sea trial results.

Many ship owners are looking into improving fleet performance, for instance by installing their ships with Energy Saving Devises (ESDs) like asymmetric stern bulbs, nozzles, or fins in front of the propeller, fins in the propeller boss cap etc. Another solution which is gathering momentum is installing Wind Assisted Ship Propulsion (WASP) devices, such as solid wings or soft sails, suction wings, Flettner rotors, or kites. In the case of wind propulsion technologies, the IMO guidelines dictate that if more than one device is installed on deck the forces produced by the devices should consider their interactions. Cape Horn Engineering have specialised in computing such interaction models, making used of advanced machine learning techniques.

For the shipping industry to adopt new WASP technology, accurate CFD assessment of potential savings and of risks involved with the technology is required. It is important that this assessment be carried out by independent, technology agnostic CFD organisations, not only to be unbiased but also to be perceived as such. Cape Horn Engineering are working with several technology developers to help optimise their systems, either by running systematic shape and condition variations to be used in Velocity Prediction Programs or by running bespoke Six-Degrees-of-Freedom (6 DOF) simulations to validate the proposed solutions. For this, they have developed a simulation workflow to directly compare the efficiency of WASP devices which allows to determine the potential power saving in a more accurate manner than by using a VPP.

This paper describes the work currently undertaken by Cape Horn Engineering to demonstrate the benefits of using CFD for EEXI/EEDI calculations, in particular when introducing ESDs and WASP devices to improve vessel performance. The requirements of demonstrating the qualification of the CFD solution by performing numerical uncertainty analysis following the ITTC and IACS recommendations is described.

The SINTEC Ocean bulk carrier benchmark was used for this purpose. The rest of this paper is organised as follows. First, open water propeller simulation results are shown, followed by results for towing resistance simulations at model and full scale, and for propulsion simulations with the virtual disk and with rotating propeller approaches. Finally, the 6 DOF simulations of the full-scale vessel in motor-sailing conditions including the WASP devices is presented. Where possible the validation with model tests by SINTEF Ocean is presented. Furthermore, our standard workflow for Validation and Verification is described. Finally, the application of Machine Learning with CFD for wing trim optimisation is presented.

2. SOBC-1 BENCHMARK

2.1 INTRODUCTION

Cape Horn Engineering is a founding member of the International Collaboration Network in Marine Technology, initiated in 2021 by SINTEF Ocean in collaboration with other research organizations, universities, classification societies and selected industry companies. The main objective of the Network is to facilitate information sharing on research and development capabilities between its members.

One of the first scientific topics for collaborative R&D activities of the Network is the SINTEF Ocean Bulk Carrier (SOBC-1) benchmark study, [1]. The project is intended to serve as "a testbed and demonstrator - physical and numerical - for various energy efficiency solutions and innovative low and zero emission technologies". SINTEF Ocean conceived the vessel to be representative of medium-size, single-screw tank/bulk carriers for short to medium range shipping, with modern hull lines, controllable pitch propeller and high-efficiency rudder. Following the design phase, an extensive model test campaign including bare hull resistance as well as both conventional and wind assisted propulsion tests was completed in their towing tank.

Cape Horn Engineering have chosen to use this benchmark for further developing their WASP modelling primarily due to the quality of the validation data available and the opportunity to exchange knowhow with the research community involved in this project. The vessel main particulars and characteristics are presented in Table 1.

Table 1: SOBC-1 main particulars

Main particulars	Value	Dimension
Length between perpendiculars , Lpp	190	(m)
Breadth, B	32.2	(m)
Height, H	23.5	(m)
Draught design, T_DWL	11	(m)
Draught at ballast, T_BWL	7.7	(m)
Displacement, Δ	50.218	(tonnes)
Block coefficient, CB	0.726	(-)
Prismatic coefficient, CP	0.730	(-)
Operation speed range	6-15(17)	(knots)
Propeller Diameter, D	6.75	(m)
Propeller type	CPP, moderate skew	
Number of blades, Z	4	
Rudder type	Spade rudder with headbox, rectangular, high-efficiency profiles	

In the following sections, we will present the different types of simulations performed and, when possible, the validation with model tests results. All simulations presented in this paper were performed with the commercial CFD package STAR_CCM+ from Siemens Digital Industries in the latest version from mid 2022.

2.2 OPEN WATER CURVE SIMULATIONS

The aim of this task was to refine Cape Horn Engineering's internal best practices to efficiently compute the Open Water Propeller (OWP) curves. Several strategies to set up the simulations were implemented and compared among each other and with cavitation tunnel results from SINTEF Ocean. The different strategies and variations were:

- Moving reference frame with one blade only and periodic boundary conditions
- Moving reference frame with all four blades
- Rotating propeller
- Full scale and model scale
- Shaft upstream or downstream
- Polyhedral mesh versus trimmed-cells mesh in the propeller rotating region
- Mesh density study, with the aim of quantifying the numerical errors of a mesh of similar resolution to the ones used for propulsion simulations with the propeller in the behind the ship condition at full scale
- Mesh topology studies, to investigate where a finer mesh is needed, e.g., in the wake or at the leading edge of blades.

The OWP curves were computed for a pitch ratio of 0.97 at $r/R = 0.7$. All different methods gave consistent results within a few percent difference to each other. This gives flexibility to use the most efficient method depending on the aims of the project.

The comparison of the simulated results to the experimental OWP curves results obtained by SINTEF Ocean shows discrepancies of less than 2% for most advance ratios (J), except for the high value of J where the differences are 8%. CFD analysis performed by SINTEF Ocean's researchers show the same level of discrepancies at the higher advance ratio. Figure 1 compares the OWP curves obtain in the experiments with those obtained from CFD using the rotating propeller approach at full scale. The mesh used was of similar resolution than the one used for running the propulsion simulations with the full ship geometry and rotating propeller. The propeller rotates two degrees per time step. Table 2 shows the differences in percent for each advance ratio and the average differences.

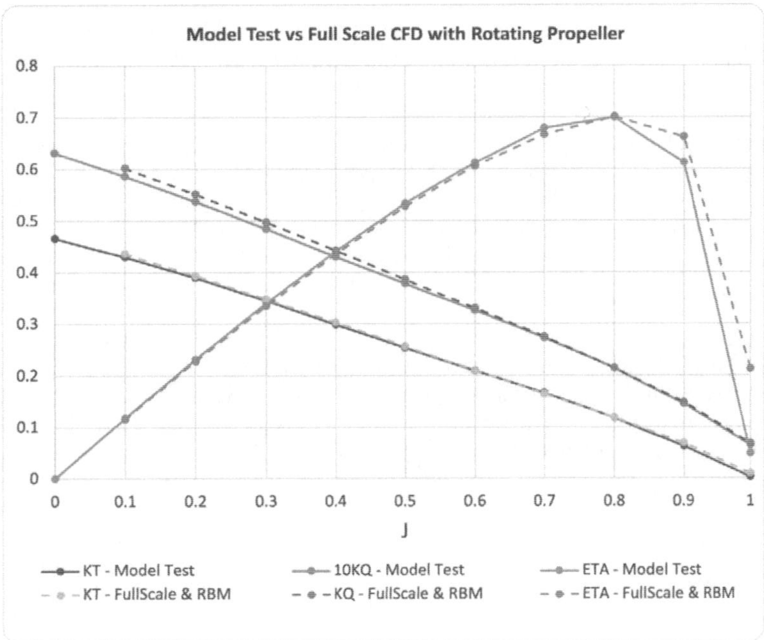

Figure 1: Open water propeller curves. Comparison CFD vs EFD

Table 2: Open water propeller curves. Deltas in %: $(CFD - EFD)/EFD$

J	KT	KQ	ETA	KT / J^2
0	-	-	-	-
0.1	1.42%	2.73%	-1.54%	1.47%
0.2	0.95%	2.61%	-1.60%	1.06%
0.3	0.90%	2.69%	-1.62%	0.83%
0.4	1.48%	2.79%	-1.22%	1.45%
0.5	1.15%	2.12%	-1.09%	1.05%
0.6	0.19%	1.23%	-1.05%	0.29%
0.7	-1.14%	0.74%	-1.77%	-0.92%
0.8	0.77%	0.00%	0.19%	0.12%
0.9	9.52%	2.08%	8.12%	10.30%
1.0	-	-	-	-
AVERAGE	1.69%	1.89%	-0.18%	1.74%

2.3 TOWING RESISTANCE SIMULATIONS

All CFD simulations presented in the following were performed for the vessel design speed of 15 knots at design water line with a draft of 11 m. Most simulations were carried out at full scale, however, to be able to validate one to one with towing tank measurements carried out at SINTEF Ocean, a set of simulations was also performed at 1:32 scale. The full-scale simulations follow our best practices and preferences for obtaining results directly for the real ship. They allow us to enter a discussion about the uncertainties in the extrapolation procedures of towing tanks. More on this below.

Several simulation parameters have been varied, such us the turbulence model (k-epsilon versus k-omega SST), the y+ value, the domain size, mesh resolution and time step size. The latter two variations will be discussed in the Verification and Validation section below.

For the full-scale simulations the y+ value was chosen to be on average 250. For the model-scale simulations three variations of y+ were investigate; with averages of 1, 4 and 40. The results of the three types of meshes were close, and only the ones for the y+=4 on average will be presented below. The results on the mesh with a y+=1 on average show worse convergence of the forces due to the higher cell aspect ratios. The mesh with a y+=40 on average was the exact same mesh used at full scale but scaled down by the model scale factor of 1:32.

As expected, there is a large variability of y+ along the hull and thus we employ the "All y+ Model" from STAR-CCM+ to better handle the areas where y+ lays in the buffer zone between 5 and 30. This formulation switches from the traditional wall-function approach to a low Reynolds number approach by means of a blending function that is a function of the Reynolds number based on wall distance. Figure 2 shows the y+ distribution for the model scale (with an average y+ of 4) and the full-scale simulations (with an average y+ of 250).

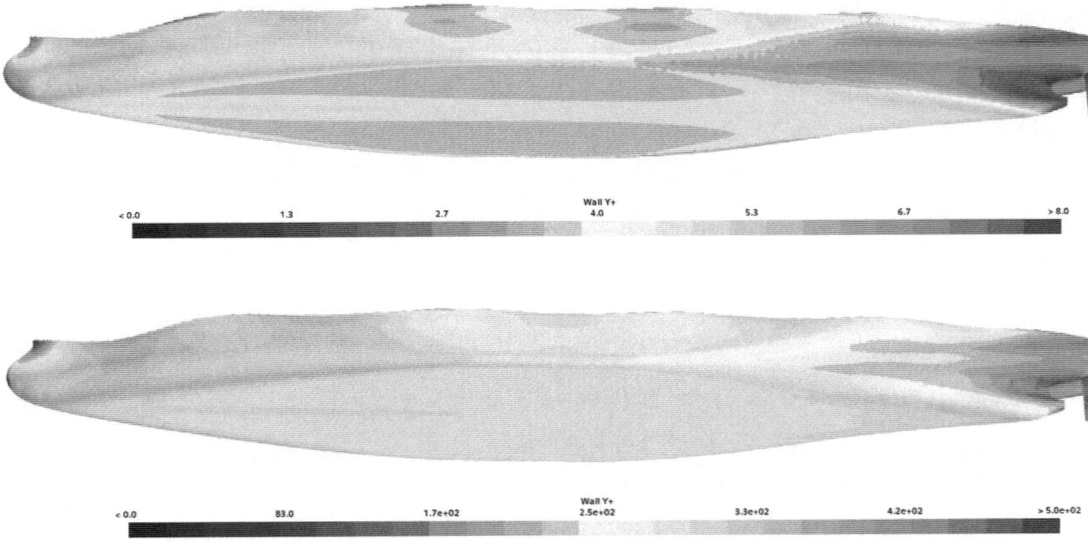

Figure 2: y+ distribution for Model Scale (above) and Full Scale (below)

For both the full-scale and the model-scale simulations the mesh count was around six million cells for half of the vessel (port side only). The wall clock time for a converged simulation was on average around seven hours on a 32-cores node (Intel Xeon Gold E5-6142 v5 CPU from 2017).

The best correlation with the measured total resistance in the tank was obtained with the k-epsilon model (0.05% difference), Table 3. Results with the k-omega SST model underpredicted the measurements by 4%. When comparing friction resistance from the simulations with the ITTC friction line, we see an underprediction with both turbulence models, with k-omega SST underpredicting by 4% more than k-epsilon. It must be noted that the values for the ITTC friction resistance in the table are computed using the wetted area and the wetted length obtained in the simulation, instead of the design wetted area and length as is used in towing tank testing. Values for dynamic sinkage and trim are also well predicted as can be seen in Table 3.

Table 3: Resistance tests – Model scale – Delta in %: (CFD-EFD)/EFD

	RTM at 15 kn		ITTC CFD	CFD Friction Resistance		Trim	Sinkage
EFD-SINTEF	30.118 N					-0.174 deg	-0.008 m
CFD - k-epsilon	30.132 N	0.0 %	24.497 N	23.887 N	-2.5 %	-0.161 deg	-0.007 m
CFD - k-omega SST	28.916 N	-4.0 %	24.497 N	22.900 N	-6.5 %	-0.164 deg	-0.007 m

At full scale we see a different picture. We compare here with results given by SINTEF Ocean following their own towing tank extrapolation procedures, Table 4. The experimental results are corrected to account for the bilge keels. The CFD results are without including the bilge keels. Sinkage and Trim are again well predicted. This time the best correlation with the tank extrapolated results were obtained with k-omega SST. We again see k-omega SST giving lower friction resistance than k-epsilon as was the case at model scale, around 3%.

It is interesting to note that at model scale both turbulence models using the low y+ approach were giving smaller values than ITTC friction line, while at full scale and with a y+ of around 250 both give higher values than the friction line results.

Table 4: Resistance tests – Full scale – Delta in %: (CFD-EFD)/EFD

	RTM at 15 kn		ITTC_CFD	CFD Friction Resistance		Trim	Sinkage
EFD-SINTEF	557690 N					-0.174 deg	-0.225 m
CFD - k-epsilon	578562 N	3.7 %	389754 N	414764 N	6.4 %	-0.177 deg	-0.273 m
CFD - k-omega SST	560879 N	0.6 %	389714 N	402827 N	3.4 %	-0.173 deg	-0.272 m

From our own experience validating CFD predictions directly with sea trials for EEXI reference velocity calculations, we are confident that using the k-epsilon turbulence model we can obtain correlation factors of less than 2% for delivered power and propeller RPM. It remains thus to be determined why in this case we see larger differences to the tank extrapolated results. We argue that this is either to the inherent inaccuracies of the tank extrapolation procedures and/or inaccuracies of the friction resistance for large y+ values. Using a low y+ value of around 4 on average as at model scale would be too difficult since the mesh count would explode exponentially if the cell aspect ratio were kept within reasonable bounds.

2.4 VERIFICATION AND VALIDATION (V&V)

We have followed our Best Practices Guidelines (BPG) developed for calculating EEXI reference velocity to conduct V&V at model and full scale. The procedures follow the International Towing Tank Conference (ITTC) [2-3] and the International Association of Classification Societies (IACS) guidelines and recommendations [4]. The ITTC guidelines set out the requirements to setup the BPG and the Demonstration of Quality of each organization. Furthermore, they set out the requirements to determine a validated correlation factor (comparison error) to sea trials or model tests results. The numerical uncertainty must be determined by systematically varying the spatial discretisation (mesh density) and the temporal discretisation (time step used to advance the simulations). This is the process known as the Verification, whereas the Validation is the direct comparison to the measurement results to determine the correlation factor.

At model scale and using the k-epsilon turbulence model results, the correlation factor is 0% as presented previously and the numerical uncertainty, which is the combination of the spatial and temporal uncertainty, is less than 1%. Table 5 shows the spatial, temporal, numerical uncertainty, the measurement uncertainty, and the validation uncertainty. Finally, the correlation factor is labelled as VALID according to the ITTC guidelines as the comparison error is smaller than the validation uncertainty.

Figure 3 shows graphically the triplet study that was completed to quantify the uncertainty for the spatial and temporal discretisation as in Table 5. The variable presented in this case is the total drag, but the same process can be applied to dynamic sinkage or trim or delivered power or propeller RPM for self-propulsion simulations. The triplets are obtained by coarsening the mesh two times by a factor of $\sqrt{2}$ or increasing the time step two times by $\sqrt{2}$. This results in 5 different simulations. The most accurate one, finest mesh and smallest time step, is the one that is compared to the measurements results. In case the convergence in space and time shows an asymptotic behaviour (convergence ratio $0<R<1$) these five simulations are enough. In case of oscillatory convergence behaviour ($R<0$) one more simulation is required to calculate uncertainty.

Table 5: Uncertainty analysis at model scale

	Total Resistance
Spatial Uncertainty	0.88%
Temporal Uncertainty	0.01%
Numerical Uncertainty	0.88%
Experimental Uncertainty	0.00%
Validation Uncertainty	0.88%
Comparison Error	0.05%
Validated	YES

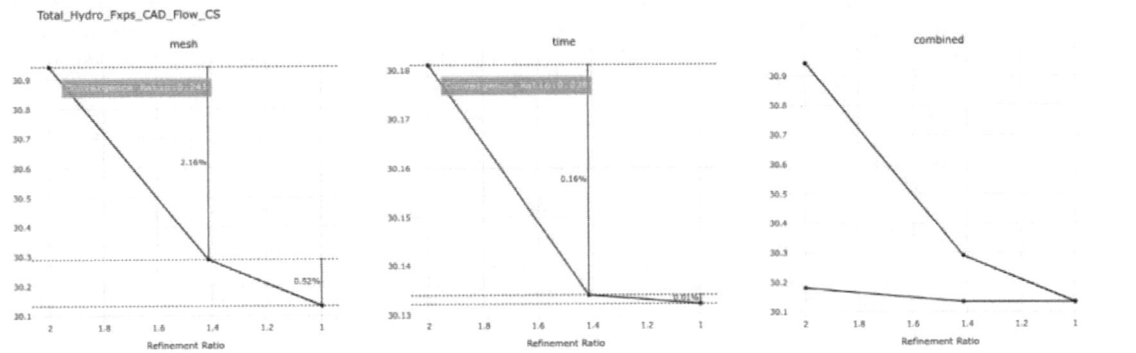

Figure 3: Verification - Triplet study for uncertainty analysis (model scale)

The validation for drag at model scale is presented graphically in Figure 4 left. In this image the blue point is the SINTEF Ocean towing tank measurement, and the blue error bar is the towing tank uncertainty which has been considered zero but a value of 1% could have been more appropriate. The red point is the CFD value, and the red error bar is the numerical uncertainty. The difference between the two dotted lines is the comparison error (or correlation factor) and the purple error bar is the validation uncertainty. As the purple error bar is larger than the difference in the dashed lines the simulations are considered validated, according to the ITTC guidelines.

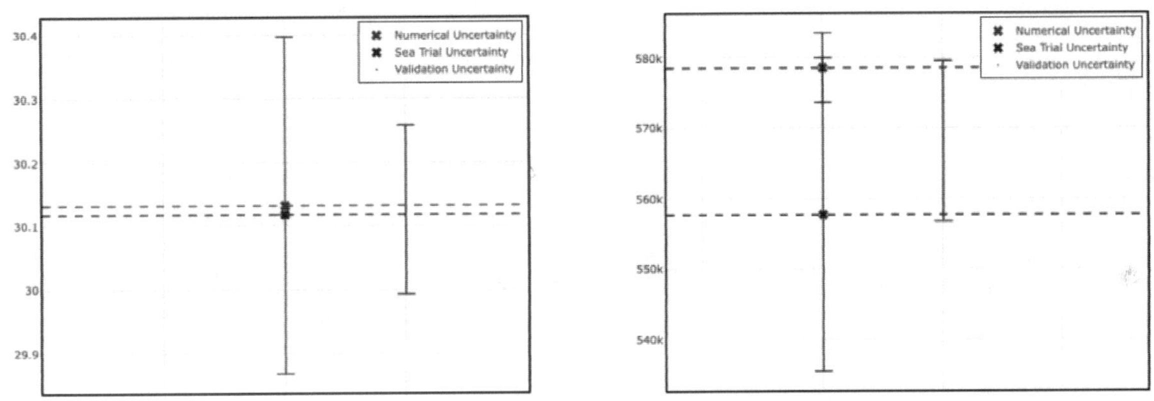

Figure 4: Validation - Graphical representation of correlation factor (left model scale, right full scale)

In a similar manner, the numerical uncertainty for the full-scale simulations has been assessed and is presented in Table 6 for total drag, for the k-epsilon turbulence model results. As for the model-scale simulations this uncertainty is less than 1%.

Table 6: Uncertainty analysis at full scale (k-epsilon)

	Total Resistance
Spatial Uncertainty	0.85%
Temporal Uncertainty	0.11%
Numerical Uncertainty	0.86%
Experimental Uncertainty	4.0%
Validation Uncertainty	3.95%
Comparison Error	3.74%
Validated	YES

For the correlation factor obtained with the k-epsilon turbulence model to be valid, an experimental uncertainty of around 4% would be needed. However, using the k-omega SST turbulence model results, an experimental uncertainty of less than 1% would give a valid correlation factor. Figure 4 right shows a graphical representation of a valid correlation factor at full scale for the k-omega SST results and 0.5% experimental uncertainty.

It is open for debate what is the uncertainty of the extrapolation procedure of each towing tank.

2.5 SELF PROPULSION SIMULATIONS

For running self-propulsion simulations with both Virtual Disk (VD) and a geometrically modelled rotating propeller (RP), both sides of the vessel, must be considered. This is due to the asymmetrical nature of the flow, with the propeller located on the centreline of the SOBC-1 vessel (for twin-screw ships the centre line plane symmetry approach could be used). Thus, the mesh size at least doubles compared to the resistance simulations. On top of that, the VD requires a refined mesh around the propeller plane, and the modelled rotating propeller requires a new region with a high-resolution mesh around the propeller. For the simulations presented here, the VD and the RP meshes have a cell count of approximately 13 and 16 million respectively.

In addition to the cell count differences, running with VD is still a very much lightweight simulation compared with RP, due to the timestep requirements. The time step size can be kept much larger for VD than with the rotating propeller. We used a time step size of approximately 0.05 seconds for the VD and 0.004 seconds for the rotating propeller, the latter corresponding to two degrees of propeller rotation per time step for the target RPM. The very small timestep is required to correctly capture the forces imparted on the flow by the moving blades.

For both the VD and RP simulations the mesh used in the far field and in the region around the hull was the same as for the resistance simulations, except in the proximity of the propeller. CPU times of around 11 hours for VD and 34 hours for RP were required to obtain converged solutions, using the 32-cores node described earlier.

The open water propeller curves needed to activate the VD model were taken from the CFD simulations using the rotating propeller at full scale already presented in Section 2.2. The pitch ratio selected for the design speed at design water line was again 0.79.

The agreement with SINTEF Ocean extrapolated prediction is acceptable. The results presented in Table 7 are with the k-epsilon turbulence model. It could be assumed that k-omega SST would shift the results lower by 3%, in which case the RP would be closer to the extrapolated prediction. At 15 knots vessel speed the split between friction and pressure drag is approximately 50-50%.

Table 7: Self-propulsion simulation compared to extrapolated tank results – Delta in %: (CFD-EFD)/EFD

	RPM		Delivered Power		Trim	Sinkage
EFD-SINTEF	77.8		5796 kW		-0.174 deg	-0.225 m
CFD – k-epsilon – Virtual Disk	77.8	0.0 %	5958 kW	2.8 %	-0.171 deg	-0.280 m
CFD – k-omega SST – Rotating Propeller	81.3	4.5 %	6376 kW	10.0 %	-0.175 deg	-0.278 m

However, we consider our results obtained with the rotating propeller model to be more accurate than with the virtual disk model, irrespectively of the comparison with the model test extrapolation. We base this assumption on V&V conducted for other similar vessels using sea trial results.

Thus, we calculate the differences in results between the VD and RP simulations based on the RP results, Table 8. The VD model is giving 4.3 % lower RPM and 2.3 % lower propeller torque. As delivered power is the product of those two values, the difference is larger, 6.6%. Dynamic sinkage and trim are again quite close.

Table 8: Self-propulsion simulation – Comparison VD vs RP – Delta in %: (VD-RP)/RP

	RPM	Torque	Delivered Power	Thrust	Trim	Sinkage
CFD – k-epsilon – Rotating Propeller	81.3	748 KNm	6376 kW	863 kN	-0.175 deg	-0.278 m
CFD – k-epsilon – Virtual Disk	77.8	731 KNm	5958 kW	674 kN	-0.171 deg	-0.280 m
Delta	-4.3%	-2.3 %	-6.6 %	-21.9 %	-2.4 %	0.7 %

Propeller thrust is a function of total hull drag, including the rudder but of course not the propeller. The difference in propeller thrust of 22 % is the same as the difference in total hull drag. Thrust and hull drag cannot be compared directly from the VD and RP simulations, since in the VD simulation the surface closing the propeller stern tube is reducing the total hull resistance with a hydrostatic pressure acting on it, while that surface is not present in the RP simulation, Figure 5. The propeller in the RP simulation is connected to the hull without a gap, and the hydrostatic pressure acting on the propeller boss cap is computed into the propeller thrust value. Thus, this hydrostatic pressure recovery is computed in one case on the hull and in another on the propeller. If we correct for the missing hydrostatic pressure on that surface the total hull drag difference from both simulations is less than 2 %.

When computing thrust deduction from bare hull resistance simulations and VD self-propulsion simulations, special care should be paid to this fact.

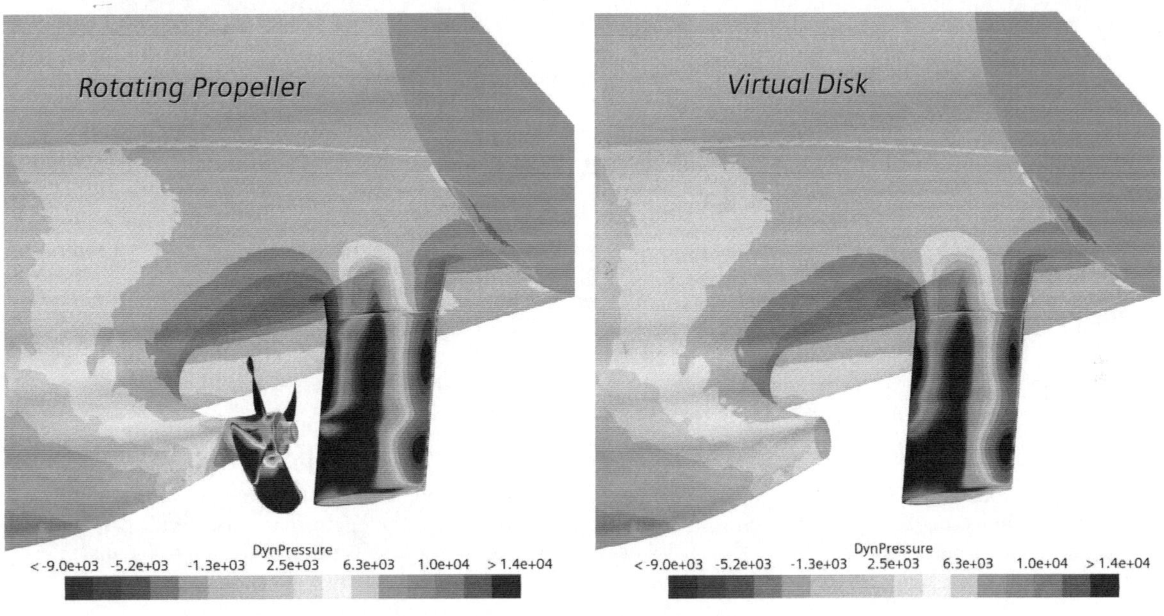

Figure 5: Dynamic pressure distribution for rotating propeller and virtual disk simulations

2.6 6 DOF SIMULATIONS INCLUDING WASP DEVICES

We will now describe the highly complex simulations developed by Cape Horn Engineering to model as realistically as possible a motor-sailing conditions with wind assisted propulsion. The high-fidelity simulations consider the most important effects of adding wings or any other type of WASP devices to a vessel. Both the water flow experienced by the hull at a given vessel speed and the air flow experienced by the wings are modelled simultaneously in a single simulation.

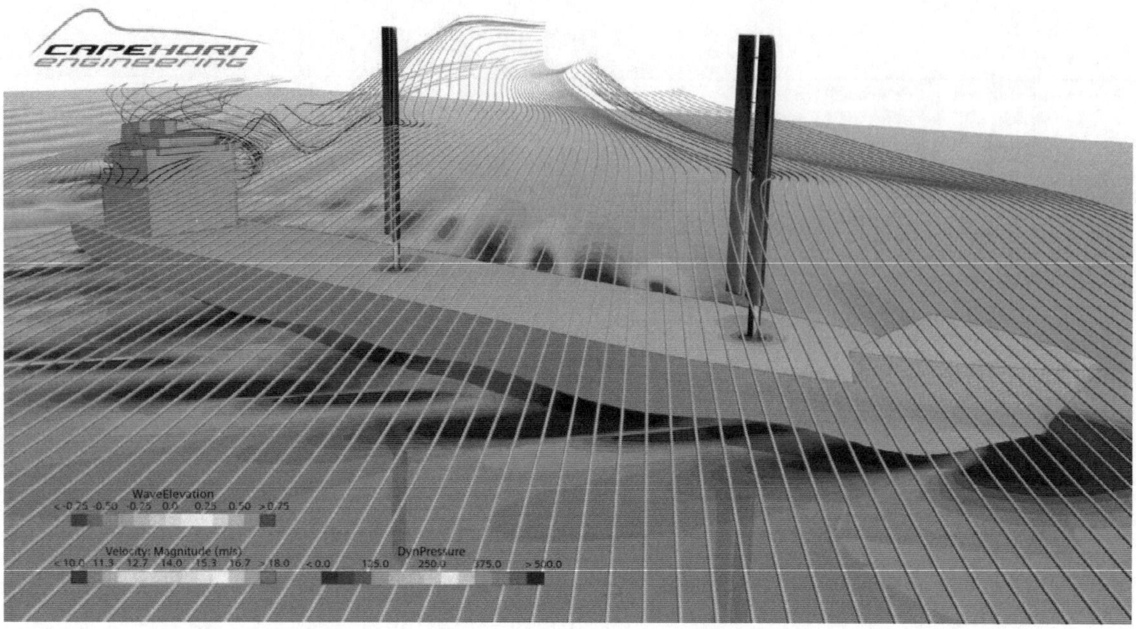

Figure 6: CFD image of bulk carrier with wings

In the simulations the vessel is sailing at constant speed, with either a virtual disk or a rotating propeller modelling the propulsion. In the cases presented here the VD approach was used which requires as input the open water propeller characteristics of the installed propeller.

The wings are mounted on deck, and aerodynamic details include the hull top sides, deck, hatches, and superstructure. In this paper we are presenting the results for the SOBC-1 vessel with two wings on deck. The wings have been modelled with simple geometries, using NACA profiles for the two elements, a main wing, and a flap. They were positioned on deck without any consideration for structure or hatches. The SOBC-1 geometry was modified closing the deck adding a generic superstructure. The underwater shape of hull and rudder is the same as presented in the previous sections for the resistance and propulsion simulations. Figure 6 shows a representation of the vessel with two wings showing CFD results.

The wind conditions above the water surface are modelled with an accurate wind profile, considering the atmospheric boundary layer wind gradient. This results in a variable apparent wind speed and direction at different heights, due to the combination of a moving ship and a wind strength that varies with height, something that sailors are very much aware of so that they need to twist the sails towards the top of the mast. In some conditions, this change of wind direction with height can be as large as 30 degrees.

Depending on the vessel resistance and the amount of thrust generated by the WASP devices, the thrust produced by the propeller balances the degree-of-freedom in the direction of travel. This thrust is used as an input to the VD model loaded with the open water curves of the propeller. The model then determines the propeller RPM and propeller torque and therefore the delivered power.

The simulation considers the deformation of the free surface, or waves generated by the vessel, the dynamic sinkage and trim, the drift or leeway angle, the heel angle and the rudder angle to keep a constant course. The wind forces on the hull, superstructure and wings induce drift and heel angles, with the rudder angle changing during the simulation via a PID controller, balancing the yaw moment of the whole system. The rudder rotation is applied using an overset mesh for the rudder region. Thus, in total all 6 degrees-of-freedom are considered. The simulations are repeated with the vessel without the wings to have an exact baseline for comparison and each simulation is stopped once they have converged to a quasi-steady solution. Then, comparing the case of the vessel with and without the wings for the same wind conditions, the reduction in delivered power to the propeller can be assessed, and with that the engine fuel and emissions reduction.

Table 9 below shows the first set of conditions investigated. A vessel speed of 15 knots, two True Wind Speeds of 15 and 25 knots each at three True Wind Angles of 70, 100 and 130 degrees were simulated. The trim angles for the wings in this first set of simulations were chosen without previous knowledge of optimum settings, just with the main wing element at 10 degrees to the apparent wind angle at 10 metres height, and the flap angle at 20 degrees wrt. the main element. The wind sheeting angles shown in the table are wrt. the vessel centreline and thus depending on the wind triangle. Simulations have the wind coming from the starboard side of the vessel.

Table 9: Motor-sailing conditions

	Vessel Speed [kn]	TWS [kn]	TWA [deg]	Wing Sheeting Angle [deg]	Flap Angle [deg]
Case 01	15	15	070	25.0	20.0
Case 02	15	15	100	40.0	20.0
Case 03	15	15	130	55.0	20.0
Case 04	15	25	070	34.9	20.0
Case 05	15	25	100	56.6	20.0
Case 06	15	25	130	83.2	20.0

Figure 7 gives the delivered power results from the simulations, shown in watts on the left graph, and as power saving percentage on the right. For the six conditions investigated, the minimum savings are 6% and the maximum 29%, with an average saving of 18%. The maximum saving occurs for the stronger wind of 25 knots from the side, and the minimum for the lighter wind from the aft quarter, since the apparent wind speed is most reduced in this case. An 18% power reduction is a considerable amount, and this average could improve even further if the wings are trimmed optimally and if a more adequate propeller is used for motor sailing conditions. Ideally, to have the greatest benefit from wind assisted propulsion the vessel would be fitted with a controllable pitch propeller to always be operating at the optimum efficiency. A hybrid power plant would also be ideal in conjunction with wind assisted technologies. At higher wind speed the vessel could even be sailing with engines just idling or switched off.

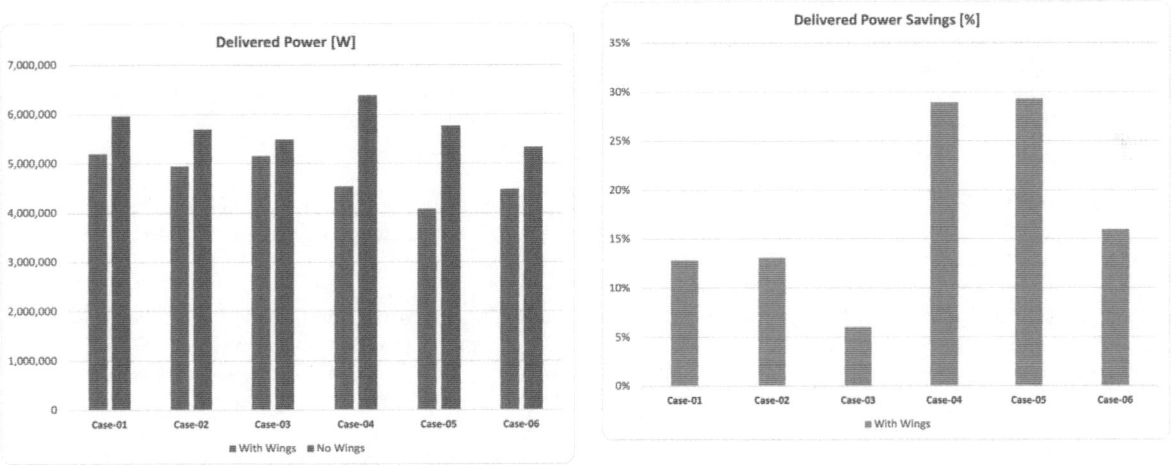

Figure 7: Delivered power and power savings

Another interesting aspect of the results is that the heel and leeway angles are very modest, Figure 8. The rudder angles needed to maintain a constant course get larger for the stronger winds from the bow quarter. However, by trimming the wings differently forward to back, the yaw moment could be reduced and an optimum in efficiency found. This remains an objective for follow up investigations.

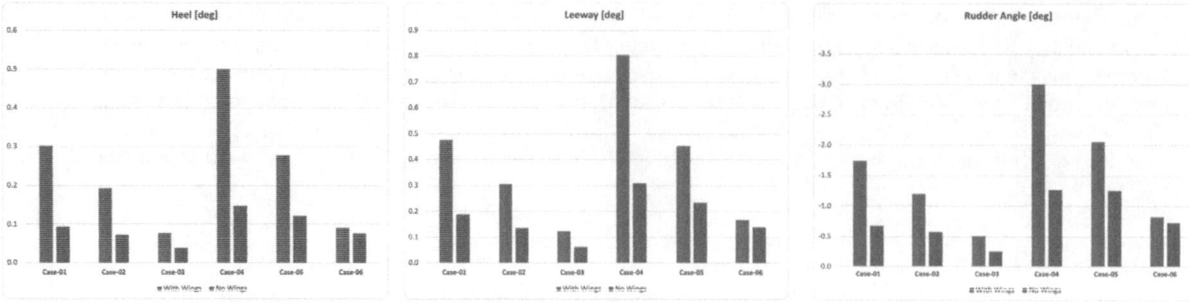

Figure 8: Heel, leeway, and rudder angles

2.7 OPTIMISATION OF WINGS' TRIM

The performance of WASP devices is highly dependent on their operating conditions. The trim angles used to evaluate the power savings in the 6-DOF simulations were determined from experience only and thus they were a rough estimate. An investigation was conducted into the use of lightweight (relative to other simulations discussed) aerodynamic CFD simulations in combination with machine learning techniques to optimise the savings gained from the wings.

The design of experiments (DOE) with surrogate modelling approach is a powerful tool for optimizing the performance of WASP devices. In this study, a DOE was conducted to systematically vary the design parameters of single phase, aerodynamic simulations of two wings. The simulations were performed for each design point to obtain the performance metrics of interest, most importantly the total thrust. The results presented are for the first-stage simplified analysis whereby we include the wing trim angle, and the flap angle for the forward and aft wings, giving four variables in total. We also kept the design space large to avoid our prior knowledge influencing the study, so many of the simulations show poor performance and actually have a negative aerodynamic thrust.

To setup the DOE, the design space must be assessed to determine the number and ranges of variables to be used. A space filling algorithm distributes the simulation points throughout the design space, such that no two simulations repeat a variable value, but the space between values is not too large as this would make it difficult for the surrogate model to be built. Furthermore, the DOE is extensible, meaning it is possible to add more points to fill more of the design space should a higher resolution be needed.

The results of the DOE were then used to train a surrogate model, which can be used to accurately predict the performance of the WASP device for any given combination of design parameters, without the need for expensive and time-consuming CFD simulations.

Once the surrogate model has been created, a global optimisation algorithm was applied to find the design point that maximizes the performance of the WASP device. The optimisation algorithm can quickly find optimum points as the surrogate model takes milliseconds to interrogate, and can include any number of constraints, or different targets. The optimisation results were then verified by performing CFD simulations.

The main focus of this study was to assess the performance of different types of surrogate models. A range of modelling methods were tested, including the relatively simple Radial Basis Function Interpolator, Kernel Ridge Regression (KRR), and also a Neural Network (Multi-layer Perceptron). It was found that KRR provided the most reliably high accuracy results. Many variations of the surrogate model were tested, with common machine learning practices employed, including hyperparameter tuning. This tuning created subsets of the data to evaluate an R^2 performance metric when different settings were used in the model. The accuracy of a kernel ridge regression model depends on the number of input samples. The model was tuned to find the best parameters for each size of the training data set. With more than 60 simulations, we get a prediction score R^2 of over 95. Above 68 input points the best surrogate models all had the same parameters. The training score is that achieved by the model during the testing phase, and the testing score is measured on a set of validation points that are not included in the training set.

With the trained surrogate model, we can find the optimum trim variables for the two wings and feed these values into a new 6-DOF simulation. From Table 10 below we can see that the initial guess was quite good, with the main difference that the whole assembly need to be 'trimmed in' more, i.e. rotated further wrt. apparent wind angle. The 'camber' or flap angles of each wing was almost spot on. There is also a small but noticeable difference between the forward and aft wings. This is due to all the interactions between the wings, and with the deck, hull, and the superstructure. In addition to the goal of maximum power savings, IMO guidelines dictate that interactions between WASP devices must be accounted for in the EEXI calculations if there is to be more than one device installed. This guideline is inherently adhered to with this optimisation technique, and we can see that interactions are an important factor to consider.

Table 10: Motor-sailing conditions

	Forward Wing Angle	Forward Flap Angle	Aft Wing Angle	Aft Flap Angle
First guess	10.0	20.0	10.0	20.0
Optimised	14.5	20.2	15.4	20.3

With the new 6-DOF simulation run, a further 4% in power reduction is observed. This value could have been much higher had the initial trim guess not been reasonably close to the optimal trims. The results of this study demonstrate that the DOE and surrogate modelling approach is an effective and efficient method for predicting the performance of WASP devices. By using this approach, the design space can be explored more thoroughly, and the optimal setup of the wings can be found with a high degree of accuracy. This will help in the development of more efficient and effective WASP propulsion systems for ships.

3. FURTHER WORK

The towing tank measurements with WASP devices was conducted at SINTEF Ocean in a different setup to the simulations just described. In the tank the investigations were run using a hybrid approach where the thrust and side force created by the WASP devices, in this case 4 Flettner rotors, were estimated beforehand and applied to the free running model via wires and actuators. No interaction between the rotors and between rotors and the hull was considered. The model did not have a deck and superstructure. Only rotors were modelled in this way but no wings.

It would be interesting in the future to have a set of towing tank measurements using a similar setup as in the simulations for validation purposes. A special tank facility with fans to create wind would be needed. This could be a good topic for a Research and Development programme in the future.

4. ACKNOWLEDGEMENTS

Special thanks are owed to Vladimir Krasilnikov from SINTEF Ocean for providing the benchmark data for the SOBC-1 vessel.

5. REFERENCES

1. RAMBECH, H.J., 'Calm Water Performance Tests, SOBC-1 (SINTEF Ocean Bulk Carrier no. 1)', *SINTEF Report, Project No. 302006033-09*, 2021.

2. 'Quality Assurance in Ship CFD Application', ITTC 7.5-03-01-02, 2021.

3. 'Uncertainty Analysis in CFD Verification and Validation Methodology and Procedures', ITTC 7.5-03-01-01, 2008

4. 'Guidelines on Numerical Calculations for the purpose of deriving the Vref in the framework of the EEXI Regulation', IACS No. 173, 2022.

6. AUTHORS BIOGRAPHY

Dr Rodrigo Azcueta is the Managing Director and founder of Cape Horn Engineering. He holds a degree of Doctor of Naval Engineering from the University of Hamburg where he specialised in Computational Fluid Dynamics applied to the maritime industry. He was head of the CFD groups during 4 America's Cup cycles and participated in the design of 3 winning Volvo Ocean Race campaigns. His company has broadened its customers base and offered services to the wider maritime and renewable energy sectors.

Hugh Ward is senior CFD Engineer at Cape Horn Engineering. He holds a Masters degree in Mechanical Engineering from the University of Southampton and has been with Cape Horn for nearly 4 years. He is responsible for the R&D development at the company.

WISP2 PROJECT ON WIND PROPULSION PERFORMANCE PREDICTION METHODS AND MANOEUVRING

R Eggers and **A S Kisjes,** Maritime Research Institute Netherlands (MARIN), The Netherlands

SUMMARY

MARIN, together with American Bureau of Shipping (ABS), started the second WiSP Joint Industry Project which has the objective to expand the knowledge base on predictions with wind propulsion and to propose changes and new standards for better rules and regulations. The present paper highlights some developments considering performance prediction before shifting to the main focus: manoeuvring. The methods and results presented in this paper are considered for later submissions to ITTC, IMO MEPC and potentially IMO MSC.

1. INTRODUCTION

MARIN became active with the resurgence of wind propulsion in commercial shipping in 2013 with background research, involvement in EU projects and organising our BlueWeek (https://blueforum.org). At that time, there were already some (potential) suppliers of wind propulsion systems active and a few design offices and developers were proposing complete ship concepts with wind propulsion. Skysails and E-ship were ahead of their time with products and one ship at operation. What was prevalent at that time is that wind propulsion was hardly acknowledged by the shipping industry. Whereas there were a lot of reasons for that, a distinct one was that technical information, more specifically promised fuel savings, was often not trusted or understood. Saving predictions were presented at conferences with a wide variety of methods and assumptions, often not documented. This may partially be understandable, because except for the EEDI guidelines (MEPC1./Circ.816 at that time) there was hardly any guidance. The same was valid for statutory and class regulations. This gap was documented in a report by CE Delft for the European Commission [1] a few years later. On this premise, the WiSP (phase 1) Joint Industry Project (JIP) started, initiated by MARIN and ABS. The project included case studies with performance predictions and gap analyses on rules and regulations. Based on the findings MARIN participated in the updated of the EEDI & EEXI guidelines in MEPC.1/Circ896 [2] through delegates Royal Institution of Naval Architects and Comoros.

The momentum on wind propulsion is now rather different. Increased ambition in the industry itself as well as adapted and new regulations have now clearly put shipping on a path towards decarbonisation. Wind propulsion is now often seen as one of the credible solutions. A small selection of trial and monitoring projects (e.g. Hurford [3] and [4]) have proven the savings. The guidelines for EEDI and EEXI were improved and most class societies came with their own wind propulsion specific guidelines and rules. So, there is progress. But there is still much to do on topics not yet addressed and improvement of existing standards, rules and regulations to equitably assess (and improve) wind propulsion performance, such as:

- Further improvement of MEPC.1/Cir.896 with guidelines on EEDI / EEXI predictions with wind propulsion
- Standards for predictions to facilitate investment decisions (higher accuracy) and standards for predictions for larger proportions of wind propulsion
- Standards for sea trials and monitoring
- Further adjustment of class rules and statutory regulations suitable for wind propulsion

Items 1, 2 and 4 have been taken op in the follow-up of WiSP, WiSP2, as described in this paper.

2. THE WiSP2 PROJECT

WiSP2 (https://www.marin.nl/en/jips/wisp-2) started in 2021 and has since attracted 23 participating organisations. Most organisations are suppliers of wind propulsion systems. WiSP2 also has a design office, technology research centres, consultancies, a flag, and class societies. MARIN and ABS are leading the project. WiSP2 is a Joint Industry Project, which means the funding is primarily brought together by the participants. As it doesn't operate under any specific thematic subsidy program there is the freedom to decide scope with very little bureaucratic overhead. Decisions in the project are confirmed in a democratic manner.

2.1 OBJECTIVES

WiSP2 aims to help further decrease barriers for increased uptake of wind propulsion by:

- Improving the knowledge base for performance predictions

- Improve rules, regulations and standards
- Present a showcase for the potential of wind propulsion

Improving rules, regulations and standards is something that of course cannot be implemented directly inside the project. The focus is rather on doing (case) studies. Recommendations are composed based on results from these case studies. These recommendations are also shared outside the project for possible implementation.

2.2 SCOPE

WiSP2 includes the following work packages:

- Case studies with performance predictions where various predictions are compared to establish suitable methods
- Manoeuvring and seakeeping predictions, first compared to experiments
- Validation of performance predictions against new monitoring data
- Update of recommended methods from WiSP(1)
- Delivery of simple prediction tool (https://blueroute.application.marin.nl/ships)
- Inventory and gap analysis of rules and regulations

3. PERFORMANCE PREDICTIONS

The focus in this paper is on manoeuvring as described in the following sections. However, of first interest with wind propulsion is of course the fuel and emission performance and savings compared to a reference ship without wind propulsion.

3.1 WiSP(1) methods

Methods for performance prediction were already compared in WiSP(1) and now in WiSP2 with the purpose to identify a good compromise between calculation or testing effort versus accuracy. The best compromise varies depending on the purpose of the performance prediction. Low accuracy is likely acceptable in early design explorations whereas high accuracy is likely required in advanced stages, e.g. when signing (build) contracts. In this respect WiSP(1) mainly focussed on an intermediate design stage where there is still some variation, but not yet the final verification for contract signing. Also, applicability in EEDI & EEXI prediction was a focus. The trade-off also depends on the relative contribution of wind propulsion. For instance, when heeling moment and side force is limited, it may be acceptable to neglect the influence of leeway, yaw balance and heel as it doesn't measurably affect performance. But that is likely not the case for wind propulsion systems with larger capacity. In fact, as illustrated in Figure 1, some guidance was provided for various topics. Tier1 in this scheme considers a rather simplistic calculation as a baseline. Depending on specific thresholds it was suggested to increase the fidelity level of the calculation. The impact of choices were established on in detailed case studies. The overall comparison between a calculation fully at

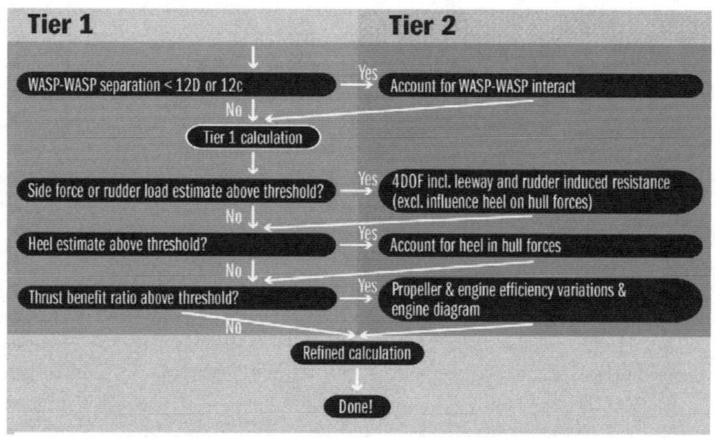

Figure 1: WiSP(1) Selection of fidelity level per topic

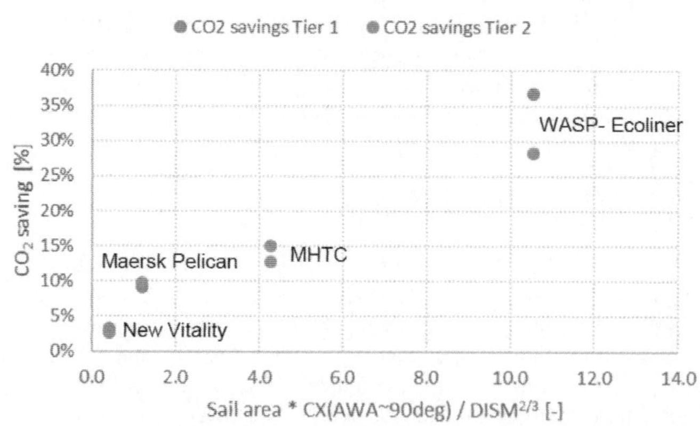

Figure 2: WiSP(1) CO_2 emission savings for Tier1 and Tier2

Tier1 or fully at Tier2 level is illustrated in Figure 2. This shows the main cases that were evaluated in WiSP(1). On the horizontal axis the ships are compared based on relative size of installation using the thrust coefficient at an apparent wind angle of 90° multiplied by the sail area and divided by the ship's displacement to the power 2/3. Then, this yields to a dimensionless number. The vertical axis gives the CO_2 emission saving in percentage relative to a reference ship without wind propulsion. It can clearly be seen that the gap between a prediction according Tier1 and Tier2 opens up as the relative size or saving of the installation increases.

3.2 WISP2 PREDICTIONS WITH WASP-ECOLINER

Additional sensitivity studies are done in WiSP2 to complement the knowledge obtained earlier. This includes a variety of topics. For performance prediction the focus is on the wind statistics as now specified in MEPC.1/Circ.896 for EEDI & EEXI, voyage simulation and the impact of maximising the wind propulsion contribution with the reference modelling methods used in the project.

The case as presented in this paper on performance prediction considers the WASP-Ecoliner concept [4] as illustrated in Figure 1 as developed by Dykstra Naval Architects and re-used with some amended/complemented specifications in WiSP(1) and WiSP2. The ship design is equipped with a diesel-direct propulsion installation. For aerodynamic modelling, wind tunnel measurements are used as generated in the NSR SAIL project [5]. Hydrodynamic performance is modelled using steady RANS computations with MARIN's CFD tool ReFRESCO. The propeller performance is taken from using the Propeller C-Series [6]. Table 1 provides the specific ship details used in the computations.

Table 1: Main particulars of WASP-Ecoliner

DESIGNATION	SYMBOL	MAGNITUDE	UNIT
Length between perpendiculars	L_{PP}	138.0	m
Breadth	B	18.0	m
Draught	T	6.5	m
Displacement mass in seawater	Δ_1	11916.3	t
Transverse metacentric height	GMt_{WET}	1.65	m
Reference ship speed	V_S	12	kn
Total sail area	A_{ref}	3859	m^2
Reference mast height	s_{ref}	60.5	m
Number of masts	-	3	-

Figure 3: Illustration of WASP-Ecoliner
(simulated ship has 3 masts)

3.2 (a) WIND STATISTICS IN MEPC.1/CIRC.896

Since the update of the EEDI/EEXI standards in 2021, the methods for selection of wind statistics have changed. From the global wind statistics now only the 50% wind conditions with best wind propulsion thrust need to be considered. The idea behind this change is that ships with wind propulsion are likely to sail in areas with more favourable wind and also that they may be use routing to guide them to better wind. The impact of this change is checked for various cases. Figure 4 shows it for WASP-Ecoliner. The fraction of wind conditions used in the overall performance prediction is shown on the horizontal scale. The green and red lines show the dimensionful CO_2 performance according the EEDI/EEXI metric for the ship with and without wind propulsion respectively. It is shown for modelling methods as documented in MEPC.1/Circ.896 itself as well as for more advanced modelling according Tier2 as defined in WiSP2. The black lines shows the saving. Firstly, the results here show again that the Circ.1/MEPC.986, which is similar to Tier1 in WiSP overestimates the savings for a design like the WASP-Ecoliner. The results also indicate that selecting 50% of the wind conditions as in MEPC.1/Circ.896 constitutes a very large improvement in performance as savings increase from about 37% (at Wind Select = 1) till 69% for the simple modelling in MEPC.1/Circ.896. This is almost a doubling of savings.

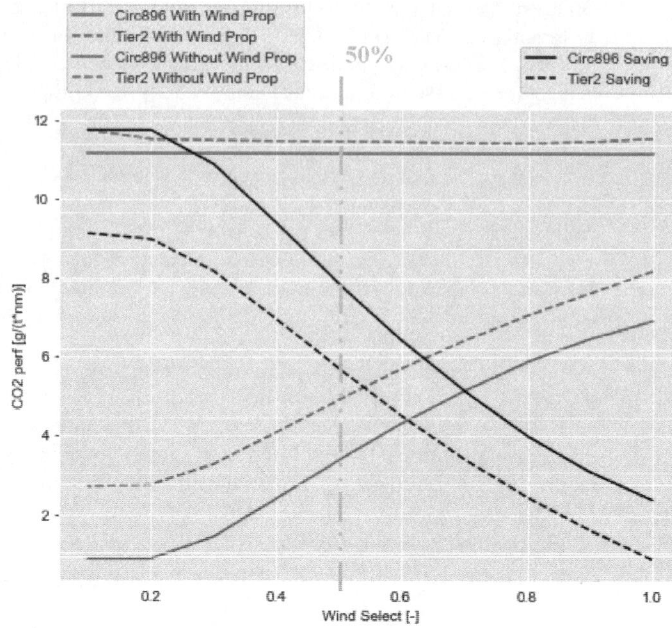

Figure 4: CO performance for variation of wind selection fraction for WASP-Ecoliner

3.2 (b) VOYAGE SIMULATIONS WITH ROUTING VERSUS SHORTEST ROUTE AND FIXED SPEED

The intention of the 50% wind selection in MEPC.1/Circ.896 was to account for better wind, in part due to routing. For WASP-Ecoliner a direct comparison is made on three routes that were also considered in previous projects. The voyage simulations were done using only the hindcast, which means the routing algorithm could work directly with the weather conditions as it would evolve. There is no uncertainty as in a forecast and thus results should be favourable. The results are depicted in Figure 5 together with the Performance Prediction Program ("PPP") results for sailing at constant speed along the shortest route. The PPP results were obtained by taking the full wind statistics, i.e. no 50% selection was done. It shows that savings can increase substantially with routing. The error bars indicate the uncertainty for individual voyages. What is interesting to see is that the error bars for the simulations with and without wind propulsion never overlap. This shows that, even if savings fluctuate, installing wind propulsion is always a good idea for this ship and these routes. Average savings increase a lot due to the routing, however not to the same extent as with the 50% approach discussed in the previous section. This implies that the benefits that are obtained with the 50% approach in MEPC.1/Circ.896 cannot be replicated with this ship and these routes.

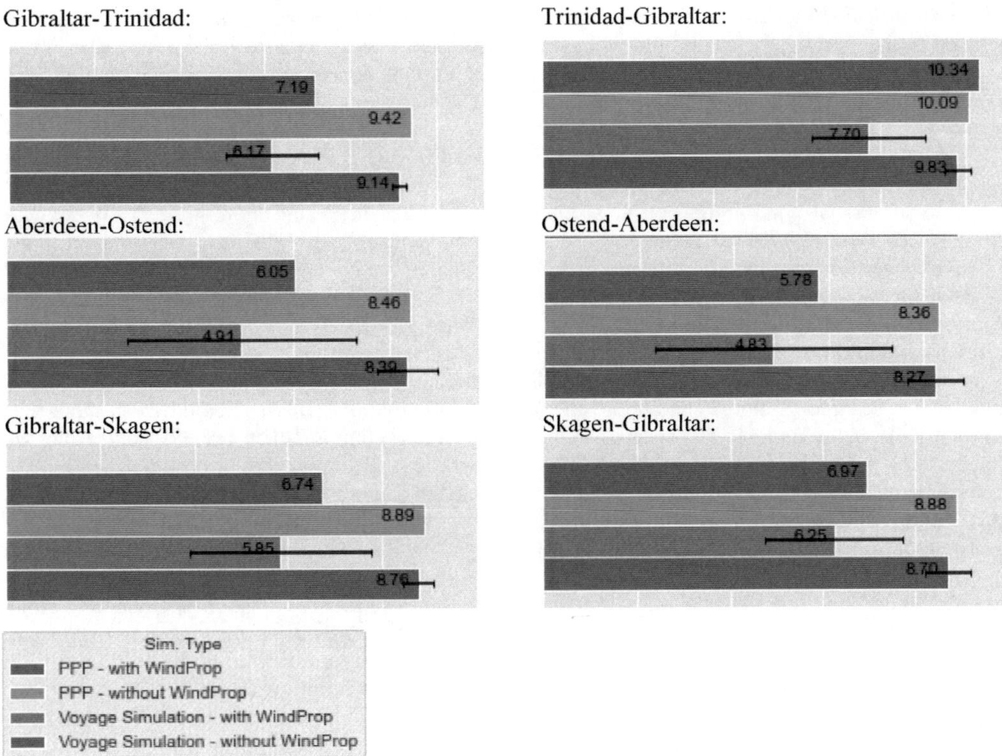

Figure 5: CO_2 performance [g/{t*nm)] for shortest route (PPP) versus routing (Voyage Simulation)

3.2 (c) TOWARDS WIND AS MAIN PROPULSION

Another sensitivity study that was done, again with the WASP-Ecoliner reference design, is to show how the prediction and methods respond when maximising the contribution of wind propulsion. This was done in two ways: by increasing the relative size of the rig (sail area) and by lowering the ship speed. In reality a substantial increase in rig size is likely not possible for this ship due to geometric, structural and stability constraints. The modelling was done by simply increasing the reference area in the calculations without changing the aerodynamic performance coefficients. Thus, this should be judged as a sensitivity study that looks at how the ship can handle such (fictitious) loads. Calculations were done using simple Tier1 methods and more advanced Tier2 methods. Further, results were obtained for the design using a Diesel-Direct propulsion line configuration ("Diesel1") as default, but also with an alternative Diesel-Electric propulsion configuration ("Electric2") using two generator sets, which should be more capable to handle the further increased variation of propulsion load. The results are shown in Figure 6. Firstly, it can be seen that an increase of 25% in rig size is less impactful than slowing down by 2 or 4 kn. At first sight, the difference between Tier1 and Tier2 seems to be relatively constant. However, with Tier2 we can actually see a cross-over where Diesel-Electric becomes

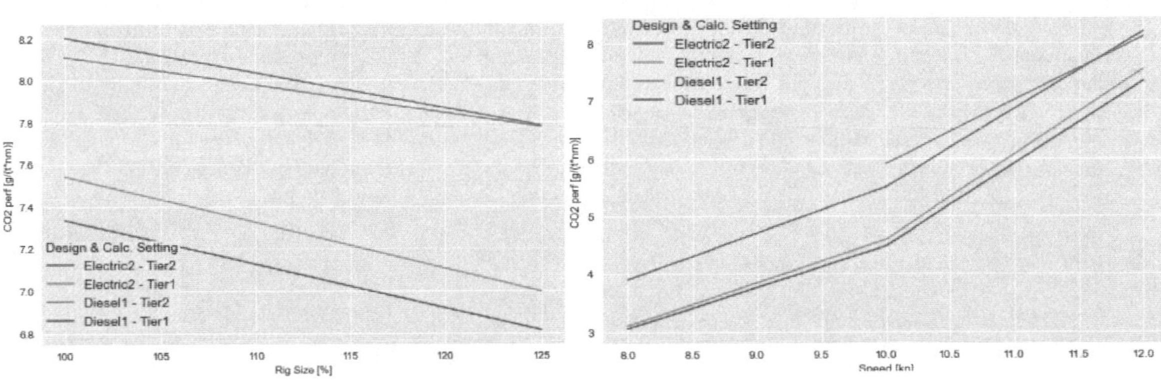

Figure 6: Towards wind as main propulsion

more efficient than Diesel-Direct. This is as expected, because with propulsion loads regularly very low, the efficiency of the diesel-electric configuration starts to decrease. But the crossover cannot be seen for Tier1. This indicates that for the depowering of the conventional propulsion, as seen in this case, modelling at Tier2 is required to draw correct conclusions on selection of the appropriate propulsion configuration.

4. MANOEUVRING CRITERIA FOR SHIPS WITH WIND PROPULSION

With more substantial wind propulsion now entering the market, more and more projects arise where the additional aerodynamic force will influence manoeuvring and course keeping. This leads to the question how manoeuvring performance may be affected and whether there is challenge to comply with existing criteria such as in IMO MSC. Resolution 137(76) [7].

The interpretation of the criteria in IMO MSC. Resolution 137(76) is not straightforward. Literally the resolution describes that the criteria should be verified in calm conditions, without wind, which would not present a problem. However, as the intention of the resolution is that ships should be able to make evasive manoeuvres at sea in operational conditions, one may interpret that the criteria should be valid in a relevant wind (and sea) condition too. In fact class societies are also interpreting it in that way. For course keeping, no official criteria exists, these are commonly defined by the owner and ship builder or design office.

For verification of the wind propulsion system, the need for a more detailed procedure and possibly re-evaluation of suitable criteria of manoeuvring (and seakeeping) characteristics when sailing with wind propulsion becomes evident. MEPC.1/Circ.896 [2] gives guidance on treatment of innovative energy efficiency technologies, such as wind-propulsion whereas section 2.4.4.4 describes that the verifier may request to demonstrate the vessel's course keeping abilities and rudder response in operational conditions: '*Where the lateral forces and yawing moment are particularly significant, the verifier may request course keeping and rudder angle demonstrations to validate conformity with the operational domain'.*

This guideline is not yet precise on what the exactly these operational conditions should be. In order to see the effect of wind propulsion, one could argue that the wind speed in which manoeuvring is verified shall not be chosen too small. This can be sailing in demanding conditions such as a rapid change of weather where the vessel experiences high winds, wind gusts or the operator attempts to maximizes the use of wind energy in such conditions. In such cases it is important to know what are operational limits on manoeuvrability and course keeping in waves. For these reasons, moderate to high wind speeds in the range from 6 to 8 Beaufort were selected in two case studies in WiSP2 to evaluate the effect of wind propulsion on the manoeuvring and seakeeping behaviour. In this paper we focus on the manoeuvring. The first case concerns the Marin Hybrid Transition Coaster (MHTC), installed with a three Flettner rotors. For the second case, the MARIN Ferry is selected which is also equipped with Flettner rotors. Both cases have a reasonable contribution of wind propulsion (average savings in the order of 20%) and are therefore interesting cases to evaluate how wind propulsion affects manoeuvring.

5. MARIN HYBRID TRANSITION COASTER (MHTC)

In an earlier publication [9], the manoeuvring response of MHTC focused on Dynarig sails. As a continuation this study expands on the manoeuvring behaviour with Flettner rotors.

5.1 SHIP DETAILS

The MHTC with the Flettner rotor configuration consists of three Flettner rotor sails as is shown in Figure 7. The Flettner positions relative from the aft perpendicular to the middle of the Flettner rotor are 19.71 m, 47.98 m and 76.25 m for the aft, mid and forward rotor respectively. Each rotor has a height of 18 m and a diameter of 3.0 and is equipped with an end plate of 5.5 m diameter. This ship represents a general cargo/coaster vessel with a length between perpendiculars of 86.76 m, a width of 14.0 m and a draught of 4.4 m even keel. The vessel has a stern gondola aft shape and was equipped with one spade rudder, one fixed pitch propeller, two lateral fins, and bilge keels, see Figure 7. The lateral fins were especially added to generate more side force at the aft to improve the directional stability. The rudder steering rate during the tests corresponds to the value of 2.32 deg/s. More ship details can be found in [9].

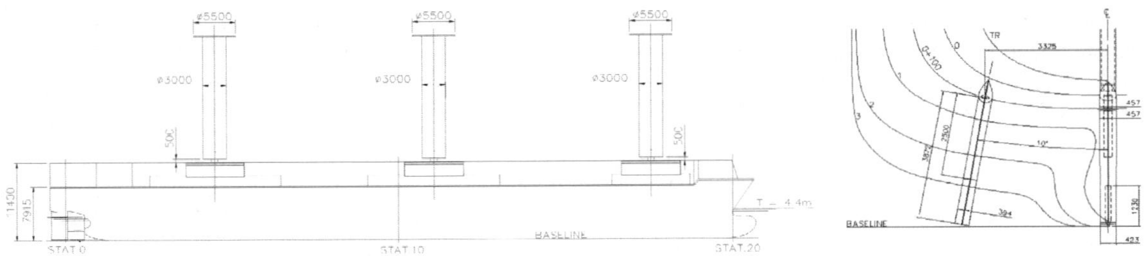

Figure 7: General arrangement of the MHTC with Flettner rotors and aft view of 2 lateral fins and rudder.

5.2 METHODOLOGY

Time domain simulations are performed with MARIN's aNySIM software. Captive CFD calculations as function of drift, rotation and heel are conducted to create a mathematical model that describe the forces on the hull and rudder in four degrees of freedom (X, Y, K and N). Two different wind models are used to calculate the Flettner rotors loads as is further described in section 5.3 (b). Superstructure loads are disregarded, but this effect is investigated for the MARIN Ferry.

5.3 RESULTS

5.3 (a) COMPARISON STUDY

A simulation programme is constructed in such a way that a comparison can be made between experiments and simulations with the focus on matching the test conditions such as wind speed, approach speed and rudder angles as good as possible.
For this comparison, zig-zags 10°/10° and 20°/20° manoeuvres are carried out for approach speeds around 12-13 knots. Only one wind condition, stern quartering wind, with a True Wind Angle of 158° from the PS bow in 7-8 Beaufort (33-35 kn) is evaluated. Due to basin restrictions, the zig-zag 10°/10° and 20°/20° were only partially possible. In these experiments, the procedure to start a manoeuvre is as follows. A few seconds prior to the start of the manoeuvre, the *'equilibrium rudder angle'* required to sail in a quasi-static equilibrium is set to the *'geometric zero rudder angle'* with respect to centre line. In this way, at the approach, t = 0 s, the rudder helm is given relative to this geometric zero rudder angle. Figure 8 shows the simulations results together with experimental results. Overall, it shows that the simulation model predicts the manoeuvring behaviour with a reasonable accuracy, considering that most of the deviations in overshoot angles vary within 0° to 5°.

5.3 (b) SENSITIVITY STUDY OF DIFFERENT VARIABLES ON MANOEUVRABILITY

The sensitivity of a number of variables are investigated for zig-zag manoeuvres. These variables are chosen because in the experiments, some test conditions were not optimal and require corrections. These variables consist of:

- *A mismatch in the intended yaw checking angle.* Therefore, the correct yaw checking angle is applied.
- *The wind modelling.* Two wind models are used: 1. *Constant apparent wind speed.* In the experiments, the apparent wind speeds was modelled as a constant, which resulted in a varying true wind speed over time when performing a zig-zag manoeuvre. 2. *Constant true wind speed.* More realistic is to model the true wind and wind angle as a constant in an earth fixed reference frame over space and time.
- *Modelling of the Flettner loads.* The first model uses *Polar coefficients.* This is a coefficient based approach for a direct comparison between simulations and experiments. Dimensionless wind coefficients are derived in four degrees of freedom using an average fitting procedure of measured Flettner forces from zig-zag model tests.
  The second model is a *Flettner model*, that describes their loads by using a relationship between spin ratio and lift and drag coefficients of a stand-alone rotor, based on a weighted average from several studies [10], [11] and [12]. Ideally, the aerodynamic model may be improved as this aerodynamic model does not incorporate rotor-rotor interaction and rotor-superstructure interactions. However, the model is regarded as a true prediction as it does not use data from the zig-zag experiments itself.

A simulation matrix is constructed such that systematically the influence of these sensitivities can be determined. All these sensitivities (yaw checking angle, wind modelling and Flettner loads) are investigated while applying the zero geometric rudder angle in the approach to the manoeuvre. These results are presented in Figure 8. Another variable, the choice for the zero steering angle is discussed separately in section 5.3 (c).

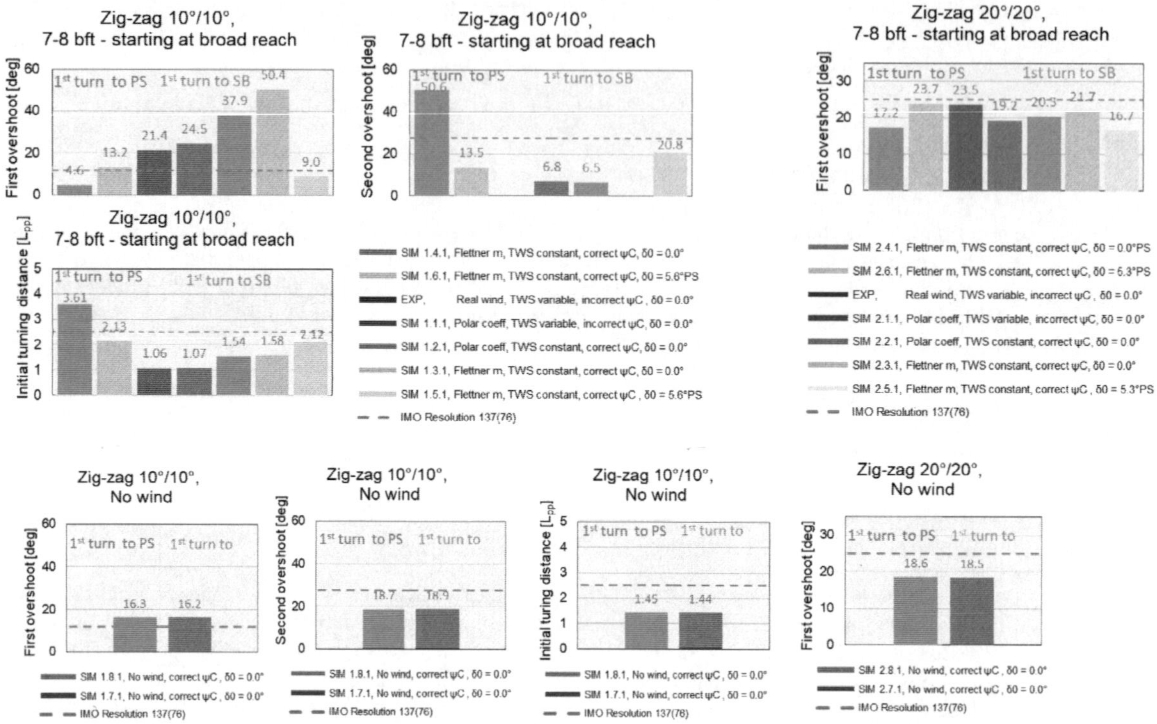

Figure 8: Simulation results compared to experiments without wind.

For the zig-zag 10°/10°, it is observed that:

- When correcting for the true wind and yaw checking angles (simulation no. 1.2.1), the first overshoot becomes 37.9°, an increase of 13.4°. The second overshoot is hardly affected and reduces with 0.3° to 6.5°. Also, the initial turning distance is affected and becomes 1.54 • L_{pp}, an increase of 0.47 • L_{pp}.
- Applying the Flettner model, the Flettner forces in longitudinal and transverse direction become larger when compared to the polar coefficients approach. The reason for the difference in forces could be the uncertainty on the actual wind velocity during the experiments and, likely to a lesser extent, the omission of aerodynamic interaction effects. The consequence is not minor. The first overshoot becomes 50.4°, which is 12.5° larger than using the polar coefficient method. After the first overshoot the vessel has changed that much of course that it finds a new quasi-static equilibrium and stays on a new straight track. It is not capable to return to its original course by applying 10° rudder to port side. The initial turning distance becomes 1.58 • L_{pp} and is still very close to the approach using polar coefficients and applying the true wind correctly.

For the zig-zag 20°/20°, it is observed that:

- When correcting for the true wind and yaw checking angles (simulation no. 2.2.1), the first overshoot becomes 20.3°, an increase of 1.1°.
- Applying Flettner model, first overshoot increases to 22.0°, which is 1.7° higher than using polar coefficients and applying the true wind correctly.

5.3 (c) CHOICE FOR ZERO RUDDER ANGLE AT THE APPROACH

Normally, zig-zag and turning circle manoeuvres are conducted in accordance to IMO Resolution 137(76), that prescribes that these manoeuvre should be carried out in calm water without external environmental forces, such as wind. In such conditions, a (near) zero rudder angle is needed to sail on a straight course.

However, the impact of wind propulsors should obviously be checked in wind from various angles. Then, the aerodynamic forces needs to be balanced with hydrodynamic forces of the ship, which is likely to require a steering angle (δ_0) to keep course. IMO resolution MSC 137(76) does not make any statement what constitutes a "zero rudder angle" that should be used at the approach. Two different implementations are investigated to quantify what effect the 'zero rudder angle' has on the characteristics of zig-zags and turning cirlces:

- *The equilibrium rudder angle.* This is the rudder angle resulting from a quasi-static equilibrium in straight ahead sailing. For instance, if the ship sails at an average rudder angle of -5° (trailing edge to port side) for a specific condition and speed, then, the zig-zag '10°/10°' is conducted with rudder execute angles of 5° and -15° whereas normally this would be 10° and -10°.
- *Zero geometric rudder angle.* This is the geometric zero rudder angle or neutral angle with respect to centreline and coincides with the resulting angle when sailing straight ahead in calm water without other external loads. When sailing with wind propulsion, using the zero geometric rudder angle can be implemented by executing, a few seconds before the manoeuvre starts, the rudder from an equilibrium rudder angle to its zero geometric position, 0°. Then a zig-zag 10°/10° is performed using 10° and -10°, this corresponds with the same rudder positions as is used for calm water manoeuvres without wind.

Figure 9 shows the time histories of the yaw and rudder angles and derived characteristics for zig-zag 10°/10° to port side and the two different methods for zero rudder angle at the approach. Figure 8 shows how much the initial turning distance and overshoot angles change for the other zig-zag manoeuvres by considering the choice for the 'zero rudder angle'. Figure 9 shows that when switching from an equilibrium rudder angle to a zero geometric zero rudder at the approach and applying this zero geometric rudder as the actual zero rudder angle gives less steering capacity in a zig-zag manoeuvre to return to its original course. Instead, by applying the equilibrium rudder at the approach as a zero rudder angle, this results in much better yaw checking behaviour and seems a better approach to perform a zig-zag manoeuvre.

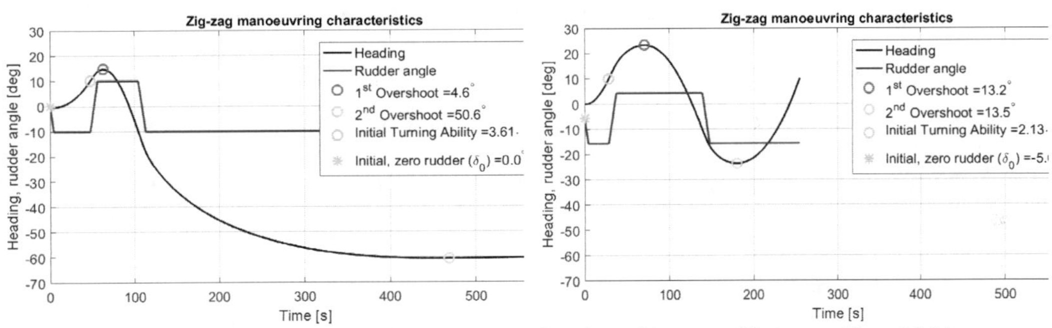

Starting with a zero geometric rudder of 0° Starting with an equilibrium rudder of 5.9°

Figure 9: Time histories of yaw and rudder angle for a zig-zag 10°/10° to port side for different choices of the 'zero rudder angle'.

It should be argued that there should be a maximum value for this equilibrium rudder angle. Typically, the maximum rudder angle applied in a turning circle according to the procedures described in IMO Resolution 137(76), is set at 35°. Often, it is seen that higher steering angles do not lead to a much higher transverse steering force, due to (partially) stalling of the rudder. Moreover, many transport ships have a mechanical rudder limit at 35°. Based on this maximum rudder angle, it is proposed to not use an equilibrium angle larger than 15°. In this way, the zig-zag 20°/20° can still be executed and not exceeding the maximum rudder of 35°. Restricting the maximum equilibrium angle at 15° is likely not problematic anyway in most applications, because induced drag is generally large at such an angle. Ships will either be designed to sail with smaller angles or the wind propulsion system will be depowered.

Similarly, as is done for the zig-zag manoeuvre, the choice for the zero rudder angle is evaluated for the turning circle manoeuvre. Again, it is assumed that the rudder is restricted not to turn further than 35° with respect to centreline. Further, a second assumption is that a certain amount of rudder angle should be applied with respect to the equilibrium rudder angle, such that is corresponds to 35° helm with respect to centre line. In these simulations, choosing the equilibrium rudder as zero rudder is implemented as follows.
Figure 10 presents manoeuvring characteristics derived from a turning circle both with and without wind. It appears that the advance becomes 4.80 • L_{pp} using the geometric zero rudder angle (SIM no. 4.3.1). When using

the equilibrium rudder angle instead (SIM no. 4.3.2), the advance reduces to 4.43 • L_{pp}. The tactical diameter is almost identical for these two approaches. The reason why the difference is larger for the advance, is that for SIM no. 4.3.1, first a rudder execute is given from equilibrium rudder 7.6° to PS to geometric zero, 0.0°. This already gives some rotation of the vessel in opposite direction, which is mainly influencing the beginning of the manoeuvre and affects the advance. The tactical diameter is derived over a long part of the manoeuvre, which is less dependent on only the initial phase.
Similar to zig-zags, the preference is also to choose the equilibrium angle as zero rudder for the turning circle manoeuvre. In addition, larger rudder angles relative to this equilibrium angle may be allowed under condition that the maximum helm of 35° with respect to geometric zero rudder should not be exceeded.

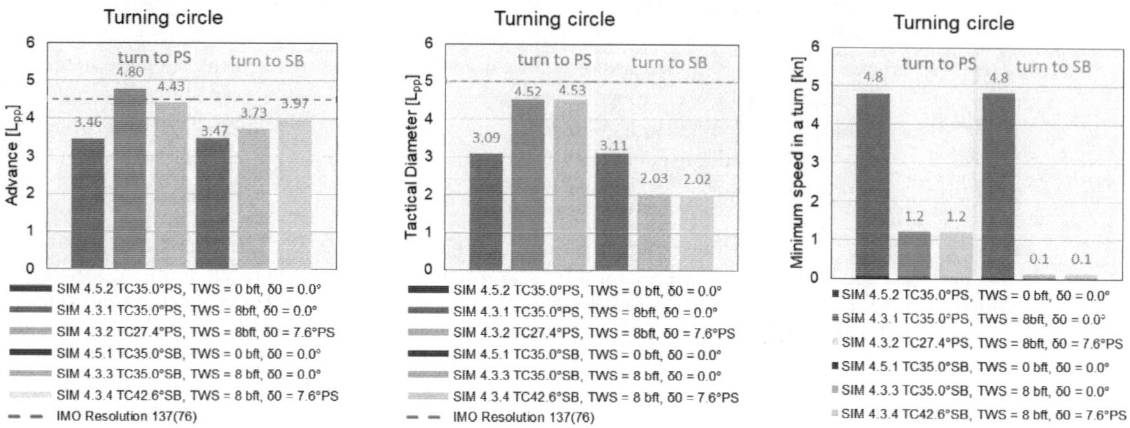

Figure 10: Manoeuvring characteristics from a turning circle 35° with and without wind either applying a zero geometric rudder or equilibrium rudder at the approach.

5.3 (d) VERIFYING COMPLIANCE TO IMO RESOLUTION 137(76)

From Figure 8 and Figure 10 also compliance to IMO criteria is verified with and without wind. For the conditions with wind, only the most relevant simulations are considered, which are the simulations with Flettner model and using the equilibrium rudder angle as zero rudder. Without wind propulsion, the vessel complies to all IMO criteria except for the zig-zag 10°/10°'s first overshoot which is exceeded with 4.5°. This already indicates that the yaw checking ability of the vessel is not satisfactory. In comparison to simulations without wind, sailing with Flettner rotors wind of 8 bft from broad reach:

- May either improve or deteriorate the yaw checking ability depending to what direction the vessel turns.
- The initial turning ability deteriorates significantly, the initial turning distance increased by 0.54 • Lpp.
- The turning ability deteriorates significantly considering the increase in advance, +0.97 • Lpp, and tactical diameter, +2.44 • Lpp.
- Again, it appears that the vessel with wind propulsion complies with all IMO criteria except for the zig-zag 10°/10° first overshoot. However, the margin for the advance, tactical diameter and most overshoot angles are small (10% or less).
- Another observation is the speed loss in a turn is high, see Figure 8. This is caused by the fact that when the vessel changes heading the apparent wind angle changes such that Flettner rotors act as a resistance. It is noted that in these simulations, the rotation rate of the Flettner rotor is kept constant, while there may be alternative measures to reduce the speed loss in a turn by reducing the Flettner rotor rotation rate prior to the turn. However, it should be questioned if an operator would always be able to react that rapidly, or alternatively, if a control algorithm exists that reduces the Flettner rotation rate in case the system recognises the action of a turning circle manoeuvre.

6. MARIN FERRY

For a second vessel, called 'MARIN Ferry', a more extended manoeuvring and seakeeping simulation programme is calculated for a wider range of wind speeds, wind angles and approach speeds.

6.1 SHIP DETAILS

In 2018, MARIN designed the MARIN Ferry as a representative modern ferry, see [13] and [14] for more details on the hull form and appendages. The MARIN Ferry has a length between perpendicular of 190 m, a draught of 7.0 m even keel and has a twin shaft-rudder arrangement. Two different skeg variants exist, a short and long skeg. For this case, the long skeg is selected, because a wind-assisted vessel favours the additional side force generation to counter balance the sail forces.

A 'substantial' sail plan is constructed such that average savings in the order of roughly 20% are achievable, depending on the route and the realised winds. As a first indication, steady sailing predictions show that a total four rotors of each 30.0 m span and 6 m diameter is sufficient to reduce the average net brake power (by also subtracting the Flettner power) by 23% at a chosen design speed of 15 kn for a wind speed of 21 kn from all directions. Two rotors are placed around midship, and two rotors at the deck house, located in longitudinal direction at 98 m and 164 m measured from the aft perpendicular respectively. The Flettner transverse offset from centreline is both 11.5 m to port side and starboard, see also Figure 11. The frontal and lateral wind area of the superstructure are 428 m^2 and 4848 m^2 respectively.

Figure 11: Render of the MARINFerry with four Flettner rotors and superstructure.

A transverse stability, GMt, of 3.0 m or more would be more representative for a ferry of this size and without a sail plan. However, when adding four rotors to the upper deck of the vessel, the centre of gravity moves upward. A rough estimation is that the corrected GMt will be 2.0 m.

6.2 METHODOLOGY

aNySIM time domain software is used to simulate the vessel in the time domain. Captive model test from [14] are used to derive a mathematical model for the hull reaction forces in 4 degrees of freedom. Flettner model as described in section 5.3 (b) is used to model the loads of the Flettner rotors. In contrast to the MHTC, an optimal Flettner rotation rate is applied, obtained from a very long speed run in which the Flettner rotation rate is varied. The optimal Flettner rotation meets predefined constraints that limit the heel angle at maximum angle of 5°. For simulations with wind, the wind loads of the superstructure are modelled as well, based on wind coefficients of a reference ferry. From the MHTC, it was found that it is more convenient to start a manoeuvre by using the equilibrium rudder angle. Consequently, this same approach is applied for the MARIN Ferry.

6.3 MANOEUVRING SIMULATIONS

The manoeuvring simulations consists of the following series of simulations and are all carried out without waves:

- *Series 1: Simulations without wind* serve as a reference to compare to simulations with wind and IMO criteria.
- *Series 2: Simulations with wind, applying only superstructure loads.* These are used to quantify the influence of the superstructure wind loads.
- *Series 3: Simulations with wind, applying superstructure and Flettner rotor loads.* With this series, the contribution of wind propulsion can be assessed by comparing to series 1 and 2.

The manoeuvring programme consists of the IMO manoeuvres: zig-zag 10°/10°, 20°/20° and turning circle 35°, all performed with the first rudder execute to starboard. Two approach speeds, 15 and 19.3 kn are used. The speed of 19.3 kn is slightly above the IMO under limit speed of 19.1 kn. For simulations with wind, two constant true wind speeds of 21 kn and 35 kn are selected with steps of 15° in direction.

6.4 RESULTS

Figure 12 shows the most interesting results, presented in a polar diagram for all wind directions. The tactical diameter for an approach speed of 15 kn is shown on the left and the maximum heel angle when approached at 19.3 kn is shown on the right.

The simulations with only superstructure loads do already reveal that the turning ability in bow quartering to beam wind from port side significantly deteriorate the turning ability. Adding Flettner rotors, shows that the general pattern is that the turning ability worsens, with some exceptions. On average the turning diameter becomes larger and is at a maximum for beam – to stern-quartering wind from port side (255°) with a maximum value of 4.40·Lpp. The manoeuvre is more demanding for an approach speed of 15 kn than 19.3 kn because it the turning ability is better 19.3 kn. At 19.3 kn, the propeller rotation rate is higher and provides a higher flow speed towards the rudder. In turn, this results in higher transverse rudder forces, increasing the turning ability.

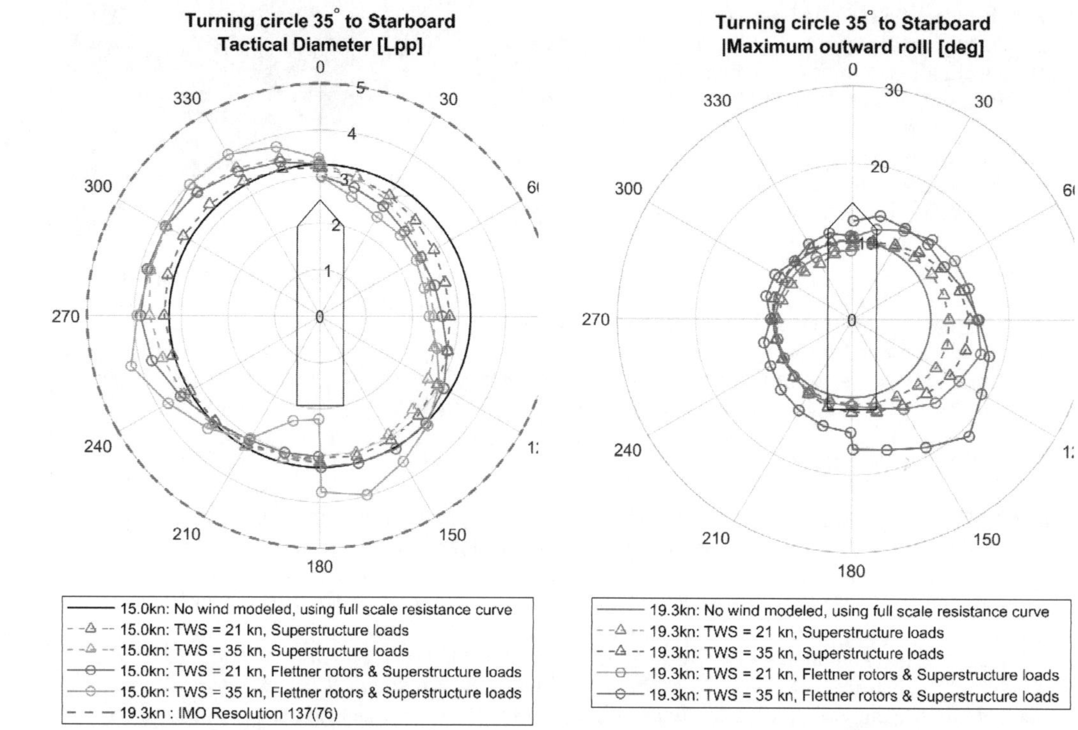

a) Tactical diameter when approached at 15 kn   b) Maximum outward heel, approached at 19.3 kn

Figure 12: Turning circle 35 manoeuvring results for a TWS of 21 kn and 35 kn presented in a polar diagram.

Table 2 gives an overview of the manoeuvring results for the condition without wind (series 1) and with wind, applying superstructure and Flettner rotor loads (series 3). This table presents the minimum and maximum values of the manoeuvring characteristics derived from the IMO manoeuvres and compares them to the IMO criteria. The number in bold are deemed most interesting.

Table 2: Simulation results for zig-zag 10°/10°, 20°/20° and turning circle 35° manoeuvres (all with 1st rudder execute to starboard), without wind and with wind and active Flettner rotors (series 3)

Series	Vs [kn]	TWS [kn]	Zig-zag 10°/10° First overshoot [deg]		Zig-zag 10°/10° Second overshoot [deg]		Zig-zag 10°/10° ITD [Lpp]		Zig-zag 20°/20° First overshoot [deg]		Turning circle 35° Advance [Lpp]		Turning circle 35° Tactical Diameter [Lpp]		Turning circle 35° |Maximum Heel angle| [deg]	
			Min	Max	Min	Max	Min	Max	Min	Max	Min	Max	Min	Max	Min	Max
1	15.0	0	3.3		4.1		1.69		8.8		3.12		3.26		5.9	
3		21	2.9	3.9	3.5	4.9	1.60	1.84	7.7	9.6	2.87	3.49	2.53	3.86	5.9	11.8
3		35	2.6	4.1	3.1	5.5	1.53	1.97	9.6	10.3	2.71	**3.91**	2.22	**4.20**	7.5	15.6
1	19.3	0	4.0		5.1		1.70		11.2		3.11		3.10		10.1	
3		21	3.6	4.8	4.5	6.3	1.62	1.80	10.0	12.9	2.90	3.29	2.60	3.39	8.8	17.0
3		35	3.3	**4.9**	4.1	**6.7**	1.56	**1.99**	9.2	**14.3**	2.79	3.67	2.45	3.54	10.4	**21.2**
-	19.3	IMO	12.4		28.6		2.50		25.0		4.50		5.00		-	

The results show that the yaw checking and initial turning ability is not that much affected as is seen that the largest increase in overshoot is only 1.6° and the initial turning ability grows by 0.29•Lpp.
The turning ability is more affected as the advance and tactical diameter increase by up to 0.79•Lpp and 0.94•Lpp respectively. Moreover, the maximum heel angles become significantly higher when sailing, especially when sailing at a ship speed of 19.3 kn.

Overall, it can be seen that the vessel complies IMO criteria without wind. And it still complies with active Flettner rotors in wind speeds of 21 kn and 35 kn. However, the margin with the criteria has reduced. Moreover in a turning circle heeling can substantially increase and the speed loss during the manoeuvre is substantial (lowest minimum speed is 2.3 kn). Even if these aspects do not influence compliance to regulations, they are likely undesired. Measures to improve the turning ability can be to automatically reduce the Flettner rotation rate at the start of a turning circle manoeuvre. The deceleration needs to be sufficiently fast to realize an significant improvement of the turning ability. Alternatively, Flettner rotors (or other wind propulsion types) could also be used to assist in steering by differentiating the control of forward and aft devices.

7. CONCLUSIONS

Related to the studies presented in this paper we draw the following conclusions.

Performance prediction with WASP-Ecoliner:

- When applying the 50% wind statistics approach as defined in MEPC.1/Circ.896 to WASP-Ecoliner, the savings increase till 69%, almost a doubling of the savings of 37% when considering the full global wind statistics.
- Voyage simulations with WASP-Ecoliner using routing on hindcast weather data yield a substantial increase in savings, but not nearly the doubling as seen with the 50% approach in MEPC.1/Circ.896.
- The Tier1 and Tier2 methods as used in WiSP2 can be used when increasing the savings for WASP-Ecoliner by increasing the rig size or decreasing the ship speed. However, only the Tier2 method predicts that Diesel-Electric propulsion becomes more efficient than Diesel-Direct at very low loads on the conventional propulsion system.

Manoeuvring with:

- MTHC in 8 bft (35 kn) wind from the stern quarter: the manoeuvring performance deteriorates when sailing with Flettner rotors for the less favourable turning direction. This concerns the turning ability, initial turning ability and yaw checking ability. In a turn, the minimum speed almost approaches 0 kn, which may be unexpected behaviour for a captain.

- MARIN Ferry for 15 kn and 19.3 kn approach speed in wind speeds of 21 kn and 35 kn from all wind directions, it is seen that the yaw checking and initial turning ability become only slightly worse with respect to no wind. The turning ability is more affected, especially for the lower ship speed of 15 kn. The speed loss in a turn is high.

Possible amendments to manoeuvring standards for ships with wind propulsion:

- To evaluate the manoeuvrability of wind propelled ships, it seems more practical to use the equilibrium rudder as zero rudder angle to start a zig-zag and turning circle manoeuvre, instead of using the geometric zero rudder angle.
- Speed loss and heeling don't have criteria associated in MSC.137(76). However, they may need attention for ships with wind propulsion.
- Measures to improve manoeuvring can be to reduce the Flettner rotor rotation rate sufficiently at the start of the manoeuvre or to use the rotors to help steer the ship. The impact of this is yet to be investigated.

Additional case studies are done in WiSP2 on the topics as described in this paper. After consultation in the project these will be used to propose improvements to existing or starting of new standards, rules and regulations at ITTC, IMO MEPC and IMO MSC.
Aside from the further development of standards, rules and regulations, the results also indicate that adaptations of conventional propulsion, control and steering need continued attention with further uptake of wind propulsion.

8. ACKNOWLEDGEMENTS

MARIN thanks the participants in the WiSP2 Joint Industry Project for their contribution to make this work possible. Moreover, this research is partly funded by the Dutch Ministry of Economic Affairs. Dykstra Naval Architects is acknowledged as original author of the WASP-Ecoliner concept.

9. REFERENCES

1. NELISSEN, D., TRAUT, M., KOEHLER, J., MAO, W., FABER, J, AND AHDOUR, S., 'Study on the Analysis of Market Potentials and Market Barriers for Wind Propulsion Technologies for Ships', *CE Delft,* 2017.

2. INTERNATIONAL MARITIME ORGANIZATION, MEPC.1/Circ.896, '2021 Guidance on Treatment of Innovative Energy Efficiency Technologies for Calculation and Verification of the Attained EEDI and EEXI', 2021.

3. HURFORD, A., 'D2.7 Vessel Performance Final Report (Q1-Q6) for HDV Marine Flettner Vessel Performance Monitoring, Data Analysis and System Modelling Project', Lloyd's Register, 2019.

4. SCHMIDT, Dr.-Ing. A., and VAHS, Prof. M., 'Entwicklung Eines Innovativen Schiffsantriebes Gekennzeichnet Durch Den Einsatz Von Magnus-Rotoren Zur Windenergienutzung in Kombination Mit Einem Hierauf Abgestimmten Antriebspropeller; Abschlussbericht', 2013

5. DYKSTRA NAVAL ARCHITECTS, 'WASP-Ecoliner', *https://www.dykstra-na.nl/designs/wasp-ecoliner/*

6. SCHWARZ-RÖHR, B., EGGERS, R., NEUMANN, D., GRIJPSTRA, J., DE VRIES, J., INSEL, M., VAN DER MEER, R., AND VOERMAN, E.J., 'Roadmap for Sail Transport: Engineering.' NSR SAIL Project, 2015

7. DANG, J., VAN DEN BOOM, H.J.J., AND LIGTELIJN, J. TH.. 'The Wageningen C- and D- Series Propellers.', *12th International Conference on Fast Sea Transportation (FAST)*. 2013

8. INTERNATIONAL MARITIME ORGANIZATION, MSC.137(76), 'Standards for Ship Manoeuvrability', 2022

9. EGGERS, R., and KISJES, A.S., 'Seakeeping and Manoeuvring for Wind Assisted Ships' *International Conference on Wind Propulsion, Royal Institution of Naval Architects in Association with International WindShip Association*, 2019.

10. JONES, L., PRINCE, M., HUDSEN, D., COCKES, AND J., 'Predicted Fuel-Savings for a Flettner Rotor Assisted Tanker Using Computational Fluid Dynamics,' *International Conference on Wind Propulsion, London, United Kingdom, Oct. 2019.*

11. TILLIG, F., RINGSBERG, W., 'Design, operation and analysis of wind-assisted cargo ships,' *Ocean Engineering, vol. 211, p. 107603, 2020.*

12. LI, D.,Q., LEER-ANDERSEN, M., ALLENSTRÖM, B., 'Performance and vortex formation of Flettner Rotors at high Reynolds numbers', 29th Symposium on Naval Hydrodynamics, *2012, vol. 29.*

13. KISJES, A.S., QUADVLIEG F., FERRARI, V., 'Uncertainty Analysis of Free running Manoeuvring Model Tests on a Modern ferry, with the emphasis on heel angles', *Proceeding of the 38th International Conference on Ocean, Offshore and Arctic Engineering (OMAE), Glasgow, United Kingdom, 2019.*

14. FERRARI, V., TONELLI, R., KISJES, A.S., HALLMAN, R., 'Manoeuvring Experiments, Mathematical Model and Sensitivity Analysis for Test-case Ferry, *Trends in Maritime Technology and Engineering, Guedes Soares, C., Santos, T. A., (Eds.), London, UK: Taylor & Francis Group, 2022.*

10. AUTHORS BIOGRAPHY

Anton Kisjes holds the position of Project Manager at the Maritime Research Institute Netherlands. He works on projects regarding manoeuvrability of ships and multi-disciplinary projects on wind-assisted ships. His master thesis was about wind propulsion and studying the suction wing in detail.

Rogier Eggers holds the position of Senior Project Manager and knowledge co-ordinator of Ships Manoeuvring at the Maritime Research Institute Netherlands. He has been leading MARIN's development of research and services to clients on wind propulsion. He co-initiated the WiSP2 project.

WASP ASSESSMENT FROM A SHIP DESIGNER POINT OF VIEW

R Prever, NAOS ship & Boat Design, ITALY

SUMMARY

Today, most of the WASP suppliers are emphasizing their pure aerodynamic performances, translating that directly into fuel saving.
Stability rules, operationally acceptable permanent heeling angles, added displacement due to device weight and ballasting requirements, are all factors equally important for a holistic and practical assessment of the WAPS on different ships type.
The paper will explore practical aspects which should be considered when comparing different systems, and show how ranking the different WAPS from that perspective, may lead to different conclusions on the appropriateness of an installation.

NOMENCLATURE

PB	Brake Power (kW)
TWA	True Wind Angle (deg)
TWS	True Wind Speed (kn)
AWA	Apparent Wind Angle (deg)
AWS	Apparent Wind Speed (kn)

1. INTRODUCTION

As ship designer we have been asked to assess different type of WAPS on specific vessels, generally RoRo type.
In this paper the practical aspect to be taken into consideration will be mentioned in a qualitative manner.

2. EFFECT OF WASP ON SHIP PROPULSION

2.1 WAPS WEIGHT

As with any energy saving device on board, a fair comparison of the effect should be done with and without it.
We often see that instead it is done with the device running or not running but installed.
That is wrong, because in all cases the installation of additional weight will have a negative effect compared to the same ship without it, due to the added displacement.
As we all know, speed performances changes with different displacement, so that any additional weight should be considered as a negative factor.

With reference to a RoRo vessel, LOA 170 m, Displacement 13.800 t, the variation of brake power (PB) per tonne of displacement is abt 0.4 kW/t.
In that case, 100 t of additional weight produces an increase of power demand of 40 kW, which should be deducted from the power supplied by the device itself.

2.2 INCREASED AERODYNAMIC DRAGWITH ZERO WIND

Since we are dealing with wind assistance to the main mechanical propulsion and not with pure sail propulsion, we have to keep in mind that the vessel still need to keep the schedule by normal sailing using its conventional propulsion.
Few of the presently available systems have the capability to not expose surfaces to the air even when not in "running mode".
The added air resistance of the ship bow section exposed area represents aproximately 3-5% of the total resistance. In bad weather or strong wind it can be more than 50% of the total resistance. In that context it appears evident that any additional exposed surface may drastically decrease the ship performance, by adding significant drag.

For example, one rotor-sail 35 x 5 in its working position, at 17 kn will generate an added resistance of abt 8 kN, which could be between 1 to 2% of the total ship resistance for a typical short sea RoRo.

2.3 UPWIND CAPABILITY

Since the apparent wind direction depends on the ship's speed which should normally be that needed to keep the schedule, when the wind speed is less than the ship's speed, the "no-thrust" upwind range becomes wider. In that sector the WAPS will make no contribution or, even worse, will reduce the performance of the vessel. Folding down mechanism will be necessary in those cases. As it can be seen in Figure 1, below 35-40 deg of true wind direction the WAPS contribution becomes negative.

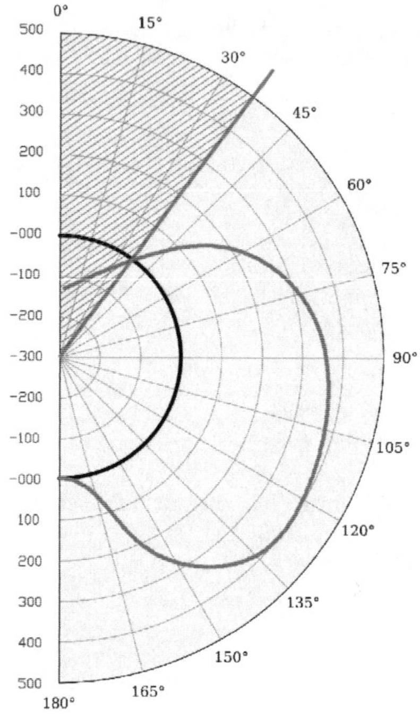

Figure 1: Power developed at different TWA of a generic WAPS

Figure 2 shows that in June, on southern Baltic sea, the probability to encounter TWS lower than 15 kn is about 80%.

For a given operational speed of 17 kn, the 80% of the time the TWA shall be not less than 40 deg to have a AWA of 20 deg. which is necessary to provide some thrust.

Only WAPS which can provide a high Lift/Drag ratio may contribute to the propulsion thrust in such a situation.

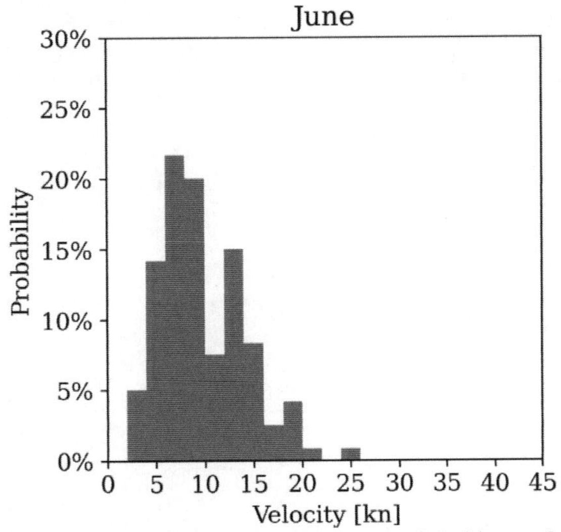

Figure 2: Statistical distribution of wind force, south Baltic sea, June (Era5)

3. EFFECT OF WASP INSTALLATION ON STABILITY

3.1 WEIGHT AND COG

Present rules for WAPS are not completely mature, but it is our understanding that any system should have a degree of safety to fulfill the IMO Weather Criterion included in the ISC 2008, with the total exposed area to the wind. A good practice is to consider the sum of the sail area, irrespective of how the sails are installed.

This is a rough assumption, since the actual pressure coefficients on specific WAPS should be considered. Indeed, different WAPSs have different lift and drag coefficients, in some cases quite different from the pressure coefficient which are correlated to the 504 Pa applied on ship side, which is approximately 1.25. Therefore, active devices like rotor or suction sails with pressure coefficients reaching very high levels, are likely to be considered differently in the near future, possibly by adding an additional WAPS induced heeling moment that more accurately reflects their physics.

Weight and center of gravity should be considered, as they deviate from the unequipped vessel, by increasing both displacement and VCG.
To compensate, additional ballast may be required to cope with the ISC, thus increasing further the displacement.

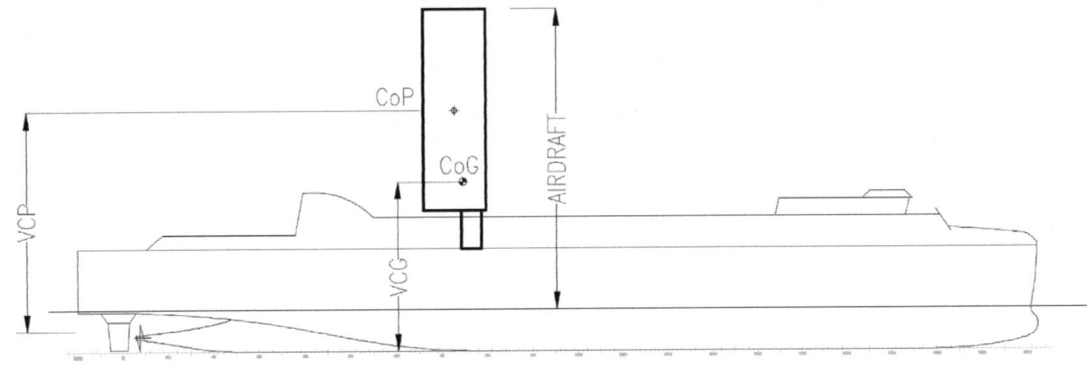

Figure 3: Generic scheme of WAPS installation on a RoRo vessel

For example, an installation consisting of 2 foldable rotor sails 35 x 5, may weigh approximately 300 tons, and be located in a relatively high position (VCG). That would require an additional 300 t of ballast to maintain the vessel within the IMO stability rules, resulting in 600 t added to the original displacement.

This is equivalent to 240 kW just to compensate the effect of weight and rise of VCG. Of course this is unnecessary on a tanker or a bulk carrier, but has to be considered on vessels carrying lighter cargo, as a RoRo or PCTC. It will be even worse if the stability criteria should be updated for "active" WAPS, due to the actual pressure coefficient when in use.

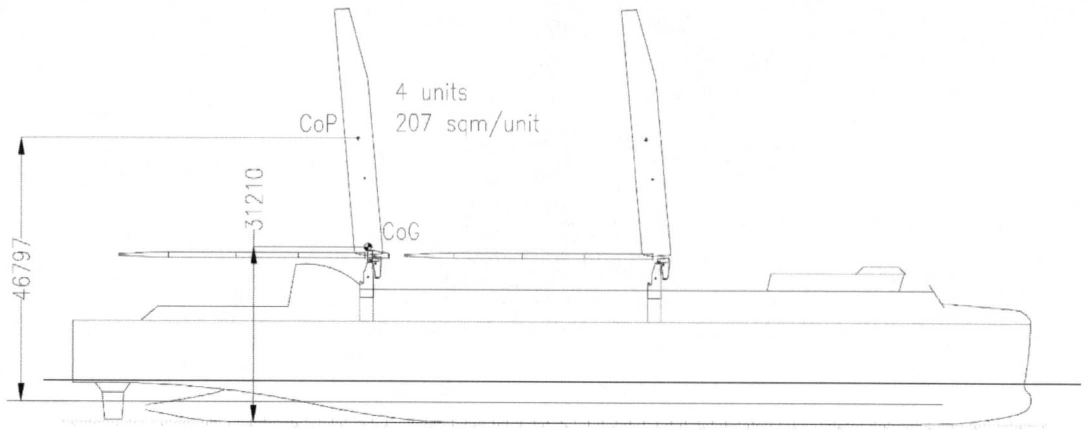

Figure 4: WSM® installation of 4 units

With this constraint well in mind, ***WSM*** ® has been designed to be as light as possible.
Four ***WSM*** ® units, delivering an equivalent thrusting power of two Rotor sails 35x5, have a total weight of abt 120 t. In the specific case we studied, this will not require additional ballast to cope with ISC.

3.2 HEELING TANK CAPACITY

All RoRo ships are normally equipped with one or more pairs of Heeling Tanks to compensate for the transversal heel during loading/unloading operations. It is generally a good design criterion to dimension the WAPS according to the maximum acceptable heeling angle achievable with the max heeling tank compensation. This may depend on the specific type of cargo, and from the level of comfort the crew need to have – a permanent heel angle during a long crossing may not be so comfortable.

This aspect is very much related to

- the sail surface area;
- the aerodynamic efficiency of the system.

On a wing type of WAPS, a high aspect ratio guarantees a high efficiency (CL/CD) which in turn provides an high ratio of Thrust/Drift.

Rotor sails have generally very high lift coefficients but also high drag coefficients which generate high side forces (drift).
In contrast, the surface area of ***WSM***® has been kept limited to abt 200 sqm to have more flexibility in the retrofit installation, potentially using multiple units.

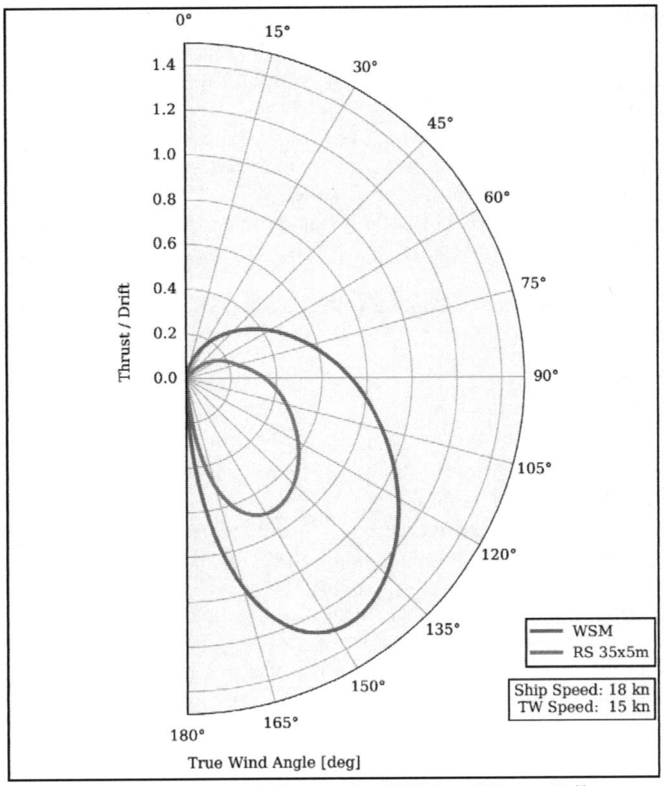

Figure 5: Thrust/drift ratio for WSM and Rotor Sail

4. WASP SHIP INTERFACE

4.1 WASP ARRANGEMENT

Different vessels types have different purposes, layouts and deck space availability.
Cargo hatches, RoRo cargo on decks, Containers in multiple tiers and the location af the accommodation block, represent challenges for the installation of any WAPS system.

Deck footprint area is an important criteria for the selection of the WAPS. In most cases a WAPS will take away some cargo space, or need additional structure to maintain the same cargo intake.

The ideal positioning of any WAPS should be as close as possible to the center of drif, to avoid yaw moments, which would require steering corrections.
Any installation which, due to unavailability of space in the right position, does not follow this principle, will end up in a poor course keeping and prone to wind gusts which easily alter the heading.

4.2 PERMANENTLY EXPOSED AREA – TILTING DOWN FEATURE

Maneuvering in port in bad weather will become more difficult with permanently exposed sail areas, mooring lines will be also more stressed.
The capability to be completed folded down when not in use is therefore a very important feature under many aspects, including the need to reduce the air draft. This capability is also necessary in extreme weather conditions...
Tilting system which can only be used to reduce air draft in good weather condition (it happen to be like that on some WAPS) will therefore be a negative factor inthe fulfillment of intact stability requirements

5. RELIABILITY

Reliability should be regarded as the most design criteria. Wind is an intermittent source of energy, so that the WASPS response should be always available.
To achieve a high degree of reliability, the number of moving parts should be minimized, to avoid potential failure. This can occur at the expense of pure aerodynamic efficiency, but a higher degree of reliability will compensate in terms of the total amount of captured energy.
Frequency of maintenance/repair work will also play an important role in the business model, not only for the costs involved, but also for the generalized lack of specific competence and crew today available on board.

Difficult access to WAPS components should also be considered. Due to their huge dimensions and vertical extension, any failures which may require the ship to be stopped or the use of cranes at the quay must be avoided.

6. CONCLUSIONS

WAPS Lightweight is fundamental for all "volume cargo" vessels, like RoRo, RoPax, PCTC, Passengers, while is much less for tankers or bulk carriers.
Simplicity of design, construction, and manning; with the increasing complexity of modern ships and the amount of paper work every crew has to do on a daily basis, WAPS should add no troubles to be welcomed by seafarers.
Reliability is more important than theoretical efficiency: the number of hours of constant service is more important than the magnitude of the peak values.
Highly efficient more than highly thrusting: high level of thrust goes together with high side forces which make the ship more difficult to handle.

7. AUTHORS BIOGRAPHY

Roberto Prever holds the current position of CEO at NAOS Ship & Boat Design, in Italy. He is responsible for the management of the group which include 4 locations in EU. His previous experience includes more than 30 years in the design of RoRo, Ro Pax vessel, sailing race and cruising mono and multihulls. In the last 10 years has promoted the development of ***WSM,*** Wing Sail Module, which is presently installed as a 1:3 prototype on board of a RoPax.

MANOEUVRE SIMULATIONS IN DESIGN PROCESS OF WIND POWERED VESSEL

S Kontos, O Lundbäck, M Kjellberg, E Wilske, S Werner, RISE Research Institutes of Sweden, Sweden

SUMMARY

Wind propulsion systems (WPS) are one of the most promising technologies for ship propulsion that can radically reduce greenhouse gas emissions. However, attention must be paid to the additional transversal forces and yaw moments connected to a wind propulsion system, as it can affect the manoeuvring and seakeeping performance of a ship, which can have a significant impact on the ship's overall performance. This paper demonstrates how time-domain simulations can be utilised to assess the manoeuvrability of a wind powered vessel to support the decision making, from the early design stage, all the way to testing the control systems, design of Human Machine Interface (HMI) and developing crew guidelines and training.

NOMENCLATURE

Beta	Leeway angle (degrees)
CE	Centre of Effort
CLR	Centre of Lateral Resistance
DOF	Degrees of Freedom
Fi	Heel angle (degrees)
IBS	Integrated Bridge System
Lpp	Length between perpendiculars
VPP	Velocity Prediction Program
WPS	Wind Propulsion System

1. INTRODUCTION

Wind propulsion has great potential in reducing shipping greenhouse gas emissions. However, wind propulsion devices introduce an additional side force and a yaw moment that needs to be balanced by the ship's hull and rudder. A vessel with poor yaw balance will require large rudder angles to keep the course which could potentially compromise it's manoeuvring performance. Therefore, informed decisions are required on the rudder size and position of the WPS to ensure a satisfactory yaw balance. Considering the lack of experience-based design rules for rudders of large wind powered vessels, efficient and reliable simulation tools are of great value. Additionally, since the position of the WPS is constrained by major design features (e.g., the deck arrangement and compartments) its effect on manoeuvrability should be assessed at an early design phase.

RISE/SSPA's software SEAMAN-Winds is capable of both steady-state (VPP) and time domain simulations. The software has a short execution time that makes it ideal for testing multiple design configurations at the early design stage. We present here a validation of the software against manoeuvring model tests of a wind powered ship, where the wind propulsion units are represented by pulling fans. We then demonstrate how complimentary time-domain simulations can reveal issues of poor yaw balance that are not always easy to detect from the results of a VPP simulation. Finally, we present how mariners operating a real-time manoeuvre simulation tool can be utilised to enhance the proof-of-concept, assess the HMI design and for crew training.

2. VALIDATION OF SEAMAN-WINDS AGAINST WIND-POWERED SHIP MODEL TESTS

2.1 SEAMAN-WINDS SIMULATION SOFTWARE

SEAMAN-Winds is a software used for dynamic simulation of ship motions in four (surge, sway, roll, yaw) or six degrees of freedom (DOF) in the time domain, or quasi-static and steady-state (VPP) simulations in up to four degrees of freedom. In this paper we are using the four DOF time-domain and the four DOF steady state VPP mode, as well as the full six DOF real-time simulation capability.

Both seakeeping and manoeuvring performance can be studied including effects of wind (incl. gust), waves (both 1st and 2nd order), current, shallow waters, canals, squat, banks, ship interaction, misc. equipment such as high efficiency rudders, bow thrusters, azimuth propulsion, moorings, anchors, etc. The calm water manoeuvring simulation model is described by [1] and later developments by [2,3]. The modelling of additional force components that SEAMAN Winds uses can be found in [4],[8], [11], among others.

The VPP module, explained in detail in [5], finds equilibrium in forces from hull, rudder, and wind propulsion system in surge, sway, roll and yaw to achieve the maximum ship speed possible. There are also modes for minimizing engine power at a minimum velocity made good and tailor-made objectives for specific projects. The hull and rudder forces modelling are based on slow-motion hydrodynamic derivatives found through CFD and/or model tests as described in [6]. For the aerodynamic modelling of rigid wing sails the highly efficient and validated lifting line based aerodynamic model CORR-SILL is used [7,8,9].

2.2 SEAKEEPING MODEL TESTS FOR A WIND POWERED VESSEL

Model tests in both calm water and waves have been performed with a model of a wind assisted ship, which has been labelled Oceanbird wPCC (wind Powered Car Carrier). It has solid wing sails targeting emission reductions in the order of 90%. This ship has been used within several research projects to develop numerical methods hand in hand with techniques for testing [4]-[11]. Part of the testing was carried out in the manoeuvring and seakeeping basin at the Maritime Dynamics Laboratory (MDL) at RISE/SSPA. These tests have been described in detail in [10]. Briefly described, a five-meter-long model of the wPCC ship was fitted with air propellers to simulate sail forces, a twin screw arrangement and transducers to measure ship motions and forces on the twin screws and air propellers. In the section below a comparison is made between model tests and VPP calculations for pure sailing course keeping, in calm water conditions.

Figure 1: Seakeeping model of research test case wPCC Oceanbird, 1:41.2 scale, [10].

2.3 SEAMAN-WINDS VALIDATION

In Figure 2, VPP calculations and model tests are compared for a bow reach, pure sailing condition. In the tests, the condition was repeated twice and showed good repeatability. The forces measured by the transducers in the air propellers were used as input to the VPP calculations. The predicted performance correlates well with the model tests. The VPP calculations have been performed in model scale. The input to the manoeuvring model in the VPP has been generated using CFD computations as described in [11].

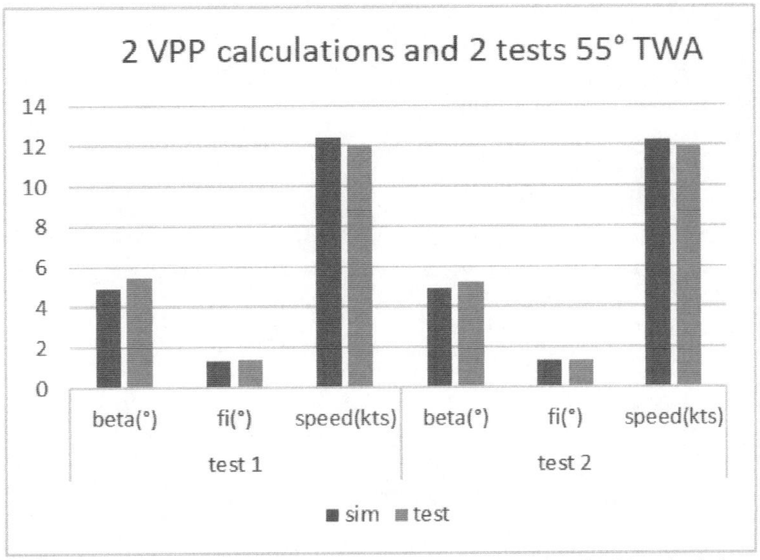

Figure 2: Comparison between free running model tests and VPP calculations

3. ASSESSING COURSE KEEPING AND MANOEUVRABILITY USING TIME-DOMAIN SIMULATIONS

The vessel considered is a generic wind powered cargo ship and is a modification of the vessel described above, designed solely for the purpose of this demonstration. Three different designs are tested. The first configuration, referred as 'Initial design' in Figures 3 and 4, positions the foremost sail 50 m forward from Lpp/2. In the second configuration the rig is shifted 50 m forward. The third design keeps the initial rig position but has a 23% larger rudder. We evaluate the performance of the different designs using VPP and time-domain simulations.

3.1 RESULTS

The VPP polars of the fully sailing vessels are presented in Figure 3. Moving the rig forward moves the centre of effort (CE) of the aerodynamic forces forward and improves the balance of the vessel. Furthermore, it increases the close-hauled sailing angle and the speed upwind. The polar plots already reveal that the initial configuration with the rig further aft has poor yaw balance since it requires large rudder angles to sail upwind. Increasing the rudder area increases the rudder's possibility to move the centre of lateral resistance of the hydrodynamic forces (CLR) aft and improves the yaw balance. The performance of the ship with the larger rudder is comparable to the initial design with the same rig position, but a smaller rudder angle is required to achieve yaw balance.

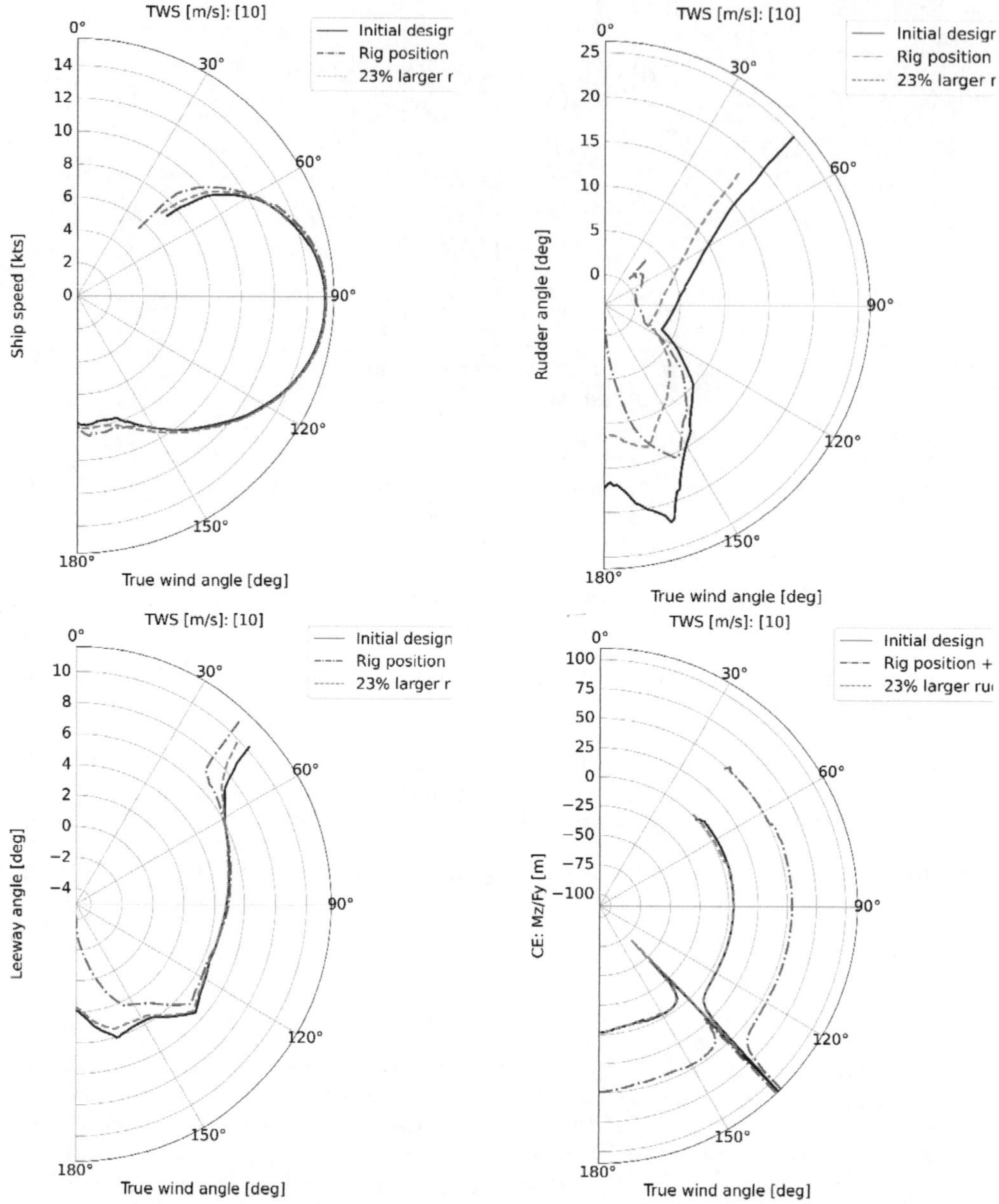

Figure 3: Initial and improved designs compared for fully sailing conditions for a generic test case ship.

Additionally, we simulated a fully sailing avoidance manoeuvre in the time-domain where the rudder is controlled by an autopilot. The track, speed, rudder angle, heading and course during the manoeuvre are plotted in Figures 4-6. With a target heading of 0° for the autopilot, the ship tries to sail on a straight course for 800 s (red) and is instructed to give way by changing the heading 40° for the next 300 s (green) and then come back to original heading (blue). The True Wind Speed (TWS) was 10 m/s and the True Wind Angle (TWA) 50°.

In Figure 4 the initial design of the ship cannot keep the target heading well and after the turn it cannot come back to the original heading. The ship with the rig shifted forward shows big improvement in keeping the heading and completes the manoeuvre in a controlled manner as shown in Figure 5. A similar effect is achieved by keeping the initial rig position but increasing the rudder area, Figure 6. This is an example where the polar performance (in terms of speed) of the ship of the initial design and the ship with the larger rudder is similar, but the manoeuvring ability and overall performance in the time domain differ significantly.

Concluding, the time-domain simulations can help evaluate the ship manoeuvring performance and overall performance during different operational scenarios and are of great importance for gaining the stakeholder's confidence in the concept and for understanding the consequences of design decisions. This also makes way for finding optimal design choices.

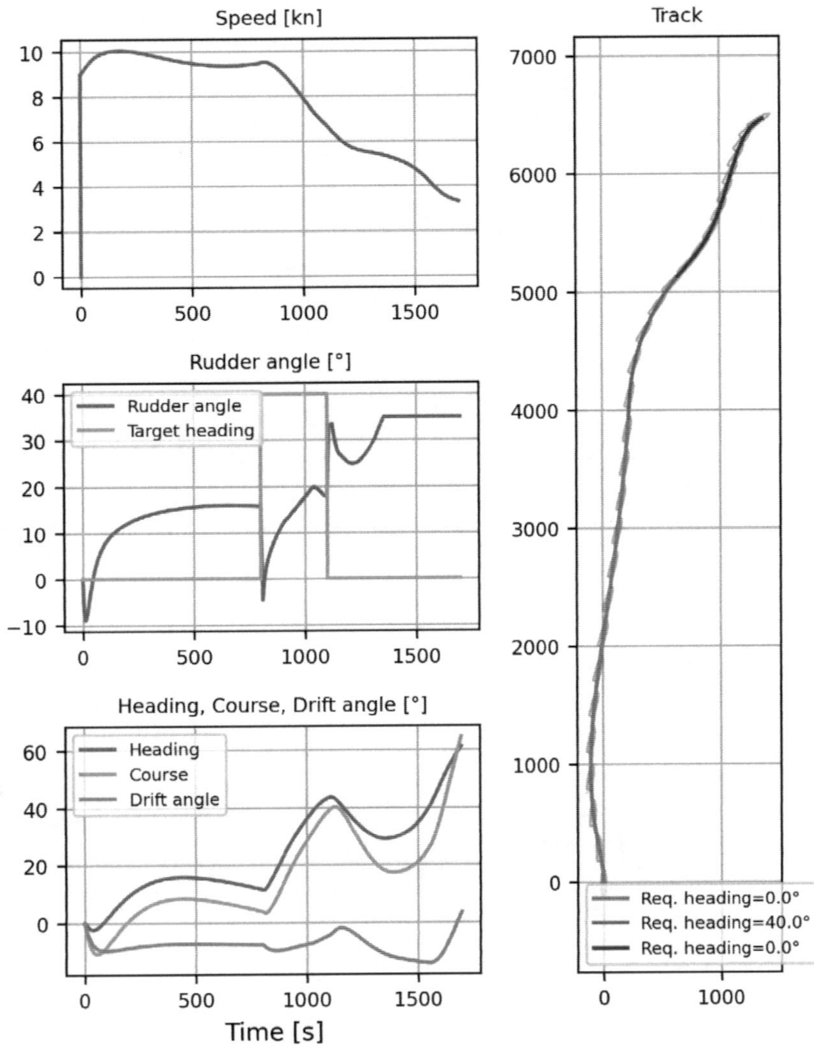

Figure 4: Initial design. TWA = 50°, TWS = 10 m/s

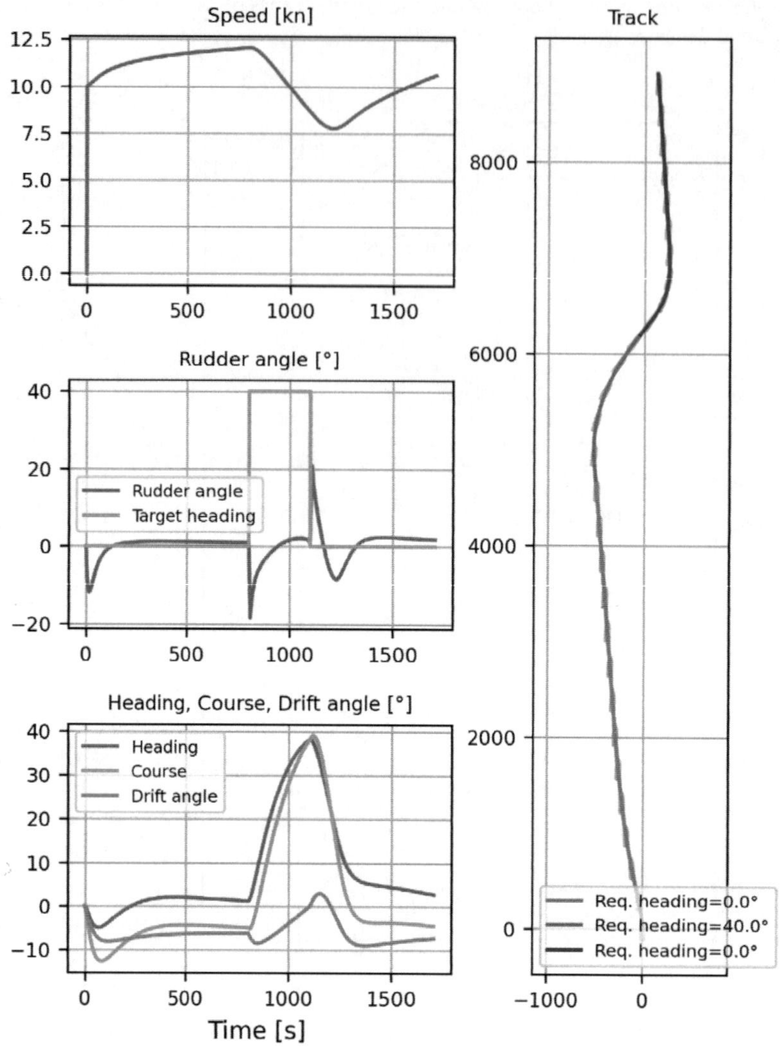

Figure 5: Improved design by adjusting the rig position. TWA = 50°, TWS = 10 m/s

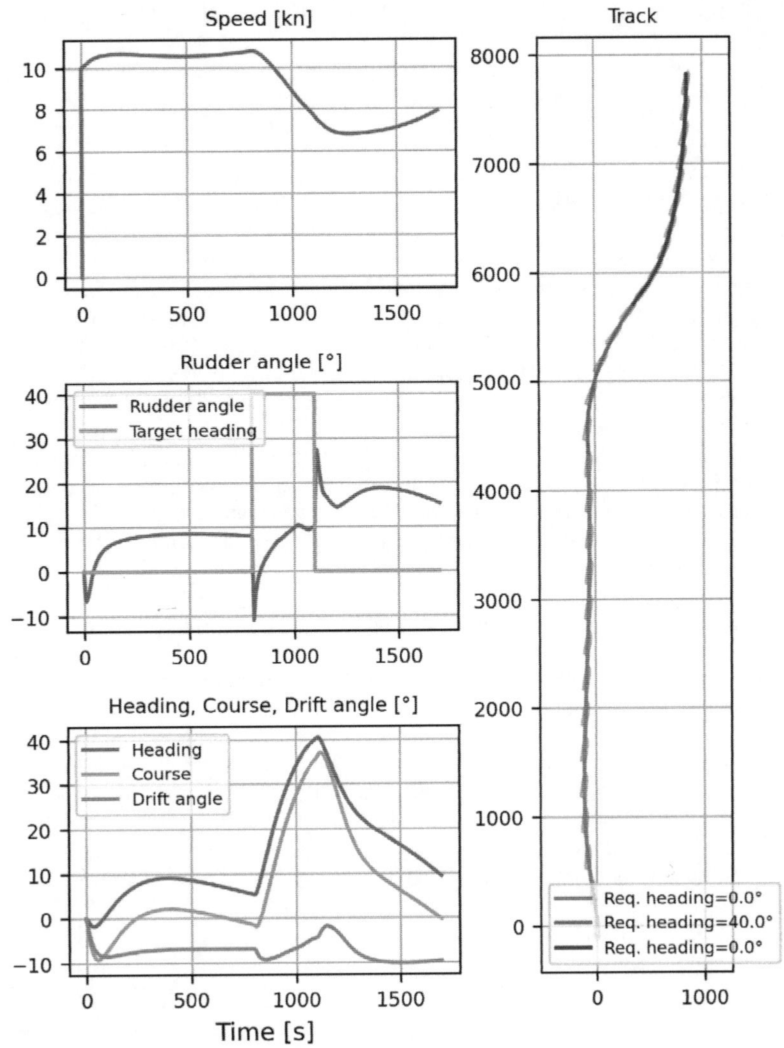

Figure 6: Initial rig position with 23% larger rudder area. TWA = 50°, TWS = 10 m/s

4. REAL-TIME SIMULATOR FOR ASSESSMENT AND TRAINING OF WPS

Wind assisted propulsion creates some operational challenges for the ship crew and in particular the bridge team. It is fair to believe that these challenges increase proportionally with the ratio of sail power compared to power from the main engine. Especially on ships where wind propulsion is the primary means of propulsion at some conditions, the bridge team needs to have a good understanding of how to operate the WPS and how it affects the ship manoeuvrability and speed performance in various conditions. The following questions need to be considered:

- In what conditions can the ship operate solely by means of wind propulsion and what readiness should the main engine have?
- What manoeuvring space is required in different wind conditions and navigable confinement?
- Will the wind propulsion require a different bridge manning compared to conventional propulsion?
- How will the WPS affect the visibility from the bridge and radar coverage and are there needs for additional support systems (e.g., cameras and additional radar antennas)?
- How should the WPS control system interface be designed and incorporated in the IBS (Integrated Bridge System)?
- What type of automation is needed for the WPS?

- What Bridge team training will be required for the following items?
  - WPS operation.
  - Route planning and execution.
  - Failure scenarios.
  - Sudden and rapid change of wind conditions.

4.1 THE TOOL FOR ASSESSMENT AND TRAINING FOR A SHIP WITH WPS

For these simulations we use the SEAMAN simulation software running in real-time. The simulation system includes all components normally found in a manoeuvring simulator. The ship evaluated in the simulation was a version the research concept Oceanbird equipped with four wings. In this phase of the research, 2D desktop simulations were utilised with 3D full mission simulations as the next planned step. The utilized 2D desktop setup include functionality for control, automation, and surveillance of the WPS. In the electronic chart, the position of the wings is indicated. Further, wind indication arrow of both true and apparent wind is displayed with two clearly distinguishable symbols.
Integrated in the conning display, the position (i.e., angle) of the wings is displayed next to the relative wind indicator. The lateral and axial forces on each wing were displayed as well as the total sum of forces.
Two operational modes were implemented for the wings. A manual and a semi-automatic mode. In the manual mode, the angle of a wing is set relative to the ship's centerline. In the semi-automatic mode, the angle of a wing is set relative to the apparent wind.

Adding the human in the loop for developing the ship design gives valuable information of what means of manoeuvring a ship needs to be equipped with for practical operation. It also gives inputs to the human-machine interface design and can be used when setting minimum requirement specifications on such interface.

It was noted that manual control of the wings quickly became exhausting for a small bridge team. Having a good monitoring of wind speed, wind direction and forces on the rig, nevertheless made manual control of the four sails possible. In the simulator tests we favoured the semi-automatic functionality that sets the wind with an angle relative to the apparent wind. This works well in most conditions. However, when tacking, the wings need at some point to be wind vaned and when on the next tack, shift the wings. Also when jibing, this automation requires manual action for shifting the wings to the new reach.

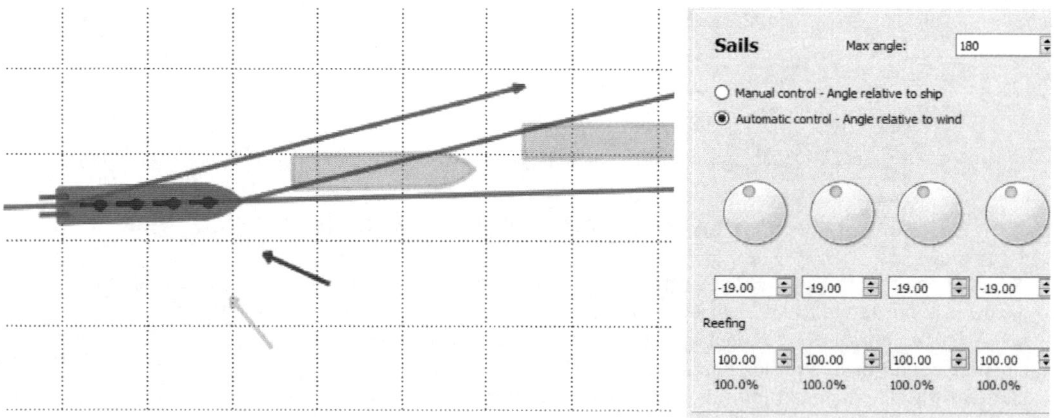

Figure 7: Part of the electronic chart and testing various types of user interface for control for the wings.

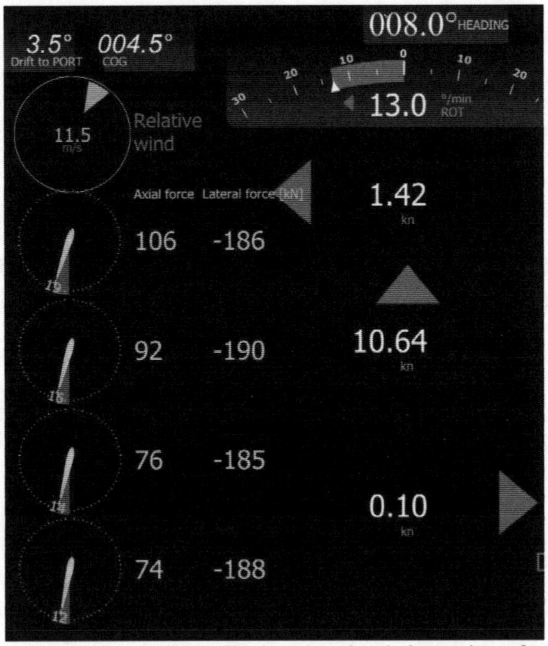

Figure 8: Test of prototype for conning display showing information of wind and wing data

5. CONCLUSIONS

The current work consists of model tests, velocity prediction program computations, time domain simulations and a real-time simulator. The results show the importance of developing the methods hand in hand for understanding the possibilities for each method and identifying where refinements can be made. The current paper also displays that the tools complement each other and that each tool is suitable to use during various stages of a ship design project. The results also demonstrate the importance of utilizing complementary tools for evaluating a design. Results also show that single and small design alterations regarding ship manoeuvring can give a significant effect on the ships overall performance. It can also be concluded that the methods and tools developed correlate well to experiments and that computational tools can efficiently be used in the various stages of a ship design project. Proposedly the usage of more refined, higher fidelity methods can be increased as a ship design project moves along, and it is likely that planning of the analyses with the right tools at each design stage improves the performance of the finished ship.

The simulator tool proves to be a valuable method for assessment of a multitude of challenges and questions that arise when operating a ship with WPS. During design and development work the importance of having a good situational awareness of the wind condition and the wing forces acting on the ship, and how they relate to forces and moments from available manoeuvring devices, became obvious.

Using a real-time ship simulator in the design process proved to be valuable when developing and testing the bridge user interface prototypes. It is also useful to understand the suitable use of automated control versus manual control. The simulator can be used in the future to achieve higher level automation strategies e.g., give maximum propulsive force ahead or wind vane all wings.

In order to operate the WPS in a safe and efficient way, the ship crew will need education and training and the simulation tool will serve as an effective way to realise such training.

6. ACKNOWLEDGEMENT

This work was made possible by financial support from the Swedish Transport Agency under grant number TRV 2018/96451 (Vinddrivet biltransportfartyg), and EU Horizon project OPTIWISE.

7. REFERENCES

1. NORRBIN, N. H., 'Ship Handling Standards - Capabilities and Requirements', *MARSIM & ICSM 90*, 1990
2. OTTOSSON, P. BYSTRÖM, L., 'Simulation of the Dynamics of a Ship Manoeuvring in Waves', *SNAME Annual Meeting 1991*, 1991
3. OTTOSSON, P., 'Mathematical models in PORTSIM', *MCMC Southampton*, 1994
4. GERHARDT F.C., WERNER S., HÖRTEBORN A., LUNDBÄCK O., NISBET J., OLSSON T., 'HORSES FOR COURSES: HOW TO SELECT THE "RIGHT" WIND PROPULSION SYSTEM AND HOW TO MAKE THE BUSINESS CASE', *in the Proceedings of the RINA Conference Wind Propulsion, London, UK*, 2021
5. OLSSON, F., MARIMON GIOVANNETTI, L., WERNER, S., FINNSGÅRD, C., 'A PERFORMANCE DEPOWERING INVESTIGATION FOR WIND POWERED CARGO SHIPS ALONG A ROUTE', *InnovSail*, 2020.
6. MARIMON GIOVANETTI, L., OLSSON, F., ALEXANDERSSON, M., WERNER, S., FINNSGÅRD, C., 'The effects of hydrodynamic forces on manoeuvrability coefficients for wind-assisted ships', *Proceedings of the ASME 2020 39th International Conference on Ocean, Offshore and Arctic Engineering*, 2020
7. A. PERSSON, D.-Q. LI, F. OLSSON AND S. WERNER, 'Performance predictions of wind propulsion systems using 3D CFD and route simulations,' in *Proceedings of the International Conference on Wind Propulsion, London*, 2019
8. MALMEK K., DHOMÉ U., LARSSON L., WERNER S., RIGNSBERG J.W., FINNSGÅRD C., 'COMPARISON OF TWO RAPID NUMERICAL METHODS FOR PREDICTING THE PERFORMANCE OF MULTIPLE RIGID WING-SAILS', *InnovSail*, 2020
9. K. MALMEK, L. LARSSON, S. WERNER, J. W. RINGSBERG, R. BENSOW, C. FINNSGÅRD, 'Rapid Aerodynamic Method for Predicting the Performance of Interacting Sails on Wind Powered Ships', Submitted to Ocean Engineering, under review.
10. F. C. GERHARDT, M. KJELLBERG, I. WIGREN, S. WERNER, M. RAZOLA, "THE HORNS OF THE TRILEMMA: SEAKEEPING MODEL TESTS FOR A WIND POWERED VESSEL"
11. MARIMON GIOVANNETTI, L., GERHARDT, F.C., KJELLBERG, M., ALEXANDERSSON, M., 'The art of model testing: Using CFD to adapt traditional tank testing techniques to a new era of wind propelled shipping,' in International Conference on Numerical Methods in Marine Engineering, MARINE, 2021

9. AUTHORS BIOGRAPHY

Stavros Kontos holds the current position of Senior Simulation Specialist at RISE. He is responsible for developing the wind propulsion module of SEAMAN-Winds and for assessing the performance of wind propulsion assisted ships. His previous experience includes developing numerical methods for hydrodynamic modelling of waves and floating structures.

Olov Lundbäck holds the current position of Senior Specialist Simulations and Algorithms at the maritime department at RISE (former SSPA). He is responsible for the development of SEAMAN simulation tools. His previous experience includes developing a simulation tool for seakeeping and manoeuvring in following and stern quartering seas.

Martin Kjellberg holds the current position of Senior Specialist Manoeuvring and Seakeeping at the maritime department at RISE. He is mainly working with seakeeping and manoeuvring and performance analysis of wind propulsion assisted ships. His previous experience includes development of numerical methods for seakeeping analysis and model tests.

Erland Wilske holds the current position of Senior Simulation Specialist at RISE. He works mainly with simulation studies for port and fairway design, assessment of ships manoeuvring- and sea-keeping performance.

Sofia Werner, Lead Researcher within Wind-Powered Ships at RISE Maritime (Formerly SSPA) is responsible for research and method development within Wind Power Ships at SSPA. Her previous experience includes towing tank testing and ship design using CFD. She currently chairs 30th ITTC Specialist Committee for Wind Assisted and Wind Powered Ships

TECHNICAL KEY PERFORMANCE INDICATORS FOR WIND-POWERED SHIPS

S Werner, RISE Maritime, Sweden
F Gerhardt, RISE Maritime, Sweden
S Kontos, RISE Maritime, Sweden

SUMMARY

The maritime wind propulsion industry evolves rapidly, and new technologies emerge on the market. This calls for standardised ways to express the technical performance of wind propulsion technologies (WPT) in a transparent manner. The wind propulsion community has not yet agreed on common key performance indicators (KPI). This complicates fair comparison, puts the level playing field at risk, and delays investment decisions.

A number of different KPIs describing the ability of a WPT to save a ships energy consumption are discussed in this paper. These have been gathered through industry workshops and interviews in co-operation with the International Wind Ship Association (IWSA), International Towing Tank Conference (ITTC) and the EU Interreg project WASP.

After an explanation of the underlying assumptions and calculation methods the investigated KPIs are exemplified for a generic ship equipped with different typical wind propulsion systems.

A set of recommended KPIs are suggested by the authors. We hope that this proposal will spark further discussions within the industry and facilitate the process of deriving useful and standardised metrics to assess wind propulsion technologies.

NOMENCLATURE

ρ_A	air density
η_D	total propulsive efficiency of ship
A	Projected area of WPU
C_D	Force coefficient from WPT aligned with to the apparent wind
C_L	Force coefficient from WPT perpendicular to the apparent wind
C_P	Coefficient of power contribution from WPT
C_X	Force coefficient from WPT in the ship's longitudinal direction
C_{X0}	Force coefficient from WPT in the ship's longitudinal direction at head wind
C_Y	Force coefficient from WPT in the ship's transversal direction
AWA	Apparent wind angle, relative course over ground (degrees)
AWS	Apparent wind speed (m/s)
EEDI, EEXI	Energy Efficiency Design Index
ESP	Energy saving potential of a WPT (J)
F_X	Force from WPT in the ship's longitudinal direction
IMO	International Maritime Organisation
ITTC	International Towing Tank Committee
KPI	Key Performance Indicator
P	Ship's propulsion power
PSP	Power saving potential of a WPT (W)
P_{in}	Power input, power consumption of WPT (W)
TWA	True wind angle, relative course over ground (degrees)
TWS	True wind speed (m/s)
V_s	Ship speed
W	wind matrix
WPT	Wind Propulsion Technology
WPU	Wind Propulsion Unit

1. INTRODUCTION

The maritime wind propulsion industry is evolving rapidly, and many new wind propulsion technologies (WPTs) emerge on the market. Such modern systems have little in common with the canvas sails of the old days and range from rotor sails over kites and suction wings to rigid sails that resemble vertical aircraft wings [1]. All these technologies have their specific strengths and weaknesses, which need to be assessed and quantified when selecting a WPT for a particular application. The wind propulsion community has, however, not yet agreed on common key performance indicators (KPI). Some technologies are described using aerodynamic coefficients, others by e.g. expected fuel savings. Percentage saving figures are commonly used, but it is often unclear what is included in the comparison. This complicates comparing technologies, puts the level playing field at risk, and delays investment decisions.

Before this background the Interreg North Sea region project WASP, the International Wind Ship Association (IWSA), and the International Towing Tank Conference (ITTC) have joined forces to develop and propose KPIs for wind-assisted ships. As part of this effort, several focus group meetings were held during the autumn of 2022. These online workshops were open to all stakeholders from the wind propulsion community and aimed at sharing ideas and discussing implications of various KPI alternatives. Figure 1 summarises and groups the participants.

In the paper we present several possible KPIs. Their advantages and drawbacks are discussed based on the industry workshops, written communication with industry representatives as well as with the authors own experience of wind propulsion applications for commercial ships.

ITTC will publish new Guidelines regarding power prediction for wind assisted ships in 2024. The present work has been conducted in close co-operation with *30th ITTC Specialist Committee for Wind Assisted and Wind Powered Ships*, and the outcome will influence new industry standard ITTC Guidelines.

The authors welcome readers who have additional comments or suggestions to contact us and continue the discussion.

Figure 1: A variety of Key Performance Indicators for wind assisted ships are used in the industry. ITTC, IWSA and the WASP project hosted focus group meetings with the industry aiming for deriving harmonised KPIs. The participants affiliation business types are show in the right graph.

2. KPI CATEGORIES

In this paper, key performance indicators are grouped into several categories and discussed separately. The categories range from A to D as shown in Figure 2. As illustrated, a Category A indicator assesses WPT performance on a general level, while a Category C indicator describes the result of a more (ship) specific assessment. Because the indicators in the four categories serve different purposes, they also need to satisfy different requirements. This is summarised in the right part of Figure 2.

Category A contains indicators that describe the characteristics of wind propulsion units without the presence of a ship. If nondimensionalized correctly, they should enable a fair comparison between technologies on a general level.

Category B consists of indicators that are ship specific and are obtained by "sailing" this ship along a pre-defined standard route. Such indices can be used for regulatory assessments, where easy, transparent, and fair definitions are more important than accurate fuel saving numbers. A typical KPI from this category is the Energy Efficiency Design Index (EEDI), where the influence of installing wind propulsion technologies is assessed by sailing the specific ship along a standard, world-wide, route.

Category C indicators are typically used in the context of analysing business cases for WPTs, or for communicating performance expectation between commercial stakeholders. KPIs in this category must therefore accurately express the expected energy or fuel savings for a particular ship trading in a specific pattern. The current paper pays particular attention to KPIs in this category because the absence of standardised Category C indicators is presently an important issue for shipowners and other stakeholders. Investment decisions often get delayed because it is difficult to compare competing technologies.

Category D comprises indicators of achieved performance during operation. These will not be discussed further in this paper.

As can be seen, indicators in these four categories serve very different purposes and it is of the utmost importance that all stakeholders in the industry understand this. KPIs of category C and D, for example, are based on ship and route specific parameters and cannot be used for comparison and ranking of WPTs in a general way.

General → Case specific

Category	Assesing	Describing	Typical usage	Requirement to a KPI: Fair "apple to apple" comparison	Ship specific	Accurate energy saving	Based on logged data
CAT A	Wind propulsion Unit (WPU) alone	Unit performance	Comparing technologies, design optimsiation	x	-	-	-
CAT B	WPU & Ship	Design index	Regulations, financial incentives	x	x	o	-
CAT C	WPU & ship & route	Energy or fuel saving (compared to same ship without WPT)	Business case, providers performance guarantee	-	x	x	-
CAT D	Performance achieved in operation	Energy or fuel saving achieved after some time of operation	Cost-saving splitting, reduced fees, rating	-	-	-	x

Figure 2: Categorisation of KPIs used in this study. To the right, requirements to possible KPIs. Here (x) denotes "required", (0) "optional", and (-) "not required".

3. TEST CASE

Throughout this paper three test cases are used to illustrate the proposed KPIs. They share the same ship hull and propeller but have different wind propulsion systems. It is important to note, that these examples are purely fictive cases. To highlight this, they are "anonymous" and will only be called WPT 1, 2 and 3.

The test cases are designed in such a manner that WPT1 and WPT3 have a rather similar fuel saving potential, whereas WPT 2 has a significantly higher saving potential.

The example ship is a 5000-dwt general cargo / bulk carrier with a length of 90m. The main parameters of the test cases are given in Table 1.

Table 1: The three test cases used throughout this paper

	WPT 1	**WPT 2**	**WPT 3**
Ship	5000 dwt general cargo, L=90m		
Max C_L	9.6	5.8	1.3
Max C_D	3.6	1.9	0.1
Active/passive	Active	Active	Passive
Projected area [m^2]	54	114	200
Deck footprint [m^2]	7	64	136
Foldable /tilting	no	yes	no

KPIs in categories C and D require a route or a trading pattern of the ship. In this paper four example routes are analysed:

- Rotterdam-Bergen (return trip)
- Copenhagen-Riga (return trip)
- New York- English Channel (return trip)
- Worldwide trade according to "EEDI matrix" [2].

The first three routes are illustrated in Figure 3 and a plot of the EEDI route can be found in IMO (2021) [2].

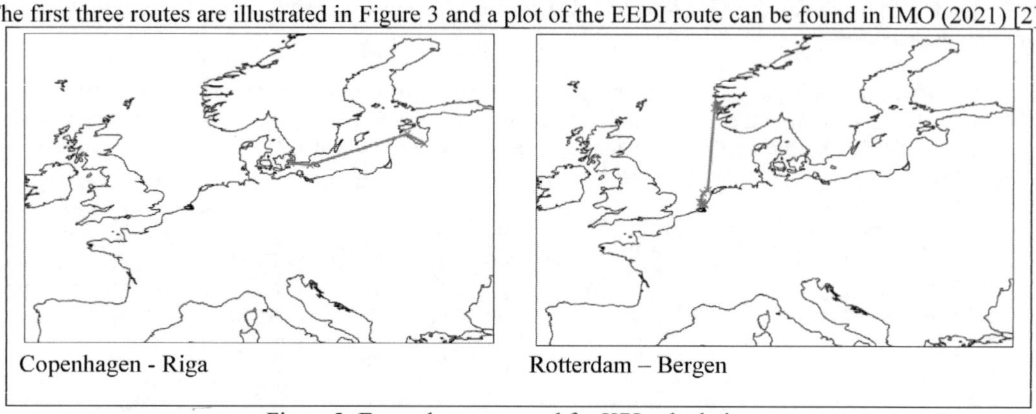

Figure 3: Example routes used for KPI calculations.

4. KPI ALTERNATIVES

This section introduces and explains possible performance indicators that have been developed and discussed in close cooperation with industry. The KPIs have also been applied to the three test cases, WPT1, WPT2 and WPT3. The results are summarised in colour-coded tables where the three design are ranked from highest performance (green) to the lowest performance (red) using the KPIs. A more extensive report, containing additional background information and additional KPIs will be published during the spring of 2023 [3].

4.1 CATEGORY A – STAND-ALONE UNITS

Indicators in this category should describe the characteristics of wind propulsion units alone, without considering a ship. "Characteristics" in this context can mean the forces they produce (thrust force, side force, lift, drag) at different wind angles or the propulsive power they are able to provide. Figure 4 provides an illustration of some typical features, which will be discussed in more details below. The results for the test cases are given in Table 2.

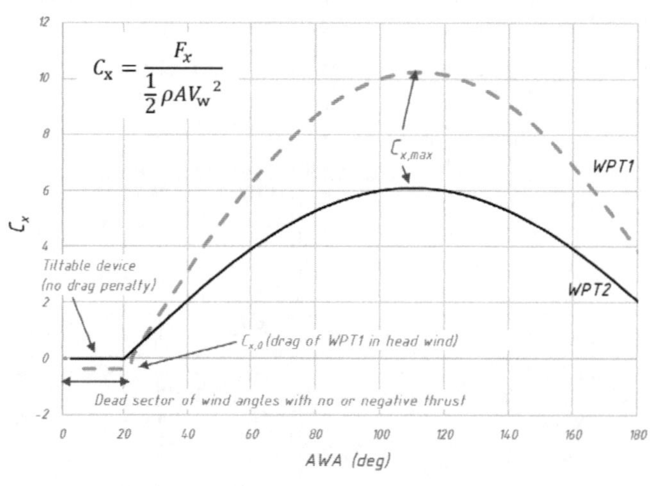

Figure 4: Example characteristics of WPTs, illustrated as
thrust force coefficient (C_x) vs. apparent wind angle curves for two different WPTs

4.1 (a) SINGLE, NONDIMENSIONAL KPIS FOR STANDALONE UNITS

Maximum lift coefficient $C_{L,max}$

The lift force is defined as the aerodynamic force component perpendicular to the flow. The maximum lift coefficient before severe stall, $C_{L,max}$, is commonly used in aeronautics to describe wings and lifting surfaces. It can be obtained by wind tunnel testing or viscous flow computations. Although the parameter has physical significance, it does not describe well how a wind propulsion device works on a ship. A similar KPI that describes the ship-specific situation better is the thrust force coefficient C_x.

Maximum thrust force coefficient $C_{x,max}$

By resolving "lift" and "drag" (aerodynamic force component in flow direction) into components parallel and perpendicular to the centreline of the ship the "thrust" and "side force" produced by the WPT can be obtained. A useful KPI could be the maximum value of the thrust force coefficient over all wind angles, $C_{x,max}$, see Figure 4.

Dead sector

The wind angles where the WPT does not contribute with power reduction can be called the "dead sector". There are both upwind and downwind dead sectors. However, the upwind sector is more critical since, for modern fast ships, small apparent wind angles occur more frequently than larger ones close to following wind. The size of the dead sector varies significant between different types of WPTs, and it varies also with the ship speed. This is therefore an important characteristic to consider. On the other hand, "weather routing" can reduce the negative effect of a large dead sector and that could possibly make this KPI less relevant. The upwind dead sector is illustrated at the left of Figure 4.

C_L/ C_D

The ratio of lift to drag gives an indication of the "pointing ability" of the WPT. High lift to drag ratios are equivalent to small upwind dead sectors and result in a ship that can sail close to the wind.

Drag of idling WPT in head wind, C_{x0}

Some WPT's can be tilted or folded on deck when the wind is not favourable. The WPTs that are not tiltable create a resistance which is often not negligible. This is an important characteristic which distinguish the differ technologies from each other. Figure 4 exemplifies this for two different WPTs.

Power requirement efficiency

Active devices, such as rotors or suction wings, require some power input (P_{in}) to function. This is of course a significant characteristic of these technologies. A suitable KPI that address this could be the ratio of required power to the power that can be delivered to the ship. Both powers (in and output) vary with the wind as well as with ship speed. A simplified expression which is independent of ship speed it could be:

$$\eta_{WPU} = \frac{P_{\text{in}}}{\left(\max\left(C_x\right)\frac{1}{2}\rho A \cdot AWS^3\right)} \tag{1}$$

where wind speed AWS is 10 m/s, and P_{in} is power consumption at the AWS=10m/s and the AWA where C_x has its maximum. P_{in} should include all losses, mechanical, aerodynamic and others i.e. this should be the power that has to be supplied by the main engine to drive an active WPT.

Looking at the η_{WPU} results from the test cases (Table 2) difference between devices 1, 2 and 3 are clearly visible, with WPT1 consuming the highest power.

KPIs related to geometrical constraints

In the absence of geometrical or financial constraints, the largest device is likely to give the highest fuel saving. In reality, the winning device will be the one that maximises power within the given geometrical limitations. Such limitations can e.g. be deck space, air draft or weight. During a concept development project, it could be relevant to use coefficients based on these parameters. As an example, Table 2 includes $C_{L,max}$ /deck-footprint (i.e the deck area required for the WPUs).

4.1 (b) COMBINATIONS OF NONDIMENSIONAL KPIS FOR STANDALONE UNITS

Combinations of single coefficients

As Table 2 shows it is not possible to rank WPTs based on a single KPI, this is because the ranking of the three devices is different for the different KPIs. One way around this could be a to use a combination of several KPIs, including at least:

[$C_{x,max}$; C_L/C_D ; C_{x0} ; η_{WPU}].

Alternatively, a graphical representation such as plotting coefficients over wind angles could be useful.

Thrust force coefficient C_x curve over wind angle

The most explanatory way to describe the aerodynamic characteristics of the WPT's is to show the full thrust force coefficient (C_x) vs. apparent wind angle curve. This can either be done in cartesian coordinates (as in Figure 5, lefthand plot) or as a polar diagram. Such curves provide a rather complete picture of WPT performance and many of the other KPIs discussed above can be obtained from the plots (e.g. width of the dead sector or $C_{X,max}$). The disadvantage of this approach obviously is that this is not a single number but a plot that requires interpretation by an expert.

Net power coefficient C_{Pnet}

A power coefficient could be one way to include the effect of power demand (PTI) and the power that the WPT can deliver to the ship. This would make it possible to compare "active" and "passive" devices in a fairer way. Ideally, since this is still a stand-alone unit coefficient, it should be independent of ship specific parameters and wind speed. However, this is not possible to derive. The ship's speed affects the ranking between high lift and high efficiency WPT's. Moreover, for a fair comparison between active and passive devices, the ship's propulsive efficiency η_D need to be included. Otherwise, the P_{in} is too large in relation to the power that can be used for propulsion. A standard value of 0.7 could be used for η_D , even though this is also unfair to active device if the real propulsion efficiency is lower.

A standard power coefficient could then be derived as

$$C_{Pnet\_V_s} = \frac{\dfrac{C_x \frac{1}{2}\rho A \cdot AWS^2 \cdot V_s}{\eta_D} - P_{in}}{\frac{1}{2}\rho S \cdot {V_s}^3} \tag{2}$$

where η_D =0.7, S is a typical ship's wetted surface.

The ship's speed should be given as a subscript in the coefficient name, for example C_{Pnet_10kt} for a ship speed of 10 knots. This coefficient is *not* independent of wind speed, hence a reference true wind speed must be given, for example 10 m/s.

In the same way as for the thrust curve, the most complete picture is given by presenting the C_{Pnet} curve over AWA or TWA. Figure 5 (right) shows the curve for the three test cases. The C_{Pnet} curve reveals a different relation between the WPT's than the C_x-curve. By including the power consumption, the relative inferior performance in upwind of the active devices WPT2 with the largest power requirement is clearer.

The plots over apparent wind angle can be either polar or cartesian plots. It is the authors opinion that the cartesian plots display the different characteristics in a clearer way.

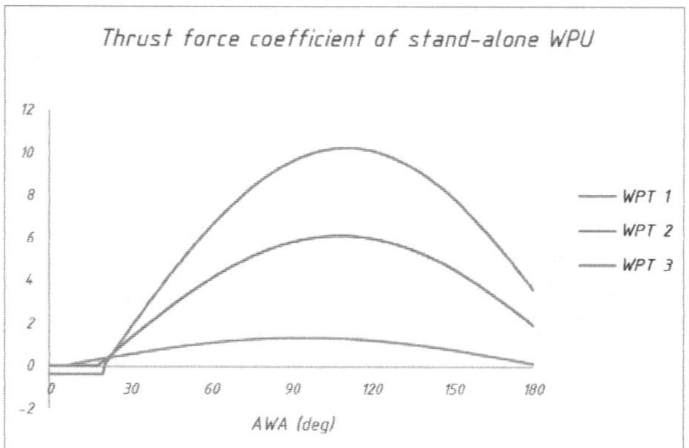

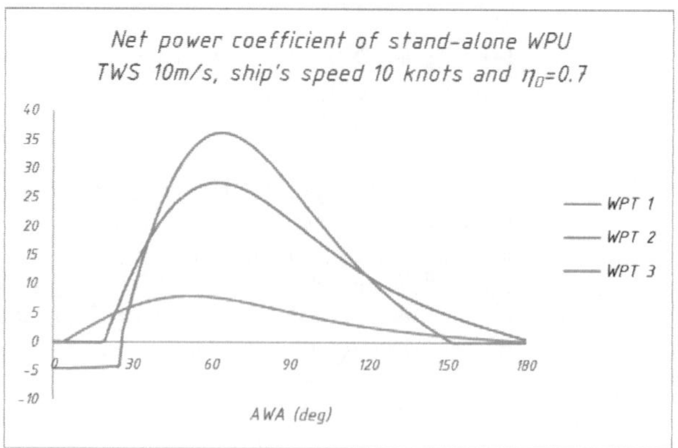

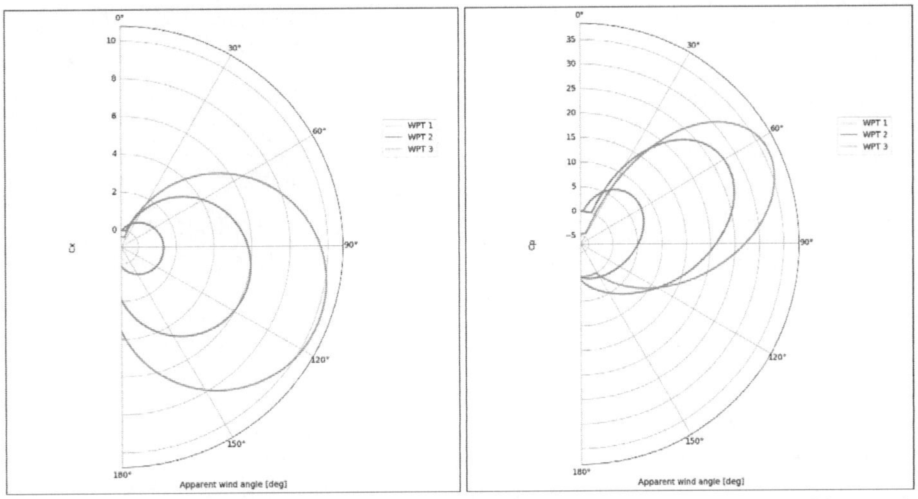

Figure 5: Thrust force coefficient (C_x) and net power coefficient (C_p) for the test cases. With cartesian plots the differences in characteristics may be clearer.

4.1 (c) DIMENSIONAL KPIS FOR STANDALONE UNITS

Rated WPU power [kW]

Wind propulsion units are usually offered in standard sizes. In the early concept phase, when scanning the market and shortlisting possible devices, it could be convenient to have easy, ship independent indicator of a unit size or its power. This could be achieved via a nominal or "rated" power derived in the same way as the "f_{eff}*P_{eff}" in MEPC.1/Circ.815 (2013) [4].:

$$Rated\ WPU\ Power_{10} = \sum_{i,j}^{n,m} \left[\frac{F_x \cdot V_s}{\eta_D} - PTI\right]_{i,j} \text{ x } [W_{i,j}] \qquad (3)$$

where

$W_{i,j}$ is the EEDI weather matrix (pre-2021)

η_D=0.7

F_x is the force matrix at the corresponding winds

V_s=10 knots

Rated power values for other standard ship speeds (e.g. 15 and 20 knots) can be worked out in a similar way This KPI is one way to describe a device while including the size. It is a theoretical number and gives an indication of a unit's theoretical potential. It should be understood that it this ignores the effects of aerodynamic side forces and that the propulsive efficiency, η_D at may be different for a specific ship. This expression "punishes" active devices if η_D is larger than the actual for a specific ship, but this flaw is at least smaller then if not including η_D at all.

Wind Route Indices

Another way to produce a dimensional "ship independent" KPI is to use a standard ship on a standard route and show the fuel saving potential. There is, however, a certain risk that such an analysis could be misinterpreted as a ship specific business case. This can lead to misleading conclusions and unfounded decisions.

Table 2: Category A KPI results for the test cases. Green means highest performance, red lowest.

Suggested KPI	unit	WPT 1	WPT 2	WPT 3
Max lift coefficient C_L		9.6	5.8	1.3
Max thrust coefficient C_x		10.3	6.1	1.3
Max C_L/C_D		2.7	3.1	11.1

C_x at AWA=0 (if not tiltable)		-0.4	0.0	0.0
Max C_L / deck footprint		1.4	0.1	0.01
Dead sector upwind, based on C_x	deg	21	19	5
Dead sector upwind, based on C_p (10 knots)	deg	27	19	5
η_{WPU}		15%	4%	0%
Max net power coefficient C_p (10 kn)		3.8	2.9	0.7
Rated WPU power (9 knots)	kW	38	63	44

4.2 CATEGORY B: INDICES – SPECIFIC SHIP CASE ON GENERIC ROUTE

Several indices in this category already exist (e.g. EEDI, EEXI [2,4]) and others are currently being developed. It is anticipated that the EEDI regulations will be revised the coming years. The current paper focuses mainly on the real-life performance indicators of the Category C type, which will be examined in the next section. Below we therefore limit the discussion of Category B indicators to the controversy around the "weather matrix". In the updated EEDI/EEDI Guidance On Treatment Of Innovative Energy Efficiency Technologies from 2021 [2], 50% of the most adverse wind conditions can be deleted from the weather matrix, i.e. ignored for EEDI rating purposes. As can be seen in Table 2, this changes the ranking between WPT1 and WPT3 (due to the different aerodynamic characteristics as described in previous section), even though the differences between them is marginal.

For this reason, some members of the industry criticise the 2021 version of the regulation for being not technologically neutral. Other voices support the change, with the argument that removing unfavourable weather conditions reflects better ship operations using weather routing.

The EEDI formula can also be criticised for not including side force effects, which in theory unable fair treatment of all WPT-types. For moderate wind assisted propulsion this is neglectable, but for more powerful installations and for ships with primary wind propulsion, this simplified equation is not suitable.

Table 3: EEDI/EEXI value for a fictive ship with three different Wind Propulsion Technologies, using the IMO regulations before and after 2021. The 2021-revision can change the ranking between devices of similar performance (like WPT1 and WPT3).

IMO regulation	**unit**	**WPT 1**	**WPT 2**	**WPT 3**
MEPC.1/Circ.896, (IMO, 2021)	g CO2/(t*nm)	10.97	10.57	11.01
MEPC.1/CIRC.815 (IMO, 2013)	g CO2/(t*nm)	11.40	11.16	11.35

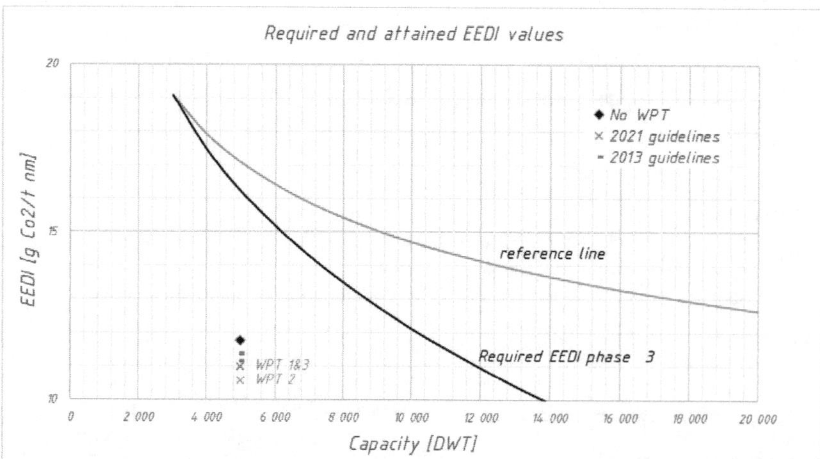

Figure 6: EEDI/EEXI value for a fictive ship with 3 different Wind Propulsion Technologies.

4.3 CATEGORY C – PERFORMANC EEXPECTATIONS FOR SPECIFIC SHIP AND ROUTE

4.3 (a) POWER SAVING, FUEL SAVING, ENERGY SAVING OR CO_2 SAVING?

The purpose of the indicators in this category is to communicate a realistic expectation of the saving potential from a WPT. They are typically used as decision support for business cases, or for agreements between commercial stakeholders

The saving due to a WPT in absolute terms can be expressed either as power, fuel, energy, or CO_2 saving. While this choice will not affect the ranking between the WPTs for the same ship it can make a difference in how far a KPI resonates with different stakeholders of the shipping industry. Owners and operators tend to think in tonnes of bunker per day, engineers are more familiar power or energy-based numbers, lawmakers will mostly focus on CO_2 savings.

If we consider ships with conventional diesel engines and power saving for constant operational speeds, then there is an almost linear relation between the predicted power, fuel, and energy savings. This allows for an easy conversion between the savings expressions by simply post-processing route simulation results. For hybrid propulsion systems, and if routing and speed optimisation are included, the relation is not that so straight forward and must be modelled within in the voyage simulations. Some of the advantages and disadvantages of the various units for savings are summarised below. These points were raised during discussions with industry stakeholders.

Fuel saving

- Ship owners can relate to fuel (tons/day and kg/h).
- Tons/miles makes it easier to scale to different routes.
- Can be used in a transition period when HFO is still the standard in shipping.
- Need modelling of machinery efficiency, or stipulate fixed specific fuel oil consumption (SFOC).
- Not easy for hybrid propulsion systems.

Energy saving

- Using propulsion energy instead of fuel allows to leave out engine efficiency.
- For specific cases ship owners can translate to fuel themselves.
- More future proof considering future fuels.

CO_2 savings/ CO_2s avoided

- Parameter assessed in EEDI, CII and emission trading schemes like EU-ETS.
- Of interest to the wider society.

Power saving

- Propulsion power is a measure that both yards, designers, ship owners and operators are familiar with.
- It does not require any modelling of engine efficiency or assumptions of fuels and hybrid propulsion.
- WPT providers do not always have information on the details of the propulsion system or engine efficiency. In this case the power makes for safe option for the saving prediction.

Considering these pro and cons, it appears that "power savings" are the most feasible way of expressing WPT performance and are also a concept that is familiar to owners and operators. In this paper we will therefore in the following focus on power-based KPIs.

4.3 (b) KPIs for power saving

A number of possible KPIs are suggested below, and again corresponding results for the test cases are given in Table 4. It is assumed that a power saving ΔP has been computed using a numerical model or formula. We will not discuss the prediction *methods* as such here but focus on how the results could be *presented.*

Table 4: KPI expressing power saving expectation for test fictive cases. EEDI means pre-2021 weather matrix.
Green means highest performance, red lowest.

	unit	Route	Ship Speed (knots)	WPT 1	WPT 2	WPT 3
Rated WPU power / Ship Propulsive power	%	EEDI	9	4.3%	7.1%	4.9%
ΔP Sea legs only, calm water	%	EEDI	9	6.0%	9.7%	6.0%
	%	Rot-Bergen	9	8.6%	13.2%	8.0%
	%	Rot-Bergen	13	4.5%	7.3%	4.4%
	%	Cop-Riga	9	6.1%	9.7%	6.0%
	%	NY- UK	9	10.2%	16.1%	9.5%
ΔP Sea legs only, incl S.M.	%	EEDI	9	5.3%	8.7%	5.4%
	%	Rot-Bergen	9	7.4%	11.6%	7.1%
ΔP vs. Total fuel consumption	%	EEDI	9	4.5%	7.2%	4.5%
	%	Rot-Bergen	9	6.2%	9.7%	5.9%
ΔP max, TWS=10m/s	%		9	27%	37%	19%
ΔP max, TWS=10m/s	kW		9	258	345	180
ΔP design	kW		9	170	233	160
ΔP most likely weather	kW	Rot-Bergen	9	92	104	100
ΔP most likely weather	kW	Cop-Riga	9	-7	11	0
ΔP yearly average, Sea legs only, incl. waves	kW	EEDI	9	57	91	56
	kW	Rot-Bergen	9	80	124	76
	kW	Rot-Bergen	13	102	165	99
	kW	Cop-Riga	9	57	91	56
	kW	NY- UK	9	95	151	90

Percentage

The percentage fuel saving is the most common KPI in communications around wind propulsion today. One could think that a percentage saving is a clear KPI that can be used for comparison between different installations, since it is nondimensional. Very often, percentage saving claims are published without any further description of the specific cases. This is, however, a problematic approach.

The first issue is to *what* the savings have been related , i.e. what number to have in the denominator. The left graph in Figure 7 shows the percentage power saving of the three test cases computed in different ways. First, the fuel saving is predicted for the sea legs in calm water. The ship's propulsion power when employing the WPT is compared to the propulsion power when there is no WPT, for the same sea leg and same speed:

$$\Delta P\% = \frac{P_{no\,WPT} - P_{with\,WPT}}{P_{no\,WPT}} \tag{4}$$

Secondly, the sea margin, or added resistance in waves, is included in the prediction. This is the standard procedures for some organisations, whereas others do not include it. It has minor effect on the predicted fuel or power saving in absolute terms (tons, kW), but it has a significant effect on the percentage, since it increases the denominator in eq (4). For WPT2, as an example, it makes the saving to decrease from 13% to 11.5%.

For the third group of numbers in Figure 7, the comparison is done against the ship's *total* fuel consumption, i.e. not only the fuel used for propulsion on the same sea legs. The denominator hence includes port manoeuvres, hotel load etc. That makes the saving to decrease further still, to 10% for WPT2. It should be noted that the trends between the WPTs are preserved: in all cases WPT 2 is still the "winner".

Finally, the %-saving is also shown for a higher ship speed, 13 knots, to illustrate how much this can affect the saving number. The saving is now down to 7%. The trends are still the same, but this is not a general conclusion since some technologies work better for higher speeds than others.

The righthand graph in Figure 7 shows that the power savings differ considerable between different routes. For WPT2, as an example, the fuel saving is 10% for the least favourable route and 16% for the most favourable. The trends between the three WPTs are preserved, although the advantage of WPT1 over WPT3 differs between the routes.

These examples illustrates that a percentage saving number, taken out if its context, may be misleading. A percentage number gives the false impression that it can be universally compared with other percentage saving predictions. For the reasons addressed here, many of the industry partners participating in the study are sceptical of using this KPI.
It could, however, be useful as limits, for example when a rule is applicable or not.

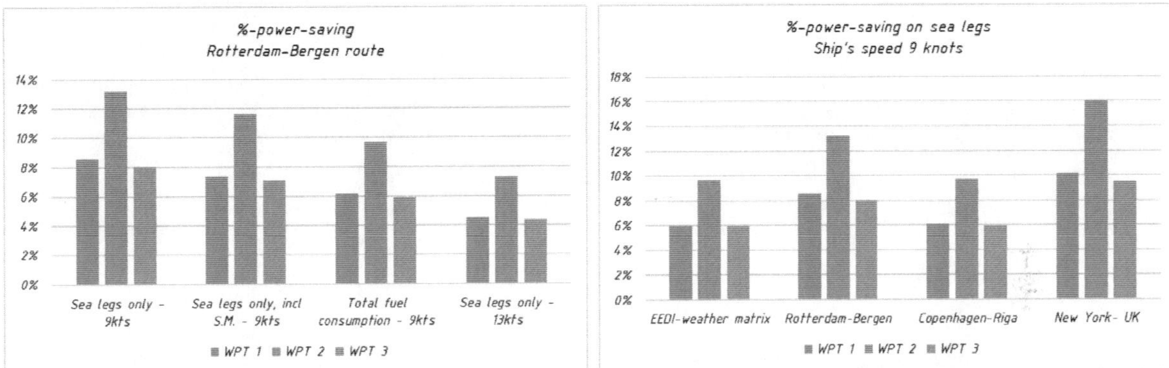

Figure 7: Percentage power saving can be computed in different ways, and for different conditions, which results in different performance expectations.

Early estimation
In the previous section on Category A KPIs, a "Rated WPU power" was defined for the purpose to be used in very early estimation of saving potential. By comparing the Rated WPU power with the ship's propulsive power at the same ship speed, we can get a rough estimate of the power saving potential. For the three test cases this indicates a saving of around 7%. The more accurately predicted average saving for the EEDI weather showed around 10%. The Rated WPU power could hence be used as a first rough approximation for fuel saving.

Power reduction
The performance of WPT could also be expressed as power reduction in kW, $\Delta P = P_{no\ WPT} - P_{with\ WPT}$. Several options for a standardised KPI related to ΔP were suggested by the industry partners involved in this study:

i. Max ΔP for TWS=10m/s (at the best wind direction)
ii. ΔP at a specified "design point" which could be for example TWS 10 m/s, TWA=60 deg
iii. ΔP at the most frequent weather
iv. ΔP on a given route, averaged over a year to include all season's weather
v. ΔP from the EEDI equation (pre or post 2021), but using weather statistics for an actual route

Figure 8 shows these options for the test cases. The advantage of option i) – iii) is that the prediction does not require any routing/voyage analysis tool. However, selecting one point where ΔP is extracted can give large over or under predictions of the power reduction (left graph in Figure 8) compared to the average saving on a route (righthand graph in Figure 8). For example, the maximum power reduction of WPT 2 is 345kw (37%), whereas the yearly average power reduction on a favourable route is about a third of that, (124kW, 12%). Communicating these two numbers would result in rather different expectations and business cases. The maximum ΔP gives also a misleading comparison between high lift versus high lift/drag devices, as explained above regarding maximum C_x. The same drawback is true for the option to derive ΔP at a specified "design point".

Selecting the most frequent weather as the design point is completely misleading. The most frequent weather could well be head wind, where WPTs cannot even be employed.

Some industry partners suggest ΔP to be derived for a given route, including route optimisation. This would give a more favourable KPI for all WPTs, but it would especially favour high lift/drag devices versus high lift devices. It can be argued that including route optimisation would reflect the real operation better. It can, on the other hand, be argued that the KPI should reflect the technical performance and not operational measures. Successful routing depends on factors like the skill of the crew, safety, logistics requirement etc. These are very hard to incorporate in a standardised KPI.
The solution could be to clearly distinguish between KPIs including and not including routing.

Power Saving Potential (PSP)
No matter how accurate we try to model the power saving on a route, it is still a theoretical value. The real saving achieved in operation will depend on many practical aspects which cannot be foreseen in the predictions, such as maintenance time, changed route and speed, changed hull efficiency due to fouling, crew skill, function of the automated WPU control system, icing and wear. For this reason, it would be wise to denote the predicted ΔP the "Power Saving Potential". This will indicate that it is an ideal number derived under certain conditions. The PSP should be possible to verify during a short, controlled sea trial. After that, it is up to the owner, operator, and crew to use this potential in the best way.

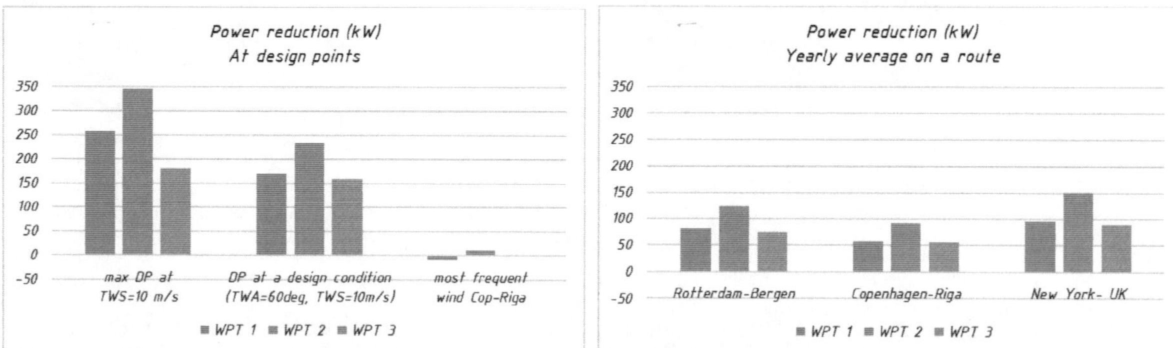

Figure 8: Power saving can be computed in different ways, and different conditions, which results in different performance expectations.

4.3 (c) OTHER KPIS

Other indicators that could be expressed in a description or offer of WPT includes:

- Operability range, such as limitation in top wind speed. Could be expressed as percentage of the EEDI matrix.
- Deck area footprint, air draft, weight and centre of gravity
- Noise and vibration
- CAPEX OPEX, LCC/LCA
- Crew training requirement

5. CONCLUSIONS AND RECOMMENDATIONS

Key performance indicators (KPI) for wind assistance ships have been discussed in a wide group of industry stakeholders. The need from the industry can be summarised as:

- It would be helpful to have a set of agreed KPIs, especially for expressing ship specific fuel saving potential in the business case and procurement phase.
- KPIs that will be communicated to ship owners and operators should be tangible to their business.
- It would be useful if the KPIs reflect the prediction method used to derive the value.
- It would be useful to have a set of KPI definitions that is consistent and can accompany the ship design process from early concept stage to operation stage, gradually adding more and more complexity.

Based on these discussions and the case study in this paper, the authors present here their recommendations. These will be further processed by ITTC Specialist Committee for Wind Assisted and Wind Powered Ships for a future inclusion in ITTC's Guidelines. Readers who have additional comments or suggestions are most welcome to contact the authors.

5.1 RECOMMENDED KPIS FOR STAND-ALONE WIND PROPULSION UNITS

The authors recommendations:

- Nondimensional stand-alone coefficients are useful to understand the characteristics of a WPT but not directly the fuel saving potential. This is relevant for experts working with the design or assessment of wind powered ships. However, it should not be the first choice when communicating performance with community in general.
- There is no single nondimensional coefficient that describes all the important characteristics of a WPT. The best demonstration of the complete picture is a power coefficient curve over apparent wind angle (eq. 2).
- The "Rated WPU power" could be a standard way to express the potential power for a single device in marketing and for short-listing suitable WPUs during early concept phase.

It is important that Standard conditions for nondimensional KPIs are defined by for example ITTC. That could be:

- Uniform/rectangular wind profile
- Area A defined as projected/planform area of WPT
- Standard air density of 1.204 kg/m3

5.2 RECOMMENDED KPIS FOR PERFORMANCE EXPECTATION AND BUSINESS CASE INPUT

The contribution of the WPT to the ship's propulsion can be expressed as power reduction, fuel saving, energy saving or CO2-saving. We recommend power saving to be used as standard. However, nothing prevents showing them all in a prediction report.

Some industry partners request that KPIs derived by different organisations should be truly comparable with each other also in terms of the derivation method. To ensure that all actors in the industry derive indicators that are truly comparable would require that some organisation could derive detailed procedures prescribing methods for CFD simulations and wind tunnel test. This is not a feasible solution. A true "apple to apple" comparison can only be achieved if the same simulation platform is used for the cases to be compared. However, we believe that it would be an improvement compared to today's situation if the industry agreed on a number of KPIs that are linked to certain levels of confidence. In this section we present a suggestion for how this could be achieved.

A Power Saving Potential (PSP) is derived by comparing the power requirement for a ship with WPT against the same ship without WPT on the same route and same speed. This can be done with a matrix multiplication of a weather matrix or by voyage simulations, in both cases using hind-cast weather covering all year around weather conditions. The power prediction should include the power consumption by the WPT if active, the drag from idling device if non-tiltable, and the operability range. The PSP is the ideal performance, assuming 100% operability within the operation range. The real saving may be lower due real life practicalities such as repairs, logistics etc. It can also be higher, of smart routing and energy management is used onboard.

The PSP can be derived using methods and input of different confidence levels. In the early stage of a concept development or feasibility study for a retro-fit, low fidelity methods and input based on estimations is sufficient. As the project advances, more accurate predictions are required for the business case support and performance expectations between provider and ship owners. Table 5 presents a system of KPIs for different confidence levels. An overview description of the corresponding power modelling methods are presented in Table 6. The details should be given in for example the coming ITTC's Guidelines.

Table 5: Recommended KPIs for Performance expectation and business case input

KPI	**Unit**	**Usage**	**Power modelling*)**	**Weather modelling**
Rated WPU power	kW	General comparing, scanning the market	Stand-alone WPU power	EEDI
PSP-I	kW	Early idea	Level I	EEDI or the ship's intended route
PSP-II	kW	Early business case assessment	Level II	The ship's intended route
PSP-III	kW	Business case & Performance expectation	Level III	The ship's intended route
ESP-IV	kW	Advanced Business case & Performance expectation	Level IV	The ship's intended route (incl. possible weather routeing, speed optimisation)

*) See table 6.

Level I-III are recommended for wind assisted ships that do not use weather routing or speed optimisation.

Level IV is recommended for ships that use extensive weather routing, primary wind powered ships, and ships with advanced hybrid propulsion systems. Modelling can include different speed profiles and different routes for the case with and without WPT. For this level, it is feasible to derive the *energy* saving rather than average power saving. For primary wind powered ships, the comparison can even be against other ship sizes and ship speeds. The energy saving should then be related to a constant cargo flow.

Table 6: Suggestion for specification of methods to derive Power Saving Potential or Energy Saving Potential at various confidence levels. Details to be given in for example the ITTC's Guidelines.

	Level I	**Level II**	**Level III**	**Level IV**
Degree of freedom	*1DOF*	*4DOF*	*4DOF*	*4-6DOF*
WPT forces	*Generic*	*Low/Mid fidelity*)*	*High fidelity**)*	*High fidelity*
Power requirement of active WPU	*Generic*	*Specific*	*Specific*	*Specific*
Ship propulsive efficiency	*fixed* η_D	*POW Propeller open water curves*	*POW +effect of leeway on prop*	*POW +effect of leeway on prop*
WPU-WPU interaction		*Low/Mid fidelity*	*High fidelity*	*High fidelity*
WPU-hull interaction		*Low/Mid fidelity*	*High fidelity*	*High fidelity*
Hydrodynamic effect of side force		*Low/Mid fidelity*	*High fidelity*	*High fidelity*
Heel		*Low/Mid fidelity*	*High fidelity*	*High fidelity*
Effects of ship motions and varying wind, incl. control systems response time				*optional*
Hybrid-propulsion (diesel electric) with energy management optimisation				*optional*

*) **Low/Mid fidelity** methods can be for example high fidelity data or regression models from similar cases, or case specific lifting line methods

) **High fidelity refers to case specific CFD, model test or full-scale test.

It is recommended not to use %-savings as KPI. If used, the standard way should be to compare propulsion power with device including power consumption, against power without WPT, both derived with the same numerical tool for the same route, same speed, and including wave added resistance or a suitable sea margin.

6. ACKNOWLEDGEMENTS

The work is made possible thanks to funding by EU Interreg North Sea Region, project WASP.
The authors wish to thank all who participated in the industry workshops or in other ways contributed with feedback and suggestions.
The members of the *30th ITTC Specialist Committee for Wind Assisted and Wind Powered Ships* contributed with valuable ideas and discussions.

7. REFERENCES

1. CHOU, T., KOSMAS, V., RENKEN, K., ACCIARO, M. (2020), New Wind Propulsion Technology - A Literature Review of Recent Adoptions, Interreg North Sea Europe Wind Assisted Ship Propulsion (WASP) report D 5.B.

2. IMO, MEPC.1/Circ.896, (2021) Guidance On Treatment Of Innovative Energy Efficiency Technologies For Calculation And Verification Of The Attained EEDI And EEXI, 2021.

3. WERNER, S. (2023) "Performance Indicators for Wind Assisted Ships", RISE report RE40201042-06-00-A.

4. IMO, MEPC.1/Circ.815 - 2013 Guidance On Treatment Of Innovative Energy Efficiency Technologies For Calculation And Verification Of The Attained EEDI (MEPC.1/CIRC.815, 2013

8. AUTHORS BIOGRAPHY

Dr Sofia Werner, Lead Researcher within Wind-Powered Ships at RISE Maritime (Formerly SSPA) is responsible for research and method development within Wind Power Ships at SSPA. Her previous experience includes towing tank testing and ship design using CFD. She currently chairs 30th ITTC Specialist Committee for Wind Assisted and Wind Powered Ships.

Dr Frederik Gerhardt holds the current position of Lead Naval Architect at RISE (formerly SSPA). He is responsible for developing RISE's business area Wind-Powered Ships. His previous experience includes seakeeping and manoeuvring testing as well as research on yacht aerodynamics. He currently chairs the 30th ITTC Seakeeping Committee.

Dr Stavros Kontos holds the current position of Senior Simulation Specialist at RISE. He is responsible for developing the wind propulsion module of SEAMAN-Winds and for assessing the performance of wind assisted ships. His previous experience includes developing numerical methods for hydrodynamic modelling of waves and floating structures.

Wind Propulsion Conference
16th – 17th February 2023
London, United Kingdom

HIGH PERFORMANCE COMPOSITE MATERIALS FOR WIND PROPULSION TECHNOLOGIES: CHALLENGES & OPPORTUNITIES

T JAMES, Hexcel, United Kingdom

SUMMARY

2022 proved to be a notable year for wind propulsion in shipping, seeing a significant gathering of momentum in both wind propulsion systems installed, and the overall visibility and perception of wind as a viable and important part of shipping's sustainable future. With the shipping industry continuing to work towards the International Maritime Organisation's (IMO) challenging Green House Gas (GHG) emissions reduction targets (40% reduction by 2030 and 50% by 2050 versus 2008 levels), and wind propulsion offering a very meaningful potential contribution towards this mission, the author is optimistic that this trend will continue. Of particular interest is the gradual slowing down of vessels as shipping lines bank fuel burn percentage savings several times the percentage drop in cruising speed needed to achieve them, thereby allowing older vessels to comply with emissions reductions targets and bringing passage times back into range of those historically achieved by sailing ships. Analysis of currently active or planned wind propulsion technologies shows that a broad range of materials are being selected today, with high-performance composite materials already playing a significant role - especially in high power systems where they are an enabling technology. This paper will look in detail at some of the experiences of other highly industrialised and commoditised composite market segments, such as wind energy, automotive and aerospace, which may provide useful input for wind propulsion manufacturers seeking to evaluate composite material solutions.

1. WIND PROPULSION TECHNOLOGIES: STRUCTURAL CONSIDERATIONS

To conveniently describe different types of wind propulsion rigs further, we can consider the following primary groups that are in existence and/or under development today. Note this list is not exhaustive; other concepts exist that are not considered here:

1.1 SOFT SAILED RIGS.

Conventional sailing rigs consist of soft sails supported by structural spars, which - along with rigging elements, if present – transfer sail loads to the ship. The spars may be supported by conventional rigging elements, or freestanding. A simpler deck layout can be achieved with freestanding rigs, but all sail loads must be reacted by the unsupported spar, cantilevered from the deck. Within this category are included the modern square rig concepts, such as the DynaRig by Southern Spars™[i], which has updated the design using multiple masts that are freestanding and rotate around bearings at deck level. Other soft sail solutions include concepts for inflatable sail plans, such as Michelin's Wing Sail Mobility (WISAMO) design which will see its first 100m^2 rig installed on a 154m Ro-Ro cargo vessel in 2023[ii], and the Aerorig type 'Solid Sail' from Chantiers de l'Atlantique. Despite its name the large Solid Sail composite panels are hinged together allowing them to be raised, lowered and flaked for storage exactly like a traditional soft sail[iii]. In most cases, the structural spars are intended to remain in position during all normal operation; depowering is achieved through reefing the sails. Retractable or folding soft sail rigs with freestanding spars is possible; where the spars are conventionally rigged this becomes impractical.

1.2 HARD WING SAIL RIGS

This classification is typically used for rigid profiled wing sails, either with or without additional trimmable flaps, which are often fully rotating and installed in series along the vessel deck. Rigid wing skins are supported by internal ribs and structural spars which transfer sail loads to the ship. Wing sail rigs are typically freestanding, with a single connection from the base of the structural spar to the ship deck. BAR Technologies WindWings[iv] by Yara Marine, Oceanwings® by Ayro[v], FastRig by Smart Green Shipping[vi], Oceanbird by Alfa Laval and Wallenius[vii], AirWing™ by GT Green Technologies[viii], and the Smart Sail by Jiangnan Shipbuilders[ix] are all examples of this type of rig. Hard wing sails may be depowered by feathering, or (if designed with the functionality) retracted by folding or telescoping to deck level.

1.3 ROTOR SAILS

Rotor sails exploit the Magnus effect to accelerate air on one side of the rotor whilst slowing it on the other, creating a pressure differential and resulting lift force. Rotor sails are freestanding. Loads from the shell of the rotor are typically transferred via a rotary bearing to the ship through an internal tower, which extends from a deck foundation up to a specified height inside the rotor. Rotor sails are one of the most mature wind propulsion technologies, with installations as far back as the 1920s and a range of modern solutions available from Anemoi[x],

Dealfeng[xi], Eco-Flettner[xii] and Magnuss[xiii]. Rotor sails cannot be reefed nor feathered, but (if designed with this functionality) may be retracted by folding to deck level.

1.4 SUCTION SAILS

Finally, suction sails are a group of wind propulsion technologies utilising wing foils with vents and internal suction that exploit boundary layer suction to increase maximum lift values. The structural configuration is typically similar to rotor sails, with a deck mounted bearing transferring loads into the ship. Suction sails can be feathered to depower and (if designed with this functionality) may be retracted by folding to deck level.

1.5 KITES

Leading the way in this category are Airseas who have developed their Seawing kite together with an integrated, automated flight control system. Fully retractable and - due to its single point connection to the bow region of a ship's deck - it can be installed with relative simplicity[xiv]. The kites are constructed from soft fabrics and inflate to form the required geometrical form. The kites are tethered to the ships deck with high-strength material.

2 MATERIAL SELECTION

2.1 THE STRUCTURAL ENGINEERING PROCESS

To better understand why wind propulsion technologies are developing such a broad range of metallic and non-metallic construction solutions it can be useful to look in more detail at some of the material selection considerations within a typical design and engineering process. Engineers developing standardised wind propulsion system will generally begin their optimisation by working within the constraints of the plan forms, shapes and sections necessary to generate the required aerodynamic and hydrodynamic performance before establishing the enabling materials and technologies that allow their concepts to work. Within those parameters, and other physical limits created by deck space or other loading and port handling requirements, engineers can then optimise again to refine the balance between material performance, weight and cost.

2.2 HISTORICAL MATERIAL DEVELOPMENT PATTERNS

As new materials become commercially available over time, manufacturing sectors have explored and adopted the most promising new options.

- Early wood and fabric aircraft were quickly superseded by aluminium airframe structures with the first resin and fibre composites arriving during WW2 as a potential solution to aluminium shortages. Only in the latter stages of the 20th century did carbon fibre composite structures start to appear in commercial aerospace with Airbus applying composites in rudders and tail fins in the early 1980s.
- Whereas steel remains highly prevalent to this day, automotive lightweighting is still key to OEMs (Original Equipment Manufacturers) being able to reach their upcoming overall range emissions targets. Aluminium has been widely adopted, whereas high-performance composites (i.e., those capable of use in primary structures) have only seen minor adoption.
- Wooden ships were only replaced by iron vessels as the Industrial Revolution unfolded in the 19th century and steam-powered iron ships began to take over. Ultimately, the availability of carbon steel in around 1880 would lead to its adoption in shipping where it still dominates to this day.
- In performance sailing vessels, weight saving, rather than cost and durability allowed the material timeline to move on from wood and steel to the point where aluminium was the recognised lightweight mast and hull material of choice until comparatively recently. Glass reinforced polyester (GRP) was followed by higher performance composites; carbon fibre reinforced polymer (CFRP) eventually prevailing in performance sailing craft and high-end luxury yachts.

2.3 THE ROLE OF HIGH-PERFORMANCE COMPOSITES

Generally, for advanced composites, the higher the performance (specific strength or stiffness of the material), the higher the cost. High performance composites must first earn their place on board as an enabling material technology, then their use must be optimised to derive the most cost-effective and structurally efficient use. These materials have established themselves as enabling technologies in several wind propulsion designs scenarios, most often where a viable manufacturing technology and capability exists and a lightweight, stiffness critical structure is necessary, such as when vessel performance (e.g., stability and roll dynamics) or the rig retraction process (e.g., folding or telescoping) demand a lighter solution than metallics can deliver. A summary of materials selected for existing and proposed wind propulsion technologies is provided within table (1) below. One can observe that whilst no "one-material-fits-all" approach exists, established steel and aluminium fabrication technologies are being joined by a significant proportion of composite and metallic-composite hybrid materials.

Table 1: Materials Selected for existing wind propulsion technologies. Source: Hexcel, January 2023

	Steel + GRP	Aluminium	CFRP	Kite fabric	GRP	Steel	Steel + Aluminium	unknown	Grand Total
France	2	1	5	5				2	15
China	7	1						4	12
Poland	9								9
Netherlands		7	1						8
Japan	3	1					1		5
Germany	3				1	1			5
UK			1					1	2
Italy			1		1				2
USA								1	1
Sweden					1				1
Singapore		1							1
Spain		1							1
Norway	1								1
Grand Total	25	12	8	5	3	1	1	8	63

2.4 LESSONS IN COMPOSITE SELECTION ACROSS INDUSTRY

Having observed the material development pathway and its movement towards composite adoption in other markets, it can also be noted that many of the lessons learned through this evolution could also be applied to wind propulsion in shipping.

2.4 (a) Automotive – not quite the high-volume composite success story promised.

Carbon composites have been touted as a viable solution for lightweighting in automotive for several decades, but the reality is that the trickle down towards high-volume composite has not yet materialised. The BMW i3 and i8 were impressively composite intensive with their carbon chassis, yet despite utilising materials from a fully consolidated supply chain whereby a BMW and SGL joint venture owned the carbon fibre producing plant, the longer-term view for the company's i cars is that carbon fibre is no longer required in such high proportions. Robert Irlinger, head of BMW's i division, has described how lower available battery capacity at the time of the BMW i3 led to a focus on taking every kilogram possible out of the car to reduce the amount of energy we needed to power it and extend range[xv]. With the latest improvements in energy capacity of batteries, it is no longer as important to search for such comprehensive lightweighting, so today BMW are instead using carbon in selected areas, primarily in premium extras including carbon roofs or trim packages. Some supercar manufacturers like McLaren Automotive continue to use carbon composites for safety cells. Tesla use composites sparingly, preferring to rely chiefly on aluminium for its vehicles. Glass composites are of growing interest where they present a more viable and sustainable solution for the future, and where the inertia in manufacturing process change can be overcome.

2.4 (b) Aerospace primary structures: cost levels currently out of reach for wind propulsion

An aircraft structure can be defined as primary, secondary, and moveable appendages. Weight reduction is the key driver for aerospace designers for all three: reducing fuel burn and expanding capacity and range with each kilogram saved. Carbon fibre composites dominate primary aerospace structures providing weight savings of around 20% versus aluminium, but this benefit does not come without a price. Quality control requirements, automation capital expenditure needed to achieve viable fibre deposition rates with automated fibre placement (AFP) and automated tape laying (ATL) machinery and a historic reliance on energy intensive autoclave (high-pressure vessel for elevated temperature and pressure curing of composites) processing all create costs an order of magnitude away from wind propulsion target levels. In terms of materials selected, there is potential for more common ground between wind propulsion and aircraft secondary structures or moveable appendages where out-of-autoclave processing and multi-material laminate solutions are more commonplace.

2.4 (c) High-performance sailing – pioneering wind propulsion technologies

The relative freedom of design restrictions and the ability to rebuild if an overly lightly built structure fails has placed yacht racing and high-performance sailing as one of the leading innovators in composite developments. Prepreg and honeycomb structures have been tested to the limit whilst early carbon fibre masts and sails were evolved quickly as breakages provided instant feedback. High performance sailing in the America's Cup has seen both solid wing sail and soft sail powered craft flying clear of the water on carbon composite hydrofoils, with both rig types and foil systems now being applied to wind ships and commercial vessels. BAR Technologies, Artemis Technologies and North Windship Technologies are prime examples of the crossover that is possible from high-speed racing to high efficiency shipping.

2.4 (d) Wind energy and wind propulsion for ships – the perfect fit?

Since the turn of the century, wind turbine blades have become much longer (for larger swept area) and the supporting towers have also grown higher to capture more energy, since winds generally increase as altitudes increase. Some of the largest units on the market now such as the GE Haliade-X offshore wind turbine featuring a maximum capacity of 14 MW from its 107-meter blades, generate similar power outputs to the main engine of a significant sized commercial vessel. However, some significant design differences exist between wind energy and wind propulsion, including:

- Wind blade tip speed typically exceeds 100 m/s (~200 knots), whereas the maximum apparent windspeed on a wind propulsion rig likely to approach 100 knots only in extreme storm conditions.
- To generate thrust, a wind blade is subjected to cyclical loading at rotational speeds of 5-10 rpm and is connected via its root end to a foundation that is fundamentally fixed and – ignoring pitch control and weather-vane capability – remains stationary despite the forces generated by the blade. Conversely, a wind propulsion rig does not rotate (to generate thrust) yet it is connected to a foundation that has six degrees of freedom. The higher the power that is generated by the wind propulsion system, the more the forces on the rig will affect the ship's movement and resulting angle of heel, pitch and yaw. Similarly, the wind propulsion rig is subjected to ship's acceleration and resultant forces.
- The close proximity of wind blades to their supporting tower means they must not flex and hit the tower, usually designers manage this by including some pre-bend away from the tower (at zero speed). This requirement is not likely to exist in wind propulsion allowing designers to proceed without the associated aerodynamic design compromise.
- The proximity of a wind propulsion rig to the ship's structure and crew presents design considerations that differ from a wind farm - typically isolated geographically and far away from people. Whilst the rig location presents higher risk due to consequential damage, it also presents an opportunity in terms of regular inspection and maintenance.

As modern wind turbines evolved, knowledge was shared between participants in the marine industry and their counterparts in wind energy, particularly in terms of material selection and manufacturing processes. Optimised methods for resin infusion and prepreg processing became dominant in modern wind turbines.

3. LEARNING FROM WIND ENERGY

Wind energy has also become fiercely price competitive, with commoditisation driving larger production runs, a reduction in customisation and a remarkably high rate of manufacture. Throughout this cost optimisation, a common structural configuration and two leading manufacturing solutions have prevailed.

3.1 STRUCTURAL ARRANGEMENT

A typical blade (as shown in Figure 1) is an assembly of two shells joined at the leading and trailing edges, supported by a structural spar positioned at or near the location of maximum chord thickness. The shells are typically of sandwich construction to provide local panel stiffness; the structural spar includes spanwise stiffening in the flanges and biaxial material within the shear webs. The leading edges are generally strengthened for damage tolerance; the trailing edges have additional stiffness for aerodynamic performance.

3.2 MATERIALS

3.2 (a) THE USE OF PRE-IMPREGNATED MATERIALS

Pre-impregnated (or prepreg) composite materials are machine made roll format products where reinforcement fibres (either unidirectional fibres or multi-axis textiles) are impregnated with a catalysed mix of (typically epoxy) resin and hardener. This factory preparation delivers precise control of the resin content and provides manufacturers with a robust product that can then be heated under vacuum to cure the final part. Setting up the mould tools for prepreg manufacturing is relatively simple with this process lending itself as well to smaller production runs of different blade types as it is for longer runs where high throughput is critical. Prepreg has historically required frozen storage, but modern advances offer the potential to avoid this whilst still benefitting from moderate & industrially achievable cure temperatures.

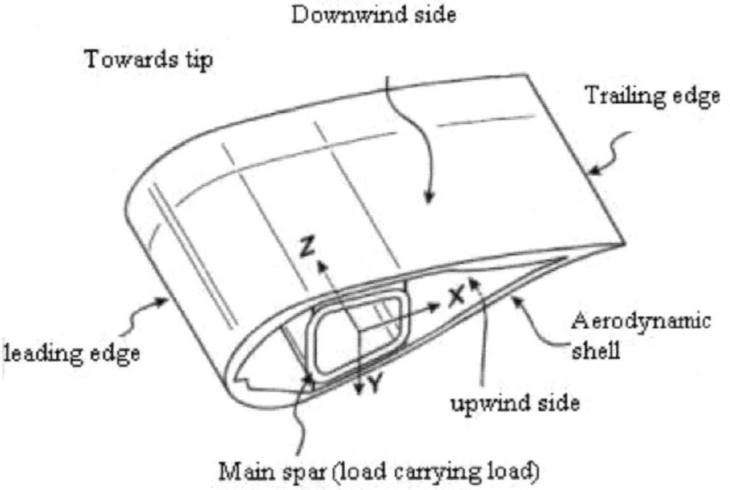

Figure 1: Typical cross sections of wind turbine blade (Sorensen et al., 2004)

3.3 (b) THE USE OF THE INFUSION PROCESS

Vacuum resin infusion processing sees the blade manufacturer take responsibility for combining liquid resins with the reinforcement fibres and fabrics. The dry reinforcements are loaded into the mould in layers and covered with a plastic film which is sealed around the mould and any air evacuated with a vacuum pump. This vacuum is then used to pull liquid resin through the dry fibre laminate stack until the part is completely wet out and the blade can be cured under vacuum consolidation. The permeability of the dry reinforcements is critical to part quality, and large-scale infusion of laminates of glass fibre (with a filament diameter of 17 microns) is considered less onerous than those of carbon fibre (7 microns) due to the higher packing ratio – and hence reduced permeability – of the latter.

3.3 (c) PULTRUSIONS

Typically, the main structural spar of an infused blade is now produced from pre-cured pultruded planks of carbon fibre composite. The pultrusion process (where unidirectional fibres are wet out with liquid resin and then pulled through a steel die to shape and cure the constant section profile in a continuous process) produces excellent fibre alignment and a high ratio of fibre to resin, delivering high mechanical performance at a cost-attractive price point.

Of the two processes, infusion has now tended to take most of the market share in wind. Whereas manufacturers using prepreg can rely heavily on their material supply chain, infusion blade builders must own and master the process. Infusion and prepreg are both well suited to wind rig construction for shipping, having been validated by the wind sector at rate and for very large structures. Prepreg's benefits of fibre alignment, guaranteed resin content/consistency and its resulting mechanical performance, together with a new capability to generate very thick (more than 100mm is possible with the latest systems) parts in one cure cycle make these materials excellent candidates for free standing mast structures where compression strength and stiffness are critical. The simplified processing (without the need for expensive, energy intensive autoclaves nor multiple cure cycles) of these wind energy derived systems enables a prepreg approach which would otherwise be unviable in shipping due to the relative complexity and cost of aerospace prepregs. In addition, infused glass fibre skins are likely to offer the optimum solution for solid wing sail, suction wing, or rotor sail mouldings where the composites must succeed in targeting a much closer position to steel or aluminium on the price/performance plot. Though each design case will differ, manufacturers have an opportunity to jump a few steps ahead with their composite learning curve by applying some of the processes and experiences of their wind energy colleagues.

3 TESTING THE CARBON FIBRE SUPPLY CHAIN

Commoditisation has fully globalised both the wind energy materials and manufacturing supply chains with an almost complete shift to low-cost manufacturing areas in East Asia, Mexico, Turkey, and India. Wind ship volumes for composite structures are not likely to trouble the suppliers of glass fibre and resin materials, with standard wind energy building blocks being well suited to the crossover in shipping and offering some of the most attractive cost/performance ratios available in composites as well extensive kitting options and global distribution channels.

The carbon fibre market is perhaps more challenging, certainly in the current market conditions of shortage and elevated pricing. Whilst the author's own estimates for wind propulsion and carbon-suited components indicating a potential loading of 6000 tons of carbon fibre per annum (around 3% of global nameplate capacity) if 100% of new build ships adopt wind power seem to allow plenty of headroom, market feedback suggests that a high level of forward planning would be required to guarantee supply throughout a programme lifespan.

With strong demand from the new hydrogen tank sector, steady pull from wind energy and aerospace having consumed excess stock generated during the pandemic, carbon fibre manufacturers are struggling to add capacity quickly. The tightness in the market seen today is likely to continue for 3-5 years, by which time it is anticipated that new capacity will be operational to service the hydrogen market needs, and the 'rebound' growth in aerospace will return to pre-pandemic levels.

4 IMPROVING THE SUSTAINABILITY OF HIGH-PERFORMANCE COMPOSITES

As a renewable energy source, the wind energy industry has had to counter many tough questions about its green credentials when so many challenges remain with end-of-life solutions for its composite heavy blades when they are retired after circa 25 years in service. The wind ship market can be expected to face similar scrutiny if engineering drivers propose tough to recycle energy intensive composites instead of easily recyclable steel. With this in mind, a brief review of recent developments follows, summarising the state of developments and available options.

4.1 LIFE CYCLE ANALYSIS – AN IMPORTANT PRECURSOR

The Life Cycle Analysis (LCA) of the materials and processes of the wind propulsion system should not be considered in isolation. Instead - to ensure appropriate focus on the overall objective – this should be assessed in relation to the positive impact of the wind propulsion technology on the LCA of the ship (as an entire system). Nonetheless, those involved in the design and engineering of wind propulsion systems need validated LCA of candidate raw materials - as well as the energy required to process them into the required geometrical forms - to help to identify the priority areas when looking for improvements in sustainability. With data in hand, the following measures may be undertaken to mitigate the environmental impact of composite materials whilst improving the sustainability benefits:

- Reduce the amount of material used and minimise the waste.
- Use of lower impact materials.
- Use less energy intensive manufacturing methods.
- Maximise the lifetime of the structure
- Use of good environmental management methods including reuse and recycling of the materials.

4.2 REDUCED MATERIAL USE

Whilst paying attention to differences in environmental impact of available materials, any optimisation of the structure to minimise its weight will help to reduce its cost and environmental impact; herein lies the value of the structural engineer. Similarly, the minimisation of in-process waste during manufacturing will reduce the bill-of-materials. As stated earlier, net-shape manufacture using AFP is likely to be limited to aerospace use, but sophisticated tools for optimisation of nesting from broadgood materials exist (thanks to Wind and Automotive) for the benefit of wind propulsion.

4.3 LOWER IMPACT MATERIALS

Focusing initially on the resin component of a composite system, several opportunities currently exist to lower environmental impact.

4.3 (a) EPOXY SYSTEMS

Some new products and some reformulated versions of established products with well characterised performance and processing data are now available with a significant percentage of their molecular carbon content derived from sustainable sources, where resin components are derived from plants that act as carbon sinks before harvesting (for example Hexcel's Hexply® Nature Range of prepregs with bio-derived resin content)[xvi] Manufacturers have been keen to demonstrate that increased sustainable content (today in the range of 20-50%, in future tending towards 100%) is available with negligible impact on structural performance. Such sustainability upgrades to the resin system have proven simple to integrate with minimal design or process adjustment necessary. Already popular in some sporting goods and leisure marine and automotive applications, certified 'biobased content' may well become a must have on composite RFQs in the future. It should be noted that these resins would still not be recyclable, and the economic impacts of the switch are also to be confirmed.

4.3 (b) RECOVERABLE RESIN SYSTEMS

Manufacturers looking for recoverable resin options will see several emerging technologies being promoted. Two that have advanced to component level evaluation include the thermoplastic Elium® liquid resin by Arkema[xvii] and the novel cleavable thermoset Recylamine range by Aditya Birla Advanced Materials[xviii]. Once again, the wind energy industry is at the forefront of this development with Elium® resin being used for a 62m blade and Siemens Gamesa also having manufactured full scale recyclable blades using Aditya Birla materials. Both options are likely to offer 'down-cycling' of the recovered component parts where the mechanical and physical properties diminish, and such circularity requires additional processing steps not necessarily executed so far, yet one imagines wind propulsion manufacturers will still be encouraged by solutions that avoid end of life landfill routes for large composite parts.

4.3 (c) FIBRES

Composite reinforcement fibre use in marine applications is predicted to remain dominated by glass and carbon fibres. Whilst basalt fibres are potentially interesting in terms of strength and modulus, they are commercially less attractive than glass and require extraction via mining - so unlikely to attract much substitution. Natural fibres such as flax and hemp are rapidly maturing in terms of their available formats for composites however, cost and availability are also an issue versus glass. For semi-structural parts that would still benefit from lightweighting, flax, with its low density and high specific stiffness, may be an alternative though care must be taken to manage the fibres' susceptibility to moisture uptake in the dry fibre form and when composite laminates are cut, drilled or trimmed. Most carbon fibres are made via the oxidation and then carbonization of polymerised acrylonitrile (PAN) derived from a petro-chemical source. Whilst such a change would be unlikely to impact the energy intensive final conversion to carbon fibre, the promising development of plant-based acrylonitrile feedstock is ongoing and would provide a more sustainable raw material feedstock, once again with the caveat that its resulting mechanical properties, as well as the economics and scale of such a supply option are yet to be confirmed.

4.3 (d) SURFACE FINISHING MATERIAL TECHNOLOGIES

Advanced composite surfacing materials technologies (for instance Hexcel's XF Surface Technology[xix]) may also provide a sustainability improvement, as well a technical benefit whereby the more consistent coating thickness contributes to improved rig balance particularly for rotor sails, when compared with traditional liquid coatings employed in marine applications. Novel machine calibrated surface films, again derived from process improvements in wind energy, can result in significant reductions in the material usage required to create a durable and uniform surface, contributing to an overall improvement in component LCA results.

4.4 REDUCE ENERGY REQUIRED TO PROCESS

In the past, high-performance composites required high temperature cures for lengthy periods (>100 degrees Celsius for 12hours), but thanks to the development in polymer chemistry and formulation in recent years, the required cure temperatures and time have reduced considerably; systems now exist with recommended cure cycles of 80 degrees Celsius for 4 hours.

4.5 MAXIMISE THE STRUCTURAL LIFE

All materials must be selected objectively in relation to both carbon footprint and service lifetime. To illustrate this point, if a newly developed materials technology exhibits twice the carbon footprint of its predecessor but increases the service life by a factor of three, the LCA would show a net-gain in favour of the new technology.

4.6 A POTENTIAL FOR RECYCLED MATERIALS

Whether obtained by pyrolysis, mechanical recycling or chemical solvolysis, a key challenge to obtaining circularity of materials lies with obtaining outputs that are viable for re-use if not delivered back to production in exactly the same format as existed before recycling. Keeping high value carbon fibre within the supply chain for as long as possible is a clear target in both wind energy and wind propulsion however product format may be a blocking point. Recycled dry fabric production waste or recovered fibre tends to be converted to short fibre products such as carbon fibre felts or mats with potential for tooling applications or less stiffness critical quasi-isotropic panels in solid sails but cannot currently be re-processed into the unidirectional prepregs, pultruded profiles or high-performance textiles that made the initial engineering and business case for carbon fibre. Industry continues to follow developments such as the latest updates in reclaiming continuous fibre whereby UK company B&M Longworth, in partnership with the National Composites Centre and Cygnet Texkimp, has successfully recovered continuous fibre from a composite pressure tank[xx]. This development, and those focused on aligning shorter recycled fibres, is more promising for re-use in highly structural parts. Additionally, reclaimed fibres may be combined with virgin fibre pultrusions or other structural to adequately support design load conditions. However, the processing of end-of-life wind ship parts or the extraction of continuous fibre from anything but pressurised gas storage cylinders still appear to be some way from becoming a reality. Nonetheless the industry continues to push for solutions: For example, Solvay offer a service of recycling CFRP into non-structural parts such as brackets and ribs[xxi], and Hexcel have recently announced a collaboration with Fairmat who aim to develop technologies and processes for CFRP recycling[xxii].

5. CONCLUSION

Fundamentally, composite materials are a good fit for the marine industry. They can be moulded into complex shapes, they are lightweight, and they do not corrode in seawater. Wind propulsion adoption in commercial shipping appears ready to enter an extremely exciting period, particularly with the number of vessels now in service, or in build, able to contribute to a much larger real-life data set for composite material selection, manufacturing and sustainability considerations.

In the short term (for a given programme) we expect structural engineers will continue to optimise within the constraints of the problem as discussed, defining the enabling material properties and manufacturing technologies required to make the concept work, before optimising to find the lowest cost and structurally efficient use of the available materials/manufacturing technologies. In the longer term (for future programmes) technology providers and leading composite material manufacturers focused on the wind propulsion sector, will add the most value by offering improvements in material and process innovations that provide engineers with the best of both worlds – lightweight, highly structural materials at low cost.

Rather than becoming a dominant material technology in wind propulsion, it seems likely that advanced composite materials will play their role in some of the most highly loaded structures, particularly when these will offer the most beneficial impact on vessel performance, long term durability and maintenance requirements. Wind rig system manufacturers currently have access to composite material technology partners who can support their process development acceleration, allowing them to focus on control systems and other key aspects of wind rig and vessel integration.

The author is optimistic about the outlook for high-performance composite materials in wind propulsion, noting that measured adoption in selected applications, matched with processing technologies rooted in wind energy, carefully optimised for rate, cost and low energy consumption would seem to be the key drivers for success.

6. REFERENCES

1. North Windships, Website, https://www.northwindships.com/, 2023

2. Michelin, Website, https://www.michelin.com.ph/auto/the-wisamo-project, 2022

3. Chantiers de L'Atlantique, Website, https://chantiers-atlantique.com/references/silenseas/, 2023

4. BAR Technologies, Website, https://www.bartechnologies.uk/project/windwings/ , 2023

5. Ayro, Website, https://ayro.fr/oceanwing/, 2022

6. Smart Green Shipping, Website, https://smartgreenshipping.com/fast-technologies, 2023

7. Alfa Laval and Wallanius, Website, https://www.theoceanbird.com/, 2023

8. GT Green Technologies, Website, https://www.gtgreentechnologies.com/, 2023

9. Offshor Energy Biz, Website, https://www.offshore-energy.biz/jiangnan-shipbuilding-unveils-smart-sail-design/, 2022

10. Anemoi, Website, https://anemoimarine.com/, 2023

11. Dealfeng, Website, http://www.dealfeng.com/news/24.html, 2022

12. Ecoflettner, Websites, https://www.ecoflettner.de/, 2023

13. Magnuss, Website, https://magnuss.com/flettner.html, 2023

14. Airseas, Website, https://www.airseas.com/seawing, 2023

15. BMWBlog.com,Website,https://www.bmwblog.com/2018/07/17/carbon-fiber-bmw-might-be-a-thing-of-the-past/#:~:text=%E2%80%9CAt%20the%20time%20of%20the%20BMW%20i3%2C%20the,need%20to%20look%20for%20the%20last%20500%20grams., 2023

16. Hexcel, Website, https://www.hexcel.com/News/News-Releases/3421/hexcel-launches-hexply-nature-range, 2022

17. Arkema,Website,https://www.arkema.com/global/en/products/product-finder/product-range/incubator/elium_resins/, 2023

18. Aditya Birla Chemicals, Website, https://www.adityabirlachemicals.com/brand-list.php?id=recyclamine, 2023

19. Hexcel,Website,https://www.hexcel.com/News/News-Releases/3298/hexcels-hexply-xf-surface-technology-revolutionizes-the-blade-surface-finishing#:~:text=Hexcel%E2%80%99s%20HexPly%20XF%20surface%20technology%20has%20been%20formulated,achieve%20the%20perfectly%20smooth%20surface%20required%20for%20painting., 2022

20. Longworth,Website,https://www.deecomlite.com/2022/05/continuous-carbon-fibre-recovery-for-hydrogen/#:~:text=Engineers%20at%20the%20National%20Composites%20Centre%2C%20the%20UK%E2%80%99s,re-used%20them%20to%20manufacture%20a%20new%20pressure%20vessel., 2022

21. Solvay,Website, https://www.solvay.com/en/chemical-categories/our-composite-materials-solutions/recycling-composites, 2023

22. Hexcel,Website,https://www.hexcel.com/News/News-Releases/3364/hexcel-and-fairmat-announce-partnership-to-recycle-carbon-fiber-prepreg-composite, 2022

7. AUTHORS BIOGRAPHY

Thomas James holds the current position of Marine Business Development Manager at Hexcel. He is responsible for development of strategy and new business opportunities within the marine industry. Having originally qualified as a Naval Architect, his previous experience includes engineering and design roles within the marine composites sector, composite structural engineering for an America's Cup yacht racing team, followed by research and commercial roles within the advanced composites materials industry. Tom is a Chartered Engineer and has been a RINA member since 1998.